FUNCTIONAL AND EVOLUTIONARY ECOLOGY OF BATS

FUNCTIONAL AND EVOLUTIONARY ECOLOGY OF BATS

EDITED BY

Akbar Zubaid, Gary F. McCracken, and Thomas H. Kunz

OXFORD UNIVERSITY PRESS

2006

OXFORD
UNIVERSITY PRESS

Oxford University Press, Inc., publishes works that further
Oxford University's objective of excellence
in research, scholarship, and education.

Oxford New York
Auckland Cape Town Dar es Salaam Hong Kong Karachi
Kuala Lumpur Madrid Melbourne Mexico City Nairobi
New Delhi Shanghai Taipei Toronto

With offices in
Argentina Austria Brazil Chile Czech Republic France Greece
Guatemala Hungary Italy Japan Poland Portugal Singapore
South Korea Switzerland Thailand Turkey Ukraine Vietnam

Published by Oxford University Press, Inc.
198 Madison Avenue, New York, New York 10016

www.oup.com

Library of Congress Cataloging-in-Publication Data
Functional and evolutionary ecology of bats / edited by Akbar Zubaid,
Gary F. McCracken, and Thomas H. Kunz.
p. cm.
Includes bibliographical references (p.).

ISBN 978-0-19-515472-6

1. Bats—Ecology. I. Zubaid Akbar. II. McCracken, G.F.
III. Kunz, Thomas H.
QL737.C5F86 2005
599.4'17—dc22 2004057541

9 8 7 6 5 4 3 2

Printed in the United States of America
on acid-free paper

Dedicated to Lim Boo Liat

Preface

Research on bats during the second half of the 20th century has had a marked influence on how we now view these so-called secretive mammals. A new generation of researchers has begun to advance our knowledge about bats in disciplines including ecophysiology, functional morphology, population biology, molecular ecology, and conservation biology, with increasing numbers of studies extending to remote corners of the Earth. Moreover, advances in molecular biology, genomics, bioinformatics, computational science, geographic information systems, sophisticated imaging systems, transponders, and the miniaturization of radio-transmitters have provided new tools that were unavailable to earlier generations of bat researchers.

The wealth of new technologies and information that has emerged from both field and laboratory studies has had an enormous impact on how research is now being conducted. Research on functional and evolutionary ecology of bats has emerged from largely descriptive studies to emerging emphases on hypothesis testing and experimentation in both the laboratory and the field. In part, this reflects an increased recognition that bats provide critical ecosystem services such as seed dispersal, pollination, and insect control that sustain healthy populations. This is especially true in light of growing concern over deforestation, habitat fragmentation, mining activities, and an increasing threat to natural habitats from air, water, and soil pollution and threats from global warming.

This volume is the outgrowth of three symposia that were convened at the 12th International Bat Research Conference, hosted by the Universiti Kebangsaan Malaysia from August 5 to 9, 2001, held in Kuala Lumpur, Malaysia. For this conference, the editors invited conveners to organize symposia based on current research on functional and evolutionary ecology. The conveners in turn selected speakers to present reviews and analyses of their respective disciplines. Chapter authors were invited because they had distinguished themselves by expanding their research beyond traditional approaches. The conveners prepared written introductions to their respective symposia so that the overall contributions could be framed in a broad perspective.

The first section on physiological ecology focuses on thermal biology of hibernation, energetics, daily heterothermy, and the evolution of basal metabolic rate. The next section on morphology includes topics ranging from form and function of dentition, ecomorphology of flower-visiting bats, form, function, and phylogeny of quadrupedal bats, wing morphology and its implications for flight performance, and the relationships between cranial morphology and feeding ecology. The third section of the book on roosting and population biology includes topics ranging from population genetic structure, life-history traits, social behavior and relatedness, mating systems, and roosting ecology.

This book would not have been possible without the dedication and enthusiasm of the conveners of each session who have helped craft the invited chapters into a synthetic whole. The convenors were not only instrumental in inviting the participants but also arranged for reviewers, assisted in the editing process, and wrote an introduction to their respective sessions. Each manuscript was thoroughly read by at least two anonymous reviewers, and we thank our colleagues who reviewed one or more of the chapter manuscripts. These individuals include Nadia Ayoub, Robert Barclay, Andy Biewener, Wieslaw Bogdanowicz, Mark Brigham, Tamsin Burland, Brock Fenton, Rod Foster, Trish Freeman, Fritz Geiser, Erin Gillim, John Hermanson, Anthony Herrel, Bruce Jayne, Kate Jones, Gerald Kerth, Barry Lovegrove, Alex Menzel, Collin O'Donnell, Steve Rossiter, Amy Russell, Randy Small, John Speakman, Tim Strickler, Carl Thulin, Don Thomas, Annie Tibbels, Peter Ungar, Maarten Vonhof, Jerry Wilkinson, Craig Willis, and John Winkelmann.

We also thank Bethany Bernasconi, Kristine Faloon, and Erin Ruppert of Boston University's Center for Ecology and Conservation Biology for assisting in the final preparation of the book manuscript. We are grateful to Kirk Jenson, who expressed enthusiasm for this project from the onset and for his patience and forbearance. We especially thank Peter Prescott who assumed responsibility for shepherding the final stages of this book to completion. We also thank Kaity Cheng of the editorial and production staff at Oxford University Press for her assistance in preparing the final product. Finally, we thank our wives, Lisa Syarma Addini, Jamey Dobbs, and Margaret Kunz, for their patience, tolerance, understanding, love, and support.

Contents

Contributors

Akbar Zubaid is Professor of Biological Sciences at the Universiti Kebangsaan Malaysia. His research focuses mostly on population biology, community ecology, and conservation biology of bats. He is the author or coauthor of over 60 publications on bats. He is the coeditor (with Zainal Abidin Abu Hasan) of *Conservation and Faunal Biodiversity in Malaysia* (Penerbit Universiti Kebangsaan Malaysia). He is a recipient of the TWAS Prize for Young Scientists in Developing Countries in the Field of Biology (for research on bats and other small mammals).

Gary F. McCracken is Professor of Ecology and Evolutionary Biology at the University of Tennessee. His research focuses mostly on behavioral ecology, population genetics, and conservation biology of bats, and he has conducted research in North America, Ecuador, and the West Indies. He is the author or coauthor of over 70 publications on bats. He serves on the scientific advisory boards of Bat Conservation International and the Lubee Bat Conservancy, and is a recipient of the Gerrit S. Miller, Jr., Award (for outstanding contributions to the study of bat biology).

Thomas H. Kunz is Professor of Biology and Director of the Center for Ecology and Conservation Biology at Boston University. His research focuses on reproductive biology, physiological ecology, behavioral ecology, and conservation biology of bats. He has conducted research in North America, the West Indies, India, Malaysia, and Ecuador. He is the author or coauthor of over 200 publications on bats, and is the editor of *Ecology of Bats* (Plenum Press, 1982), *Ecological and Behavioral Methods for the Study of Bats* (Smithsonian Institution Press, 1988), coeditor (with P.A. Racey) of *Bat Biology and Conservation* (Smithsonian Institution Press, 1998), and coeditor (with M.B. Fenton) of *Bat Ecology* (University of Chicago Press, 2003). He serves on the scientific advisory boards of Bat Conservation International and the Lubee Bat Conservancy. He is a recipient of the Gerrit S. Miller, Jr., Award (for outstanding contributions to the study of bat biology), and the C. Hart Merriam Award (for outstanding contributions to the study of mammals).

Maryem-Fama Ismael Aguirre
Department of Ecology and Evolutionary Biology
Box G-B206, Brown University
Providence, RI 02912
USA

Johnson Balasingh
St. John's College
Palayamkottai 627006
Tamilnadu
South India

Elizabeth M. Barratt
3 Rutland Road, Hazel Grove
Stockport, Cheshire SK7 6JD
United Kingdom

Andrew F. Bennett
School of Ecology and Environment
Deaking University
Victoria 3125
Australia

Hari R. Bhat
107, Awanti Appartments
Opp. Kamala Nehru Park
Erandawana
Pune 411004, India

Kristin Bishop
Department of Ecology and Evolutionary Biology
Box G-B206
Brown University
Providence, RI 02912
USA

Tamsin M. Burland
School of Biological Sciences
Queen Mary & Westfield College
University of London
Mile End Road
London E1 4NS
United Kingdom

Ariovaldo P. Cruz-Neto
Departmento de Zoologia
IB, UNESP, Rio Claro
Brazil

Elizabeth R. Dumont
Department of Biology
University of Massachusetts
Amherst, MA 01003-9297
USA

Abigail C. Entwistle
Fauna & Flora International
Great Eastern House
Tension Road
Cambridge CBI 2TT
United Kingdom

Alistair R. Evans
Evolution and Development Unit
Institute of Biotechnology
University of Helsinki
Helsinki
Finland

Fritz Geiser
Department of Zoology
University of New England
Armidale, NSW 2351
Australia

Gerald Heckel
Computational and Molecular Population Genetics Lab
Zoologisches Institut
Universität Bern
Baltzerstrasse 6
3012 Bern
Switzerland

Otto von Helversen
Institut für Zoologie II
Universität Erlangen
Erlangen University
Staudtstrasse 5
91058 Erlangen
Germany

Murray M. Humphries
Natural Resource Sciences
McGill University,
McDonald Campus
21, 111 Lakeshore Road
Ste-Anne-de-Bellpuue
Quebec H9X 3V9
Canada

Gareth Jones
School of Biological Sciences
University of Bristol
Woodland Road
Bristol BS8 1UG
United Kingdom

Kate E. Jones
Center for Environmental Research and Conservation
Columbia University
New York, New York 10027
USA

Gerald Kerth
Zoologisches Institut
Abteilung Verhaltensbiologie
Universität Zürich-Irchel
Winterthurerstrasse 190
8057 Zurich
Switzerland

Thomas H. Kunz
Center for Ecology and Conservation Biology
Department of Biology
Boston University
Boston, MA 02215
USA

Linda F. Lumsden
Department of Natural Resources
Arthur Rylah Institute
123 Brown Street
Heidelberg 3084
Australia

Gary F. McCracken
Department of Ecology and Evolutionary Biology
University of Tennessee
Knoxville, TN 37996
USA

P. Thiruchenthil Nathan
St. John's College
Palayamkottai 627006
Tamilnadu
South India

Christopher W. Nicolay
Department of Biology
University of North Carolina at Asheville
One University Heights
Asheville, NC 28804-8511
USA

Colin F.J. O'Donnell
Southern Regional Science Centre
Department of Conservation
PO Box 13049, Christchurch
New Zealand

Paul A. Racey
Department of Zoology
University of Aberdeen
Aberdeen, Scotland AB9 2TN
United Kingdom

Roger D. Ransome
School of Biological Sciences
Queen Mary & Westfield College
University of London
Mile End Road
London E1 4NS
United Kingdom

Stephen J. Rossiter
School of Biological Sciences
Queen Mary & Westfield College
University of London

Mile End Road
London E1 4NS
United Kingdom

Amy L. Russell
Department of Ecology and Evolutionary Biology
Yale University
New Haven, CT 06520-8105
USA

William A. Schutt, Jr.
Natural Science Division
Southampton College of Long Island University
236 Montauk Highway
Southampton, NY 11968
USA

Jane A. Sedgeley
Department of Zoology
University of Otago
Dunedin
New Zealand

Nancy B. Simmons
Department of Mammalogy
Division of Vertebrate Zoology
American Museum of Natural History
New York, NY 10024
USA

John R. Speakman
Department of Zoology
University of Aberdeen
Aberdeen, Scotland AB24 3TZ
United Kingdom

Jay F. Storz
School of Biological Sciences
University of Nebraska
Lincoln, NE 68588
USA

Sharon M. Swartz
Department of Ecology and Evolutionary Biology
Box G-B206
Brown University
Providence, RI 02912
USA

Donald W. Thomas
Département de Biologie
Université de Sherbrooke
Sherbrooke, Québec
Canada

Christian C. Voigt
Institute for Zoo and Wildlife Research
RG Evolutionary Ecology
Alfred-Kowalke-Strasse 17
10315 Berlin
Germany

Craig K.R. Willis
Department of Biology
University of Regina
Regina, Saskatchewan S4S 0A2
Canada

York Winter
Department Biologie
Universität München
Luisenstrasse 14
80333 Munich
Germany

Akbar Zubaid
Department of Zoology
Universiti Kebangsaan Malaysia
Bangi, Selengor
Malaysia

1

Physiological Ecology

Donald W. Thomas & John R. Speakman

Body temperature and body size have such profound effects on how animals function that no studies in behavioral, physiological, and evolutionary ecology can escape taking both variables into account either explicitly or implicitly (Speakman and Thomas, 2003). Body temperature affects the rate of metabolic and thermal processes and so determines the rates of food ingestion and energy extraction, metabolism and the depletion of stored energy, reproduction and somatic growth, and heat flux between an animal and its environment. Because the effect of temperature on metabolism is exponential, even small changes in body temperature have profound effects on the rate of catabolic and anabolic processes.

The choice of a temperature setpoint by an individual and the regulation of body temperature around this setpoint by endogenous heat production can be thought of in the context of a cost–benefit trade-off. The cost of a high body temperature is that it necessitates a continuous channeling of assimilated energy into heat production, which may at times seem wasteful. Endotherms have far lower production efficiencies than ectotherms (roughly 3% vs. 40%; Brafield and Lewellyn, 1982) simply because they invest so much ingested energy in the maintenance of an elevated body temperature. The benefit of a high body temperature, however, is that when food is abundant the rate at which it can be ingested, processed to extract energy and nutrients, and finally invested in growth and reproduction is greatly enhanced. Thus, 3% of a huge food processing capacity is greater than 40% of a small capacity, allowing homeotherms to build tissue and grow far faster than ectotherms.

When food is plentiful or when animals can permit themselves to carry a substantial on-board energy store (fat) to buffer variation in food supply, the benefits of high body temperature outweigh the costs. However, when either food supply or on-board energy is limiting, the cost of homeothermy may outweigh the benefit. The most extreme response to energy limitation is death, but a common strategy for limiting the impact of energy constraints is the lowering of the body temperature setpoint, with the effect of slowing metabolic processes, decreasing heat flux, and finally bringing energy expenditure into balance with intake.

Size affects the amount of food and energy required to build bodies and sustain cellular integrity: large animals obviously require more food than small animals. However, the importance of size in biology lies not just in its effect on the absolute requirements of energy and nutrients, but even more so in its all-pervasive effect on scaling (Peters, 1983). As body size decreases, both thermal conductance and metabolic rate increase on a per gram basis, increasing mass-specific food requirements and profoundly reducing the fasting endurance of small versus large homeotherms. Consequently, even small declines in ambient temperature pose a great challenge for small endotherms, forcing them to seek metabolic flexibility by making use of multiple setpoints for body temperature.

Although size and temperature affect all animals in terms of their physiological ecology, bats are particularly interesting for a variety of reasons. Although they span three orders of magnitude in body mass, ranging from less than 2 g for *Crasseonycteris thonglonglyai* to over 1,200 g for the largest *Pteropus* species, bats generally can be classed as small. Most bat species are strictly nocturnal and spend 50% or more of their time in day roosts, usually without access to food or water. Bats are well represented in almost all climatic zones and thus are able to adapt to an impressive array of environmental temperature and rainfall conditions. Bats exhibit a remarkable diversity of feeding and foraging strategies, including frugivores, nectarivores, carnivores, piscivores, sanguinivores, as well as aerial and gleaning insectivores. And, of course, bats fly.

These factors combine to make the physiological ecology profile of bats unique among mammals. Bats must routinely reconcile high mass-specific metabolic rates with extended roosting periods when food is not available. These fasting periods may last for single days under normal conditions, for short multi-day periods when weather conditions may make foraging unprofitable, and even for periods exceeding 6 months during winter hibernation. Diet and foraging mode determine how variable food availability is, both within and between seasons, and so affect the length of fasting. For example, surface-gleaning insectivores may not experience the same reduction of prey availability as aerial insectivores when night temperatures are low. As an energetically expensive locomotory mode (per unit time), flight not only requires a high metabolic capacity but also imposes constraints on the size of the fat reserve that can be transported economically. These considerations and constraints explain the long-standing interest in the physiological ecology of bats.

Although the four chapters that comprise part I do not span the entire scope of current studies on the physiological ecology and energetics of bats, they do present many of the aspects that have oriented considerable research over recent years. Namely, these include regulation of body temperature and use of daily torpor, hibernation strategies, impact of body size on ecology, and the evolution of metabolic rate.

In chapter 1, Geiser examines whether the thermal and metabolic patterns exhibited by the Australian bat fauna are different from those shown by their continental counterparts. Because the Australian bat fauna has been separated from the Asian and Oriental faunas for considerable time, giving rise to a high degree of endemism in the Microchiroptera, and because the Australian climate can be highly variable, there is reason to question whether bats have evolved a characteristic metabolic profile in response to these conditions. Geiser's analyses indicate that small species have lower-than-expected basal metabolic rates and that most species, including small Megachiroptera, exhibit a high propensity for the use of torpor in response to food restriction. Again we see the importance of body size and body temperature in the ecology of bats.

In chapter 2, Humphries, Speakman, and Thomas rely on the large knowledge-base concerning the physiological ecology of hibernating little brown myotis, *Myotis lucifugus*, to develop a mechanistic model of overwinter energy use. This model serves to demonstrate how a wide variety of behavioral and metabolic data can be combined in an energetic analysis. However, it has even greater value by offering a functional explanation for small-scale (within-hibernaculum) and large-scale (geographic) patterns in the distribution of *M. lucifugus* and by allowing predictions in responses to climate change. This analysis suggests that the northern distribution of hibernating *M. lucifugus* is constrained by the size of pre-hibernation fat reserves and a rapid increase in fat requirements at high latitudes due to declines in cave temperatures below 2 °C. Because climate change will have a strong warming effect in the northeastern Arctic, this model allows us to predict major northward shifts in the distribution of this species. Curiously, this analysis also shows that energy requirements for successful hibernation increase rather than decrease toward the southern part of their range due to increasing cave temperatures.

In chapter 3, Willis focuses on the expression of torpor by nonhibernating bats and the importance of roost microclimates. Recent telemetry studies have shown that the expression of short bouts of torpor on a daily basis is a common metabolic response of male and female bats to food restriction. Willis also points out that heterothermy may be an essential physiological trait of bats because of their high metabolic rates and lengthy daily fast. Reproductive constraints shape the expression of torpor, with males showing more and deeper torpor bouts than females and pregnant females expressing more torpor than lactating females. If the primary aim of torpor is to reduce metabolic costs to bring them into line with energy income and reserves, then the selection of roosts that reduce rewarming costs can be viewed as a related strategy. Species that roost in rock crevices and among foliage choose specific sites that are non-random selections of available roosts. Thus, thermoregulatory and metabolic constraints shape the roosting ecology and distribution of bats.

Finally, in chapter 4, Cruz-Neto and Jones apply a modern statistical approach to examine evolutionary patterns in basal metabolic rate in bats and to test several current hypotheses. For more than two decades now, studies have sought to explain interspecific variation in basal metabolic rate in mammals and birds by differences in body size, diet, distribution, and ecology (e.g., Hayssen and Lacy, 1985; Lovegrove, 2000; McNab, 1986, 1987, 1988). The data presented by Cruz-Neto and Jones show that there is a distinct phylogenetic signal in basal metabolic rate, with closely related clades showing similar rates and distant clades differing more than would be expected by chance. When phylogenetic effects have been controlled for, both body mass and body temperature explain 90% of the variation in basal metabolic rate, underlining the pervasive effect of these two variables. However, distribution and resource limitation also affect basal metabolic rate, and island species have lower metabolic rates than widely distributed continental species.

We believe that these four chapters testify to the important contributions that research in physiological ecology can bring to our understanding of the functional and evolutionary ecology of bats.

LITERATURE CITED

Brafield, A.E., and M.J. Lewellyn. 1982. Animal Energetics. Chapman and Hall, New York.

Hayssen, V., and R.C. Lacy. 1985. Basal metabolic rates in mammals: taxonomic differences in the allometry of BMR and body mass. Comparative Biochemistry and Physiology A, 81: 741–754.

Lovegrove, B. 2000. The zoogeography of mammalian basal metabolic rate. The American Naturalist, 156: 201–219.

McNab, B.K. 1986. The influence of food habits on the energetics of eutherian mammals. Ecological Monographs, 56: 1–19.

McNab, B.K. 1987. Basal rate and phylogeny. Functional Ecology, 51: 159–167.

McNab, B.K. 1988. Food habits and the basal rate of metabolism in birds. Oecologia, 77: 343–349.

Peters, R.H. 1983. The Ecological Implications of Body Size. Cambridge University Press, New York.

Speakman, J.R., and D.W. Thomas. 2003. Physiological ecology and energetics of bats. Pp. 430–492. Bat Ecology (T.H. Kunz and M.B. Fenton, eds.). University of Chicago Press, Chicago.

1

Energetics, Thermal Biology, and Torpor in Australian Bats

Fritz Geiser

Although most Australian bats have been isolated from bat species in other parts of the world for prolonged periods and may functionally differ, little detailed research has been conducted to determine how Australian bats cope with seasonal and short-term food shortages and adverse environmental conditions. This chapter provides a comparative summary about the limited information on the thermal biology and energetics of Australian bats. The data suggest that, in general, Australian bats are similar in their thermal characteristics and energy use to other bats. Thermal conductance of Australian bats is almost identical to what has been observed in other bat species, although conductance in some tropical taxa is higher than predicted. The basal metabolic rate (BMR) of Australian bats tends to be somewhat below that predicted from allometric equations for bats and, in general, is well below that of placental mammals. However, BMRs of the insectivorous/carnivorous microbats (Microchiroptera) do not appear to differ from those of frugivorous/nectarivorous megabats (Megachiroptera). Torpor appears to be common in Australian bats and has been observed in six of seven families: Pteropodidae (blossom-bats and tube-nosed bats), Emballonuridae (sheathtail bats), Rhinolophidae (horseshoe bats), Hipposideridae (leaf-nosed bats), Vespertilionidae (long-eared and bentwing bats and others), and Molossidae (free-tailed bats). Australian vespertilionids (and likely members of other families) have the ability to enter deep and prolonged torpor in winter (i.e., hibernate) and members of the genus *Nyctophilus* have been observed entering brief bouts of torpor in the field on every day during the resting phase, even in summer. The body temperature (T_b) in some vespertilionids falls to minima between 2 and 5 °C and the metabolic rate (MR) during torpor can be as low as 3–4% of BMR. Small megabats (e.g., blossom-bats) enter daily torpor, their T_b falls to a minimum of 17–23 °C and their MR to about 50% of BMR. Unlike many other species, torpor in blossom-bats is more pronounced in summer than in winter, likely due to the low supply of nectar during the warm season. The low BMR and the high proclivity of Australian bats for using torpor suggest that they are constrained by limited energy availability and that heterothermy plays a key role in their life history. However, more research on families that have received little scientific attention and more fieldwork are needed to establish how Australian bats are functionally adapted to the specific challenges of their local environment.

INTRODUCTION

Bats and rodents are the only terrestrial placental mammals native to Australia. In contrast to monotremes and marsupials, Australian bats apparently do not share a Gondwanan origin and thus probably migrated to Australia from Southeast Asia. All seven families of extant Australian bats appear to be derived from the Oriental bat fauna and reached Australia by island-hopping via Cape York and perhaps the Northern Territory (Hall, 1984). Although about 60% of the approximately 75 Australian bat species are considered to be endemic, endemism is mainly restricted to the Microchiroptera (microbats). Only two of 12 Megachiroptera (megabats) are currently considered to be endemic (Churchill, 1998; Hall, 1984).

Because of the long separation of most Australian bats from those in other parts of the world, some of their physiological characteristics may differ, stemming from specific adaptations to the Australian climate. Although Australia is often (incorrectly) considered to have a mild climate, aridity, unpredictable rainfall, cool to cold winter temperatures, and infertile soils are characteristic of much of the continent. Thus, we might expect Australian bats to exhibit some of the same energetic and thermoregulatory traits that have allowed the endemic marsupials to diversify and persist under the specific constraints of the Australian climate (Lovegrove, 1996). It is known, for example, that Australian microbats as well as small megabats use torpor extensively to reduce energy expenditure (Bartels et al., 1998; Coburn and Geiser, 1998; Geiser et al., 1996; Geiser and Brigham, 2000; Hosken, 1997; Hosken and Withers, 1997; Kulzer et al., 1970; Morrison, 1959, Turbill et al., 2003a). This is most likely due to their large relative surface area, high costs of thermoregulation at low air temperatures (T_a), and fluctuating, unpredictable, or temperature-dependent food supply, as observed in bats from Eurasia and North America (Brigham, 1987; Barclay et al., 2001; Fenton 1983; Henshaw, 1970; Hock, 1951; Kunz and Fenton, 2003; Lyman 1970; Pohl, 1961; Speakman et al., 1991; Speakman and Thomas, 2003; Thomas, 1995; Thomas et al., 1990; Willis and Brigham, 2003). Although many Australian bats live in a relatively warm climate for much of the year, they too must conserve energy because they must cope with unpredictable weather and food resources, often pronounced daily temperature fluctuations, aridity, and other environmental challenges.

In this chapter, available information on basal metabolic rate (BMR), thermal conductance, roosting behavior in relation to energetics, and torpor patterns in Australian bats is summarized and compared with that on bats from other continents, as well as mammals in general.

Relevant data were assembled from the literature and are summarized in tables 1.1 and 1.2. For data on metabolic rate (MR) and body temperature

Table 1.1 Body Temperatures, BMR, and Thermal Conductance in Australian Bats

	Body Mass (g)	T_b (°C)	BMR (mL O_2 $g^{-1}h^{-1}$)	Minimum C (mL O_2 g^{-1} h^{-1} $°C^{-1}$)	References
Pteropodidae					
Common blossom-bat *Syconycteris australis*	18	34.9	1.44	0.121	Geiser et al., 1996; Coburn and Geiser, 1998
Northern blossom-bat *Macroglossus minimus*	16	35.3	1.29	0.17	Bartels et al., 1998
Little red flying-fox *Pteropus scapulatus*	362	37	0.67	0.054	Bartholomew et al., 1964
Grey-headed flying-fox *Pteropus poliocephalus*	598	36	0.53	0.044	Bartholomew et al., 1964
Megadermatidae					
Ghost bat *Macroderma gigas*	107	35.6	0.88	0.114	Baudinette et al., 2000
Hipposideridae					
Orange leaf-nosed bat *Rhinonicteris aurantius*	8	36.1	1.96	0.466	Baudinette et al., 2000
Vespertilionidae					
Gould's wattled bat *Chalinolobus gouldii*	17.5	34	1.44	0.186	Hosken and Withers, 1997
Large bentwing bat *Miniopterus schreibersii*	11	37.7		0.446	Baudinette et al., 2000
Lesser long-eared bat *Nyctophilus geoffroyi*	7	35.7	1.36	0.35	Geiser and Brigham, 2000
Gould's long-eared bat *Nyctophilus gouldi*	10	36	1.22	0.26	Geiser and Brigham, 2000
Greater long-eared bat *Nyctophilus timoriensis*	14	33	1.5	0.187	Hosken, 1997

T_b, body temperature; BMR, basal metabolic rate; C, thermal conductance.

Table 1.2 Torpor in Australian Bats

	Body Mass (g)	Minimum T_b (°C)	Torpor Duration (days)	Minimum TMR (mL O_2 g^{-1} h^{-1})	TMR/BMR	References
Pteropodidae						
Common blossom-bat *Syconycteris australis*	18	17.4	0.34	0.58	0.403	Geiser et al., 1996; Coburn and Geiser, 1998
Northern blossom-bat *Macroglossus minimus*	16	23.1	0.34	0.7	0.543	Bartels et al., 1998
Eastern tube-nosed bat *Nyctimene robinsoni*	50		<1			Hall and Pettigrew, 1995
Emballonuridae						
Coastal sheathtail bat *Taphozous australis*	23	16[a]				Kulzer et al., 1970
Rhinolophidae						
Eastern horseshoe bat *Rhinolophus megaphyllus*	8	16[a]				Kulzer et al., 1970
Hipposideridae						
Orange leaf-nosed bat *Rhinonicteris aurantius*	7	23.6[a]				Kulzer et al., 1970
Vespertilionidae						
Gould's wattled bat *Chalinolobus gouldii*	17.5	5		~0.06	0.042	Hosken and Withers, 1997

Little pied bat *Chalinolobus picatus*	6	12.8[a]				Kulzer et al., 1970
Large bentwing bat *Miniopterus schreibersii*	15	10.5[a]	12			Kulzer et al., 1970; Hall, 1982
Large-footed myotis *Myotis adversus*	8	8[a]	8			Kulzer et al., 1970
Lesser long-eared bat *Nyctophilus geoffroyi*	7	1.4	13	0.037	0.027	Geiser and Brigham, 2000; Turbill et al., 2003c
Gould's long-eared bat *Nyctophilus gouldi*	10	2.3	13	0.052	0.042	Geiser and Brigham, 2000; Turbill et al., 2003c
Greater long-eared bat *Nyctophilus timoriensis*	14	6[a]		0.05	0.033	Hosken, 1997
Inland broad-nosed bat *Scotorepens balstoni*	7	3.2		0.044		Geiser and Brigham, 2000
Eastern forest bat *Vespadelus pumilus*	4	14.5[a,b]	0.25[c]			Turbill et al., 2003a
Inland cave bat *Vespadelus finlaysoni*	5	9.5[a]				Geiser, unpublished
Molossidae						
Little northern freetail bat *Mormopterus loriae*	8.5	10[a]				Kulzer et al., 1970

T_b, body temperature; BMR, basal metabolic rate; TMR, metabolic rate during torpor.
[a]Likely not a regulated minimum body temperature.
[b]T_{skin}.
[c]Measured in summer.

(T_b) during torpor, only values from undisturbed bats, preferably recorded in steady-state torpor are included. Some unpublished information is also provided and the methods used for these measurements were essentially similar to those that were previously described (Geiser and Brigham, 2000; Körtner and Geiser, 2000). Linear regressions were performed using the method of least squares and *t*-tests were performed on residuals to compare physiological variables of micro- and megabats.

BASAL METABOLIC RATE (BMR) AND FIELD METABOLIC RATE (FMR)

The maintenance of vital body functions, such as breathing, heart contractions, cell division, and biosynthesis requires energy. This maintenance metabolic rate (often termed basal metabolic rate, BMR), when measured under specific standardized conditions that exclude energy costs of thermoregulation, movement, growth, and digestion, represents the minimum energy use during normothermia (i.e., at high T_b). Because selection pressures probably set the level of BMR, and because data for more than 10% of living mammals are available in the literature (Lovegrove, 2000), BMR is commonly used for comparative studies among mammals.

The slope and elevation of the regression of BMR on body mass of most Australian bats appears to differ from that of other bats and of placental mammals in general (figure 1.1). Although large Australian bats (body mass >100 g) have BMRs that are almost identical to those predicted from the allometric equation for bats in general (Hayssen and Lacy, 1985), most small Australian bats (body mass 7–18 g) have BMRs that are lower

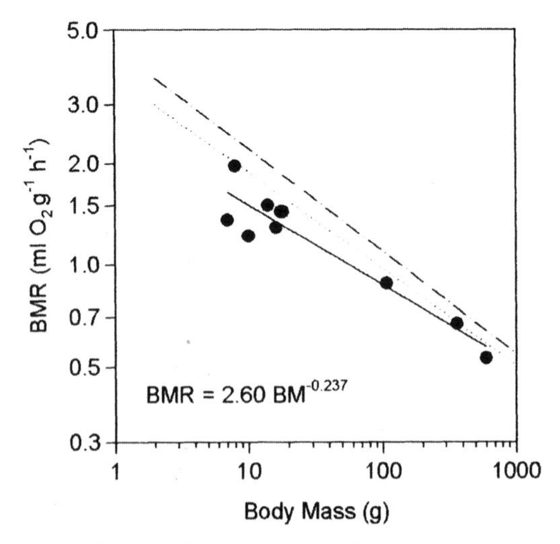

Figure 1.1 Basal metabolic rates (BMR) of Australian bats (continuous lines and filled circles) as a function of body mass (BM) in comparison with all bats (dotted line) and placental mammals (broken line). Regressions for all bats and other placentals are from Hayssen and Lacy (1985).

than predicted. At a body mass of 10 g, the allometric equation for Australian bats predicts a BMR that is only 80% of that for bats in general. There is, however, considerable variation in BMR at a given body size, and some small Australian bats have BMRs that range from that predicted by the allometric equation for bats, to only 65% of the predicted value. These differences do not appear to be caused by diet as suggested by McNab (1969), because the residuals of species calculated from the regression of BMR versus body mass did not differ between the insectivorous/carnivorous microbats and the frugivorous/nectarivorous megabats (t-test, $P = 0.97$). However, compared with other placental mammals, Australian bats had BMRs that were about 68% (10 g) and 88% (500 g) of that predicted by the allometric equation for placentals (Hayssen and Lacy, 1985) and their normothermic resting T_b values (table 1.1) were lower than those of many placentals (Withers, 1992).

Although low BMR may reduce energy expenditure when bats are at rest within the thermoneutral zone (TNZ), BMR is of little consequence to bats that face low ambient temperatures (T_a), especially if they have a high proclivity to enter torpor when not active (see below). Thus, BMR alone should not be used to make predictions about energy expenditure of bats in the wild.

Field metabolic rate (FMR), unlike BMR, provides an integration of all energy expenditures of free-ranging organisms over time, including costs for thermoregulation and locomotion. FMR can be quantified using doubly labeled water turnover, which provides a measure of CO_2 production and thus an estimate of energy expenditure (Kunz and Nagy, 1988; Speakman, 1997). Unfortunately, data on FMR in Australian bats are currently restricted to a single species, the blossom-bat *Syconycteris australis* (18 g). The FMR of *S. australis* is about 7 times BMR (Geiser and Coburn, 1999), representing one of the highest values reported for endotherms to date and being roughly double the value predicted by allometric equations (3.5 times BMR; Degen and Kam, 1995). The likely reason for the high FMR of this blossom-bat is its prolonged periods of activity and flight at night (Law, 1993). This further emphasizes that BMR is a poor predictor for energy expenditure of wild animals, as a low BMR of a species does not preclude a high energy expenditure in the field.

THERMAL CONDUCTANCE AND ROOSTING

Since all bats are relatively small and their surface area/volume ratio is large, heat loss during cold exposure should be substantial. Nevertheless, minimum thermal conductance at rest (the inverse of insulation and a measure of the ease of heat flux; Speakman and Thomas, 2003) of Australian bats is similar to that of similar-sized terrestrial mammals (figure 1.2; Bradley and Deavers, 1980). Thermal conductance also appears to be similar between Australian micro- and megabats, as the residuals of species from the regression did not differ between the two groups (t-test; $P > 0.1$).

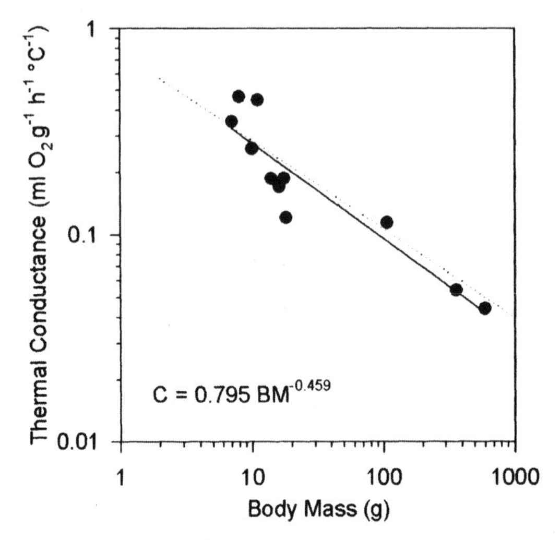

Figure 1.2 Thermal conductance (C) of Australian bats (continuous line and filled circles) as a function of body mass (BM) in comparison with that of mammals in general (dotted line; Bradley and Deavers, 1980).

However, the thermal conductance of a 10 g microbat is about 6 times that of a 500 g megabat simply because of the difference in body size, and thermal conductance of species from the tropics tended to be higher than those from temperate areas. Because of the small size of bats and the resulting heat loss, selection for thermal environments that limit excess heat loss should be important. Many bats show thermal preferences that reflect energy use during normothermia (high T_a) or torpor (low T_a) (Nagel and Nagel, 1991; Brown and Bernard, 1994), although much of this information on bats from other continents has been derived from cave bats.

Australian blossom-bats, *Syconycteris australis*, roost solitarily under leaves and apparently also select roosts according to the local thermal environment (Law, 1993). In autumn and spring, *S. australis* prefers to roost near the center of rainforest patches, a location that is thermally buffered from the high T_as encountered at the edges of patches. In winter, *S. australis* shifts its roosting sites closer to the edge of the rainforest where T_as are warmer than in the center (Law, 1993). Thus, *S. australis* appears to avoid T_a extremes and prefers T_as that are between 18 and 24 °C (Law, 1993). Although these T_as are below the TNZ, they offer bats the ability to exploit the energetic benefits of torpor while not facing T_as that fall below the minimum T_b (Coburn and Geiser, 1998). This thermal selectivity allows them to minimize energy expenditure.

Nyctophilus geoffroyi, which often roosts individually or in small groups, selects exposed positions under bark or shallow fissures in summer (Turbill et al., 2003b). Although these sites appear to provide little buffering of the bats' thermal environment (and little protection from potential predators), roost selection appears to be related to the thermal biology of this species.

Even in summer, *N. geoffroyi* enters torpor on 100% of roost days, but usually arouses near midday, apparently partially offsetting arousal costs by relying on rewarming by the rising T_a (Geiser et al., 2000; Turbill et al., 2003b). Passive rewarming appears more common in mammals than previously thought and has been shown to substantially reduce energy costs of rewarming from torpor (Geiser et al., 2002; Geiser and Drury, 2003; Lovegrove et al., 1999; Mzilikazi et al., 2002). Since temperatures of bat roosts under bark were warmed to several degrees above ambient T_a during the day (Hosken, 1996) it seems likely that the roosts used by *Nyctophilus* play an important part in minimizing energy expenditure, partly by allowing them to exploit the metabolic benefits of torpor while at the same time minimizing arousal costs (Lovegrove et al., 1999).

Miniopterus schreibersii blepotis (syn. *Miniopterus oceanensis*; Reinhold et al., 2000) is a cave bat that may roost in large colonies on the east coast of Australia. In spring and summer, pregnant females form large clusters of several thousand individuals on domed ceilings of caves. Heat flow from these clusters raises the T_a of their immediate surroundings by about 8 °C above that of the cave T_a of 14–15 °C (cluster T_a = 19–23 °C; Dwyer and Harris, 1972). This 8 °C rise in T_a reduces heat loss and thus thermoregulatory costs by about one third (Baudinette et al., 2000). These energy savings may be important especially toward the end of pregnancy (Thompson, 1992). In autumn, during the time of feeding and fattening, the species roosts at the highest T_a available of 19.5 °C near the rear of the cave (Hall, 1982). In winter, when they are fat and undergo multiday torpor bouts of up to 12 days, the bats select cooler areas with T_a of 9.5–11 °C toward the front of the cave (Hall, 1982). Although heat loss and energy expenditure are no doubt important for the selection of specific roosting sites in these bats, the selection of roosting sites may also be based on high humidity, which acts to reduce evaporative water loss (Baudinette et al., 2000; Thomas and Cloutier, 1992; Thomas and Geiser, 1997).

While Australian bats seem to be well equipped to deal with low T_a there is no detailed published information to my knowledge on how Australian bats deal with extreme heat. However, exposure to a T_a of 40 °C substantially increases T_b, resting MR, thermal conductance, and evaporative water loss of Australian bats (Bartholomew et al., 1964; Baudinette et al., 2000). In some bats from other continents, the short-term tolerance of much higher T_a (up to 60 °C) is substantial (Maloney et al., 1999) and it is likely that Australian bat species in hot areas also must be able to survive similar temperatures.

TORPOR

Unlike during normothermia, when endotherms can regulate a more or less constant high T_b by a proportional increase in heat production to compensate for heat loss, torpor involves a controlled reduction of T_b and MR (Speakman and Thomas, 2003). Torpor is widely used by small mammals and birds to reduce energy expenditure during adverse environmental conditions or during food shortage, but also to balance energy use under more favorable conditions or to enhance fat storage for future energy bottlenecks. Torpid animals do not only have reduced costs for thermoregulation, but as MR during torpor is usually well below BMR because of the lowered T_b and, in some species, metabolic inhibition, their maintenance costs are also substantially reduced (Geiser, 1988; Geiser and Brigham, 2000; Humphries et al., 2003). Nevertheless, thermoregulation during torpor is not abandoned, but T_b is regulated at or above a species- or population-specific minimum by an increase in MR (Heller and Hammel, 1972). Torpor has been observed in both microbats and small megabats, and occurs in six of the seven bat families in Australia. Reduced activity in winter (Brigham and Geiser, 1998; Lumsden and Bennett, 1995) suggests that torpor is used extensively by many Australian microbats, especially in the southern part of the continent.

Megabats

Contrary to a widely held view that megabats are strictly homeothermic, torpor has now been observed in three small Australian megabats with a body mass up to about 50 g. Blossom-bats (*Syconycteris australis* and *Macroglossus minimus*) and the tube-nosed bat (*Nyctimene robinsoni*) from tropical and subtropical regions have been observed to enter daily torpor (table 1.2). To date, information on prolonged torpor has not been published for any Australian or other megabat (Bartels et al., 1998; Bartholomew et al., 1970; Geiser et al., 1996); however, it is not unlikely that some are capable of prolonged torpor. Large megabats appear to be homeothermic when studied in the laboratory (Kulzer et al., 1970; Morrison, 1959); however, this does not preclude use of torpor during energy emergencies in the wild.

Both species of the largely nectarivorous Australian blossom-bats (Law, 1994) appear to exclusively use daily torpor, and their T_b and MR are similar to those of many other daily heterotherms (Geiser and Ruf, 1995). In captivity, torpor usually commenced in the morning soon after lights on, and lasted for several hours (figure 1.3). Both species entered induced torpor when food was withheld even at T_a as high as 25 °C, and spontaneous torpor (food ad libitum) was frequently observed at T_a 20 °C. Arousal from torpor in undisturbed bats usually occurred between midday and the afternoon (Geiser et al., 1996; Bartels et al., 1998). Torpor in *S. australis* (18 g) from a subtropical region was deeper than in *M. minimus* (16 g) from a tropical region (figure 1.4). The T_b of torpid *S. australis* fell to a minimum of about 18 °C where it was regulated by an increase in MR, and the minimum MR was about 40% of the BMR. In contrast, the minimum T_b of *M. minimus* was about 23 °C and the minimum MR was about 55% of BMR (Bartels et al., 1998; Geiser et al., 1996). These observations suggest that the depth of torpor and specifically minimum T_b are affected by the climate that a species or population normally encounters through its range, as has been suggested for birds (Wolf and Hainsworth, 1972).

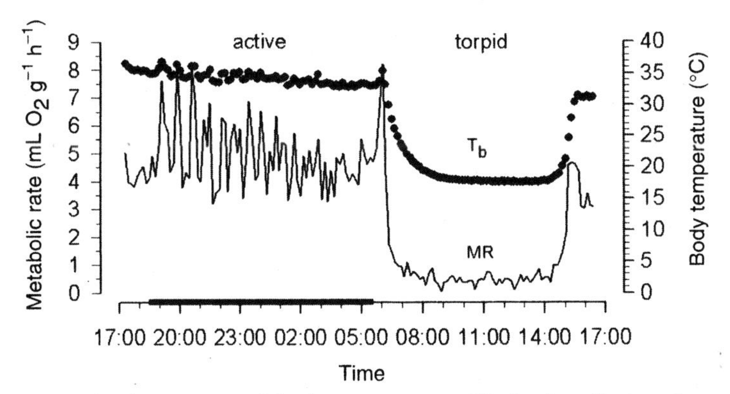

Figure 1.3 Daily fluctuations of body temperature (T_b, broken line) and metabolic rate (MR, continuous line) of a common blossom-bat, *Syconycteris australis*. Torpor entry occurred after lights on in the morning and spontaneous arousal in the afternoon.

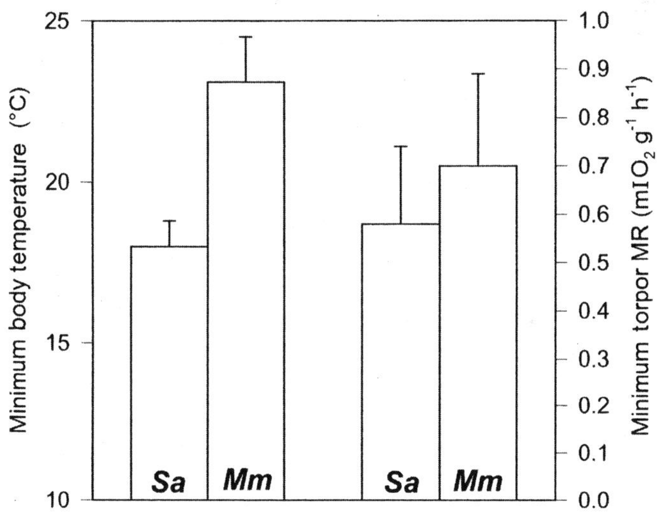

Figure 1.4 A comparison of the minimum body temperature (left) and the minimum metabolic rate (right) during torpor of the common blossom-bat *Syconycteris australis* (*Sa*) from a subtropical habitat and the northern blossom-bat *Macroglossus minimus* (*Mm*) from a tropical habitat.

The mean monthly minimum T_as in the habitat of both bat species is similar to or slightly below their minimum T_b. By matching minimum T_b to minimum T_a, bats can minimize metabolic costs by avoiding the necessity of increasing heat production to stabilize T_b by thermoregulatory heat production when T_a falls below a minimum T_b (Geiser and Kenagy, 1988; Song et al., 2000; Speakman and Thomas, 2003).

Seasonal changes in the use of torpor by *Syconycteris australis*, from the subtropical east coast of New South Wales, were the opposite of those observed for heterothermic mammals from cool regions of the Northern

Hemisphere, which generally show more pronounced and more frequent torpor in winter than in summer (Wang, 1989). In *S. australis* captured in winter, average torpor bout duration was only 5.5 hours and minimum MR during torpor about 55% of BMR. In contrast, in *S. australis* captured in summer, torpor bouts lasted for an average of 7.3 hours and minimum MR was about 35% of the BMR (Coburn and Geiser, 1998). While these findings may seem counterintuitive at first glance, different day length and food availability in summer and winter appear to explain the unusual seasonal response. In winter, bats can forage for prolonged periods during long nights and have access to an abundance of flowering plants. Even in winter, T_a on the New South Wales north coast is relatively mild. In summer, nights and thus foraging times are brief and the availability of nectar is substantially reduced (Coburn and Geiser, 1998). Thus, the unusual seasonal pattern of torpor use in *S. australis* appears to be an appropriate physiological adaptation to ecological constraints of their subtropical habitat. Currently there is no information on the seasonal use of torpor in other subtropical species and it would be of interest to know whether the reversed seasonal pattern observed in *S. australis* occurs in other species and whether it is related to diet.

Microbats

Of six families of Australian microbats only the ghost bats (Megadermatidae), represented by the single species *Macroderma gigas*, appear to be strict homeotherms (Leitner and Nelson, 1967). However, as for large microbats, no field study has been conducted to confirm that ghost bats enter torpor in the wild. Detailed information on torpor is available only for vespertilionids (table 1.2). Much of the information on the other families stems from a comparative study on T_b, bout length, and other observations on torpor by Kulzer et al. (1970). Thus, a substantial amount of research in both the laboratory and the field is required before we can gain some understanding about the occurrence and patterns of torpor of the approximately 65 species of Australian microbats.

Some Australian microbats have the ability to enter deep and prolonged torpor (hibernation). The longest torpor bouts that have been observed are 12 days for *Miniopterus schreibersii*, 8 days for *Myotis adversus* (Hall, 1982; Kulzer et al., 1970), and 2 days in summer and up to 2 weeks in winter for *Nyctophilus* spp. (Turbill et al., 2003b, c). Observations of reduced trap success of bats (Brigham and Geiser, 1998; Lumsden and Bennett, 1995) and reduced activity in winter (Ellis et al., 1991) suggest that prolonged torpor is common in bats living in southern Australia.

The T_b of torpid Australian microbats fell to values between 1.4 and 23.4 °C (table 1.1). However, many of these values are unlikely to represent regulated T_b minima because bats were exposed to mild T_a or were not in steady-state torpor. The minimum T_b of 1–3 °C observed for *Nyctophilus* spp. and *Scotorepens balstoni* were defended during torpor by an increase in MR and these values are similar to those of many microbats from other continents

(Geiser and Brigham, 2000). Similar to the T_b of torpid Australian microbats, the MR during torpor fell to very low values. Minimum MRs during torpor at about T_a 5 °C were 0.5% of that of normothermic bats at the same T_a and about 3–4% of the BMR within the thermoneutral zone (table 1.2). These values are near the mean for hibernating endotherms (Geiser and Ruf, 1995), suggesting that both minimum T_b and MR of torpid Australian microbats are similar to those of Northern Hemisphere microbats and other hibernators in general.

Considering that Australia is perceived to be a warm continent, the low T_b of torpid bats may appear surprising. Why should Australian bats living on a warm continent let T_b fall to such low temperatures? Although most of Australia is warm in summer, more than 80% of the continent experiences regular frost in winter, and the average minimum T_a in winter for most of the southern half of Australia is below 6 °C (Colls and Whitaker, 1993). As tree roosts provide limited thermal buffering, and tree-dwelling bats usually are torpid when T_a is low, it is likely that Australian tree-dwelling bats in many regions regularly experience T_b between 1 and 5 °C. For a 10 g normothermic, resting bat this would require a 7–9 times increase of MR in comparison with BMR.

Although *Nyctophilus* spp. are capable of prolonged torpor, individuals investigated in the laboratory frequently displayed short bouts of torpor and aroused almost daily after sunset. The abrupt signal of lights off in the laboratory at the beginning of the dark phase appears to be a strong trigger for arousals even at low T_a. Nevertheless, although most torpor bouts were shorter than 24 hours, the MR in torpid *Nyctophilus* spp. was similar to that of hibernating species, and only about 10% of that in species that enter daily torpor exclusively and appear unable to undergo prolonged torpor (Geiser and Ruf, 1995). Thus, it appears that although *Nyctophilus* spp. and other Australian (and likely non-Australian) microbats may frequently arouse on a daily basis, their short bouts of torpor do not resemble daily torpor, but functionally appear to be short bouts of hibernation.

In the field during summer, tree-dwelling Australian microbats (*Nyctophilus* and *Vespadelus*) display a high proclivity to re-enter torpor in the late afternoon after daily arousals around midday and before their nocturnal activity phase (Geiser et al., 2000; Turbill et al., 2003a). This may seem especially surprising for *Vespadelus pumilus*, as this species was studied in a subtropical region when food appeared to be abundant and T_a was mild (Turbill et al., 2003a). The exposed roosts allow bats to re-enter torpor and reduce energy expenditure in the afternoon as soon as T_a begins to decline (Turbill et al., 2003b,c). Nevertheless, if bats remained torpid throughout the day, energy expenditure would be substantially lower than when arousing near midday. This suggests that normothermic periods around noon may be related to some other important physiological function, such as digestion of food that was captured in the previous night, or passive rewarming and high T_a for several hours that minimize the energetic costs associated with the process.

Seasonal changes in MR of torpid *N. geoffroyi* were not pronounced (Geiser and Brigham, 2000) and information on this topic from other microbats is not available. As summer data on *Nyctophilus* spp. were similar to those obtained in Northern Hemisphere bats during hibernation in winter, it seems likely that MR of torpid bats shows only small seasonal changes and that the known seasonal changes in torpor patterns and bout duration in bats are to a large extent determined by changes in environmental conditions.

CONCLUSIONS

The present comparison shows that Australian bats are conservative with energy use, especially during their resting phase. As torpor is used frequently and widely, and seems to reduce energy expenditure substantially, it appears that it plays a central role in the biology of these bats and that energy is a limited commodity. However, much of the available information on the ecological physiology of Australian bats is based on data from a few species from essentially two families and field data on physiology are scarce. Thus, it is paramount to study members of other families and to put more emphasis on physiological studies in the field to obtain a better understanding about how bats have adapted to the environmental challenges of the Australian continent.

ACKNOWLEDGMENTS

I wish to thank Michael Barritt, Mark Brigham, Dionne Coburn, Rebecca Drury, Nicola Goodship, Sandy Hamdorf, Gerhard Körtner, Brad Law, Karen May, Bronwyn McAllan, Chris Pavey, Chris Turbill, and Wendy Westman for their help with various aspects of this work, and Don Thomas for constructive comments on the manuscript. The work was supported by a grant from the Australian Research Council to the author.

LITERATURE CITED

Barclay, R.M.R., C.L. Lausen, and L. Hollis. 2001. What's hot and what's not: defining torpor in free-ranging birds and mammals. Canadian Journal of Zoology, 79: 1885–1890.

Bartels, W., B.S. Law, and F. Geiser. 1998. Daily torpor and energetics in a tropical mammal, the northern blossom-bat *Macroglossus minimus* (Megachiroptera). Journal of Comparative Physiology B, 168: 233–239.

Bartholomew, G.A., P. Leitner, and J.E. Nelson. 1964. Body temperature, oxygen consumption, and heart rate in three species of Australian flying foxes. Physiological Zoology, 37: 179–198.

Bartholomew, G.A., W.R. Dawson, and R.C. Lasiewski. 1970. Thermoregulation and heterothermy in some of the smaller flying foxes (Megachiroptera) of New Guinea. Zeitschrift für vergleichende Physiologie, 70: 196–209.

Baudinette, R.V., S.K. Churchill, K.A. Christian, J.E. Nelson, and P.J. Hudson. 2000. Energy, water balance, and roost environment in three Australian cave-dwelling bats. Journal of Comparative Physiology B, 170: 439–446.

Bradley, S.R., and D.R. Deavers. 1980. A re-examination of the relationship between thermal conductance and body weight in mammals. Comparative Biochemistry and Physiology A, 65: 465–476.

Brigham, R.M. 1987. The significance of winter activity by the big brown bat (*Eptesicus fuscus*): the influence of energy reserves. Canadian Journal of Zoology, 65: 1240–1242.

Brigham, R.M., and F. Geiser. 1998. Seasonal activity patterns of *Nyctophilus* bats based on mist-net captures. Australian Mammalogy, 20: 349–352.

Brown, C.R. and R.T.F. Bernard. 1994. Thermal preference of Schreiber's long-fingered (*Miniopterus schreibersii*) and Cape horseshoe (*Rhinilophus capensis*) bats. Comparative Biochemistry and Physiology A, 107: 439–449.

Churchill, S. 1998. Australian Bats. Reed, Sydney.

Coburn, D.K., and F. Geiser. 1998. Seasonal changes in energetics and torpor patterns in the subtropical blossom-bat *Syconycteris australis* (Megachiroptera). Oecologia, 113: 467–473.

Colls, K., and R. Whitaker. 1993. The Australian Weather Book. National Book Distributers, Brookvale, Australia.

Degen, A.A., and M. Kam. 1995. Scaling of field metabolic rate to basal metabolic rate ratio in homeotherms. Ecoscience, 2: 48–54.

Dwyer, P.D., and J.A. Harris. 1972. Behavioral acclimatization to temperature by pregnant *Miniopterus* (Chiroptera). Physiological Zoology, 45: 14–21.

Fenton, M.B. 1983. Just Bats. University of Toronto Press, Toronto.

Ellis, W.A., T.G. Marples, and W.R. Phillips. 1991. The effect of a temperature-determined food supply on the annual activity cycle of the lesser long-eared bat, *Nyctophilus geoffroyi* Leach, 1921 (Microchiroptera: Vespertilionidae). Australian Journal of Zoology, 39: 263–271.

Geiser, F. 1988. Reduction of metabolism during hibernation and daily torpor in mammals and birds: temperature effects or physiological inhibition? Journal of Comparative Physiology B, 158: 25–37.

Geiser, F., and G.J. Kenagy. 1988. Torpor duration in relation to temperature and metabolism in hibernating ground squirrels. Physiological Zoology, 61: 442–449.

Geiser, F., and T. Ruf. 1995. Hibernation versus daily torpor in mammals and birds: physiological variables and classification of torpor patterns. Physiological Zoology, 68: 935–966.

Geiser, F., and D.K. Coburn. 1999. Field metabolic rates and water uptake in the blossom-bat *Syconycteris australis* (Megachiroptera). Journal of Comparative Physiology B, 169: 133–138.

Geiser, F., and R.M. Brigham. 2000. Torpor, thermal biology, and energetics in Australian long-eared bats (*Nyctophilus*). Journal of Comparative Physiology B, 170: 153–162.

Geiser, F., and R.L. Drury. 2003. Radiant heat affects thermoregulation and energy expenditure during rewarming from torpor. Journal of Comparative Physiology B, 173: 55–60.

Geiser, F., N. Goodship, and C.R. Pavey. 2002. Was basking important in the evolution of mammalian endothermy? Naturwissenschaften, 89: 412–414.

Geiser, F., D.K. Coburn, G. Körtner, and B.S. Law. 1996. Thermoregulation, energy metabolism, and torpor in blossom-bats, *Syconycteris australis* (Megachiroptera). Journal of Zoology (London), 239: 583–590.

Geiser, F., J.C. Holloway, G. Körtner, T.A. Maddocks, C. Turbill, and R.M. Brigham. 2000. Do patterns of torpor differ between free-ranging and captive mammals and birds? Pp. 95–102. In: Life in the Cold: 11th International Hibernation Symposium (G. Heldmaier and M. Klingenspor, eds.). Springer, Berlin.

Hall, L.S. 1982. The effect of cave microclimate on winter roosting behaviour in the bat, *Miniopterus schreibersii blepotis*. Australian Journal of Ecology, 7: 129–136.

Hall, L.S. 1984. And then there were bats. Pp. 837–852. In: Vertebrate Zoogeography and Evolution in Australasia (M. Archer and G. Clayton, eds.). Hesperian Press, Carlisle, Western Australia.

Hall, L.S., and J. Pettigrew. 1995. The bat with the stereo nose. Australian Natural History, 24: 26–28.

Hayssen, V., and R.C. Lacy. 1985. Basal metabolic rates in mammals: taxonomic differences in the allometry of BMR and body mass. Comparative Biochemistry and Physiology A, 81: 741–754.

Heller, H.C., and H.T. Hammel. 1972. CNS control of body temperature during hibernation. Comparative Biochemistry and Physiology A, 41: 349–359.

Henshaw, R.E. 1970. Thermoregulation in bats. Pp. 188–232. In: About Bats (B.H. Slaughter and D.W. Walton, eds.). Southern Methodist University Press, Dallas, TX.

Hock, R.J. 1951. The metabolic rates and body temperatures of bats. Biological Bulletin, 101: 289–299.

Hosken, D.J. 1996. Roost selection by the lesser long-eared bat, *Nyctophilus geoffroyi*, and the greater long-eared bat, *N. major* (Chiroptera: Vespertilionidae), in *Banksia* woodlands. Journal of the Royal Society of Western Australia, 79: 211–216.

Hosken, D.J. 1997. Thermal biology and metabolism of the greater long-eared bat, *Nyctophilus major* (Chiroptera: Vespertilionidae). Australian Journal of Zoology, 45: 145–156.

Hosken, D.J., and P.C. Withers. 1997. Temperature regulation and metabolism of an Australian bat, *Chalinolobus gouldii* (Chiroptera: Vespertilionidae), when euthermic and torpid. Journal of Comparative Physiology B, 167: 71–80.

Humphries, M.M., D.W. Thomas, and D.L. Kramer. 2003. The role of energy availability in mammalian hibernation: a cost-benefit approach. Physiological and Biochemical Zoology, 76: 165–179.

Körtner, G., and F. Geiser. 2000. Torpor and activity patterns in free-ranging sugar gliders *Petaurus breviceps* (Marsupialia). Oecologia, 123: 350–357.

Kulzer, E., J.E. Nelson, J. McKean, and F.P. Möhres. 1970. Untersuchungen über die Temperaturregulation australischer Fledermäuse (Microchiroptera). Zeitschrift für vergleichende Physiologie, 69: 426–451.

Kunz, T.H., and K.A. Nagy. 1988. Methods of energy budget analysis. Pp. 277–302. In: Ecological and Behavioral Methods for the Study of Bats (T.H. Kunz, ed.). Smithsonian Institution Press, Washington, DC.

Kunz, T.H., and M.B. Fenton (eds.). 2003. Bat Ecology. University of Chicago Press, Chicago.

Law, B.S. 1993. Roosting and foraging ecology of the Queensland blossom bat (*Syconycteris australis*) in north-eastern New South Wales: flexibility in response to seasonal variation. Wildlife Research, 20: 419–431.

Law, B.S. 1994. *Banksia* nectar and pollen: dietary items affecting the abundance of the common blossom bat, *Syconycteris australis*, in south-eastern Australia. Australian Journal of Ecology, 19: 425–434.

Leitner, P., and J.E. Nelson. 1967. Body temperature, oxygen consumption and heart rate in the Australian false vampire bat, *Macroderma gigas*. Comparative Biochemistry and Physiology, 21: 65–74.

Lovegrove, B.G. 1996. The low basal metabolic rates of marsupials: the influence of torpor and zoogeography. Pp. 141–151. In: Adaptations to the Cold: 10th International Hibernation Symposium (F. Geiser, A.J. Hulbert, and S.C. Nicol, eds.). University of New England Press, Armidale, Australia.

Lovegrove, B.G. 2000. The zoogeography of mammalian basal metabolic rate. The American Naturalist, 156: 201–219.

Lovegrove, B.G., G. Körtner, and F. Geiser. 1999. The energetic cost of arousal from torpor in the marsupial *Sminthopsis macroura*: benefits of summer ambient temperature cycles. Journal of Comparative Physiology B, 169: 11–18.

Lumsden, L.F., and A.F. Bennett. 1995. Bats of a semi-arid environment in south-eastern Australia: biogeography, ecology and conservation. Wildlife Research, 22: 217–240.

Lyman, C.P. 1970. Thermoregulation and metabolism in bats. Pp. 301–330. In: Biology of Bats, Volume I (W.A. Wimsatt, ed.). Academic Press, New York.

Maloney, S.K., G.N. Bronner, and R. Buffenstein. 1999. Thermoregulation in the Angolan free-tailed bat *Mops condylurus*: a small mammal that uses hot roosts. Physiological and Biochemical Zoology, 72: 385–396.

McNab, B.K. 1969. The economics of temperature regulation in neotropical bats. Comparative Biochemistry and Physiology, 31: 227–268

Morrison, P. 1959. Body temperatures in some Australian mammals. I. Chiroptera. Biological Bulletin, 116: 484–497.

Mzilikazi, N., B.G. Lovegrove, and D.O. Ribble. 2002. Exogenous passive heating during torpor arousal in free-ranging elephant shrews, *Elephantulus myurus*. Oecologia, 133: 307–314.

Nagel, A., and R. Nagel. 1991. How do bats choose optimal temperatures for hibernation? Comparative Biochemistry and Physiology A, 99: 323–326.

Pohl, H. 1961. Temperaturregulation und Tagesperiodik des Stoffwechsels bei Winterschläfern. Zeitschrift für vergleichende Physiologie, 45: 109–153.

Reinhold, L., T.B. Reardon, and M. Lara. 2000. Molecular and morphological systematics of the Australo-Papuan *Miniopterus* (Chiroptera: Vespertilionidae). 9th Australasian Bat Conference, Abstract, p. 60.

Song, X., G. Körtner, and F. Geiser. 2000. Temperature selection and energy expenditure in the marsupial hibernator *Cercartetus nanus*. Pp. 119–126. In: Life in the Cold: 11th International Hibernation Symposium (G. Heldmaier and M. Klingenspor, eds.). Springer, Berlin.

Speakman, J.R. 1997. Doubly Labelled Water: Theory and Practice. Chapman and Hall, London.

Speakman, J.R., P.I. Webb, and P.A. Racey. 1991. Effects of disturbance on the energy expenditure of hibernating bats. Journal of Applied Ecology, 28: 1087–1104.

Speakman, J.R., and D.W. Thomas. 2003. Physiological ecology and energetics of bats. Pp. 430–490. In: Bat Ecology (T.H. Kunz and M.B. Fenton, eds.). University of Chicago Press, Chicago.
Thomas, D.W. 1995. The physiological ecology of hibernation in vespertilionid bats. Symposium of the Zoological Society of London 67: 233–244.
Thomas, D.W., and D. Cloutier. 1992. Evaporative water loss by hibernating little brown bats, *Myotis lucifugus*. Physiological Zoology, 65: 443–456.
Thomas, D.W., and F. Geiser. 1997. Periodic arousals in hibernating mammals: is evaporative water loss involved? Functional Ecology, 11: 585–591.
Thomas, D.W., D. Cloutier, and D. Gagné. 1990. Arrhythmic breathing, apnea, and non-steady-state oxygen uptake in hibernating little brown bats (*Myotis lucifugus*). Journal of Experimental Biology, 149: 395–406.
Thompson, S.D. 1992. Gestation and lactation in small mammals: basal metabolic rate and the limits of energy use. Pp. 213–259. In: Mammalian Energetics (T.H. Tomasi and T.H. Horton, eds.). Cornell University Press, Ithaca, NY.
Turbill, C., B.S. Law, and F. Geiser. 2003a. Summer torpor in a free-ranging bat from subtropical Australia. Journal of Thermal Biology, 28: 223–226.
Turbill, C., G. Körtner, and F. Geiser. 2003b. Natural use of torpor by a small, tree-roosting bat during summer. Physiological and Biochemical Zoology, 78: 868–876.
Turbill, C., G. Körtner, and F. Geiser. 2003c. Daily and annual patterns of torpor by tree-roosting microbats. Comparative Biochemistry and Physiology A, 134, Suppl. 1: S93.
Wang, L.C.H. 1989. Ecological, physiological, and biochemical aspects of torpor in mammals and birds. Pp. 361–401. In: Advances in Comparative and Environmental Physiology, Volume 4 (L.C.H. Wang, ed.). Springer, Berlin.
Willis, C.K.R., and R.M. Brigham. 2003. Defining torpor in free-ranging bats: experimental evaluation of external temperature-sensitive radiotransmitters and the concept of active temperature. Journal of Comparative Physiology B, 173: 379–389.
Withers, P.C. 1992. Comparative Animal Physiology. W.B. Saunders, Fort Worth, TX.
Wolf, L.L., and F.R. Hainsworth. 1972. Environmental influence on regulated body temperature in torpid hummingbirds. Comparative Biochemistry and Physiology A, 41: 167–173.

2

Temperature, Hibernation Energetics, and the Cave and Continental Distributions of Little Brown Myotis

Murray M. Humphries, John R. Speakman, & Donald W. Thomas

Hibernation permits endotherms to survive prolonged periods of cold temperatures and reduced food supply through a combination of energy storage, microhabitat selection, and metabolic reduction. For many mammals, predictable thermal relationships define both the length of the hibernation period and the level of energy expenditure during hibernation, facilitating the ability to make precise predictions about the energetic consequences of microhabitat and latitudinal temperature gradients. Here we develop a quantitative model predicting the effect of ambient temperature on the hibernation energetics of little brown myotis (*Myotis lucifugus*) and compare predictions of the model with the observed distribution of bats across thermal gradients within caves and across the North American continent. Our model predicts pronounced effects of ambient temperature on total winter energy requirements and a relatively narrow combination of hibernaculum temperatures and lengths of winter that permit successful hibernation. Empirical distributional patterns of *M. lucifugus* correspond closely to these predictions, suggesting that temperature effects on hibernation energetics severely constrain the distribution of bats within hibernacula at the northern limit of their range.

INTRODUCTION

Endothermic organisms generally maintain a constant, elevated body temperature (T_b) independent of ambient temperature (T_a), but their activity and energy use remain highly temperature-dependent (Speakman, 2000). Small endotherms in temperate and arctic zones face a dual problem in winter when increased thermoregulatory requirements coincide with reduced food availability. The adaptations that permit persistence during these periods generally involve some form of avoidance of the most extreme energetic stresses (King and Murphy, 1985). Migration (Gwinner, 1990) and occupation of thermally buffered microhabitats (Walsberg, 1985) can be thought of as spatial avoidance, whereas energy storage (Blem, 1990) and metabolic depression (Hochachka and Guppy, 1987) represent temporal avoidance. Hibernation is arguably the most extreme energetic adaptation used by endotherms (Lyman et al., 1982; Nedergaard and Cannon, 1990) because it combines at least three and sometimes all four of the avoidance strategies: substantial energy storage, prolonged dormancy, precise microhabitat selection, and in some populations, migration to locations suitable for hibernation (McNab, 1982).

The energy savings associated with hibernation are sufficiently pronounced that small endotherms, capable of storing enough fat to support euthermic energy requirements for only a few days, can survive an entire winter solely on stored energy (Lyman et al., 1982). This is accomplished by allowing T_b to drop to near-ambient levels, which reduces metabolic rate to less than 10% of euthermic levels (Geiser, 1988; Heldmaier and Ruf, 1992). Even during hibernation, however, endotherms actively regulate metabolism and T_b, alternating between brief bouts of temperature regulation at elevated, euthermic levels and prolonged bouts of regulation while in torpor (Lyman et al., 1982; Nedergaard and Cannon, 1990). As a result, both activity (in terms of the proportion of time spent euthermic) and energy requirements (while euthermic and in torpor) remain highly temperature-dependent.

The severity of the energetic constraints faced by hibernators and the simplicity of the thermal relationships that define winter energy consumption, facilitate the prediction of what microhabitats and geographical localities should permit successful hibernation. The considerable complexity normally involved in modeling animal energetics (e.g., Moen et al., 1997) is reduced to three simple factors: the size of the energy store at the onset of hibernation, the rate at which the energy store is depleted during winter, and the length of the winter. If reserve size is less than depletion rate times length of winter, the hibernator will not survive. In this chapter, we construct a model that predicts the quantitative relationship between ambient temperature and total hibernation expenditure, then combine this with estimates of length of winter and the maximum size of fat stores to predict under what thermal conditions successful hibernation should be possible.

Our modeling approach could be applied to any hibernating species, but we focus on the common North American little brown myotis, *Myotis lucifugus*. This small (*c*. 8 g), insectivorous species hibernates in caves, where a range of temperatures is available, and has a geographic range that extends from the northern tropics to the Arctic (Fenton and Barclay, 1980). During winter, *M. lucifugus* does not feed (Whitaker and Rissler, 1993) but relies completely on stored body fat (Kunz et al., 1998). In combination with small body size, an aerial mode of locomotion, and very long winters in the northern part of its range, this reliance on fat stores to fuel metabolism results in severe constraints on the amount of energy available for hibernation (Blem, 1990). We base our model on the well-quantified hibernation energetics of this species, and compare the model's predictions with empirical data on the microhabitat and biogeographical distribution of hibernating *M. lucifugus*. Specifically, we evaluate whether the hibernaculum temperatures that minimize winter energy requirements are selected, and the possibility that hibernation energetics constrain the geographical distribution of this species.

THE MODEL

The energy expenditure of a euthermic endotherm, E_{eu}, varies with T_a, according to a well-described metabolic response curve (figure 2.1A, continuous line; Scholander et al., 1950; Speakman and Thomas, 2003). Within the range of thermoneutrality, an inactive animal expresses basal metabolic rate, BMR. When ambient temperature declines below the lower critical temperature, T_{lc}, expenditure increases according to

$$E_{eu} = \text{BMR} + (T_{lc} - T_a)C_{eu} \quad (1)$$

where C_{eu}, is euthermic thermal conductance, a measure of the ease with which heat flows between the animal and surrounding air.

However, when T_a drops far below T_{lc}, many endotherms are capable of expressing adaptive hypothermia (Lyman et al., 1982; Geiser and Ruf, 1995). The resulting torpid state is associated with substantial metabolic savings related to Q_{10} effects and/or active metabolic inhibition (Geiser, 1988; Heldmaier and Ruf, 1992). During torpor, metabolic rate, TMR, and body temperature, $T_{b\text{-}tor}$, decline with T_a until a lower setpoint temperature, $T_{tor\text{-}min}$, is reached, after which torpor T_b is defended (i.e., remains constant) and consequently TMR increases (figure 2.1A, dotted line). Thus, TMR varies with temperature according to

$$\begin{aligned} E_{tor} &= \text{TMR}_{min} Q_{10}^{(T_a - T_{tor\text{-}min})/10}, \quad \text{if } T_a > T_{tor\text{-}min} \\ E_{tor} &= \text{TMR}_{min} + (T_{tor\text{-}min} - T_a)\, C_t, \quad \text{if } T_a \leq T_{tor\text{-}min} \end{aligned} \quad (2)$$

where Q_{10} represents the change in the metabolism of torpid bats resulting from a 10 °C change in T_a, and C_t represents conductance below $T_{tor\text{-}min}$.

Hibernating mammals remain inactive for a prolonged winter period during which no energy is acquired through foraging (Lyman et al., 1982; Nedergaard and Cannon, 1990). Despite this, hibernators arouse from torpor at regular intervals during winter, remaining euthermic for a short time before re-entering torpor. The reasons for these brief arousals are poorly understood but they are characteristic of all species studied to date and account for a disproportionate amount of total winter energy expenditure. The energetic cost of arousal, E_{ar}, is primarily a function of the required increase in T_b from $T_{b\text{-}tor}$ to euthermic levels, $T_{b\text{-}eu}$, and the specific heat

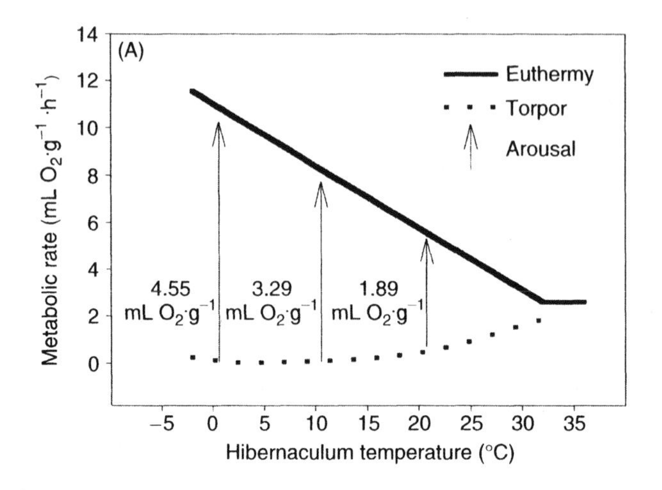

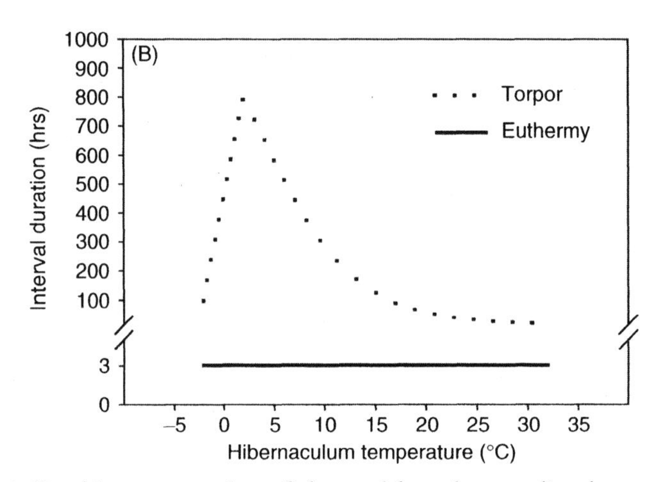

Figure 2.1 Graphic representation of the model used to predict the consequences of ambient temperature on energy requirements of *M. lucifugus* during hibernation. (A) The effect of temperature on metabolism while in torpor, during arousals, and when euthermic. (B) The effect of temperature on the duration of torpor and euthermic intervals.

capacity, S, of the hibernator's tissues (Speakman and Thomas, 2003; Thomas et al., 1990; but see Geiser and Baudinette, 1990):

$$E_{ar} = (T_{b\text{-}eu} - T_{b\text{-}tor})S \tag{3}$$

Thus, the energetic cost of a complete torpor–arousal cycle, E_{bout}, is

$$E_{bout} = E_{eu}(t_{eu}) + E_{tor}(t_{tor}) + E_{ar} \tag{4}$$

where t_{eu} and t_{tor} are the times spent euthermic and torpid. E_{bout} could plausibly increase or decrease with a decrease in ambient temperature.

Low ambient temperatures will reduce the energy expended while in torpor but at the same time increase the energetic costs of arousals and euthermy.

The length of torpor bouts varies dramatically with ambient temperature (figure 2.1B), presumably because of the thermal dependency of TMR (Brack and Twente, 1985; Twente et al., 1985; Geiser and Kenagy, 1988). Thus, t_{tor} falls from maximum levels, $t_{tor\text{-}max}$, above and below $T_{tor\text{-}min}$ in accordance with increases in E_{tor} (equation 2) such that

$$t_{tor} = t_{tor\text{-}max} Q_{10}^{-(t_a - t_{tor\text{-}min}/10)}, \quad \text{if } T_a > T_{tor\text{-}min}$$
$$t_{tor} = t_{tor\text{-}max} - (T_a - T_{tor\text{-}min})k, \quad \text{if } T_a < T_{tor\text{-}min} \qquad (5)$$

where k is an analytical constant that yields equal values of t_{tor} for a given TMR above and below $T_{tor\text{-}min}$.

Finally, total energy requirements for hibernation, E_{winter}, can be predicted for a given winter length, t_{winter}, according to

$$E_{winter} = (t_{winter}/t_{bout})E_{bout} \qquad (6)$$

where t_{bout} is the temperature-dependent duration of a torpor–arousal cycle.

Although our model is applicable to mammalian hibernators in general, for the purposes of this chapter we apply it only to *M. lucifugus*. Individual bats of this species hibernate in loose aggregations in natural or artificial caves for 4–7 months per year. Although they do not feed during winter arousals (Whitaker and Rissler, 1992), some mating activity may occur (Thomas et al., 1979). Fat stores for hibernation are developed during late summer and autumn, particularly during a pre-hibernation swarming period that consists of active foraging and intermittent occupancy of hibernation caves (Kunz et al., 1998). The size of fat stores at the onset of hibernation differs widely between individuals, but generally does not exceed 3.5 g (Kunz et al., 1998).

Model values for hibernation energetics in *M. lucifugus* were taken from the literature (table 2.1). For cave-dwelling, insectivorous bats, relatively simple thermal relationships determine the length of the hibernation period and the maximum cave temperatures available for hibernation. Because *M. lucifugus* is an obligate, nocturnal insectivore (Fenton and Barclay, 1980) and insect activity is inhibited by freezing air temperatures (Rainey, 1976), the minimum length of the hibernation period should approximate the period when average nightly minimum temperatures are less than 0 °C. Thus, climatological data from weather stations (Canadian Meteorological Centre, Climate and Water Information, Environment Canada; National Climatic Data Center, National Oceanic and Atmospheric Administration, USA) can be used to predict how the duration of the hibernation period for *M. lucifugus* will vary across its wide geographic distribution. Similarly, although the temperature profiles of caves vary according to internal geometry, deep cave temperatures generally approximate average annual temperature while regions nearer to the entrance track winter temperatures more closely (Dwyer, 1971; Richter et al., 1993). As a result, although the coldest sites available for hibernation are much less than annual average temperatures, the warmest sites exceed the annual average by only a few degrees, excepting the case of geothermally heated caves (Bell et al., 1986; Parker et al., 1997). We calculated average annual temperature for each region, and assumed that maximum cave temperatures available in a given region could exceed this average by 2 °C.

Table 2.1 Parameter Values Used to Predict the Effect of Ambient Temperature on Hibernation Energy Requirements for *Myotis lucifugus*

Parameter	**Value**	**Units**	**References**
Resting metabolic rate (RMR)	2.6	$mLO_2g^{-1}h^{-1}$	Stones and Wiebers, 1967
Minimum torpor metabolic rate (TMR_{min})	0.03	$mLO_2g^{-1}h^{-1}$	Hock, 1951
Euthermic body temperature (T_{eu})	35	°C	Thomas et al., 1990
Lower critical temperature (T_{lc})	32	°C	Stones and Wiebers, 1967
Torpor setpoint temperature ($T_{tor\text{-}min}$)	2	°C	Hock, 1951
Euthermic conductance (C_{eu})	0.2638	$mLO_2g^{-1}h^{-1}$	Stones and Wiebers, 1967
Q_{10}	$1.6 + 0.26T_a - 0.006T_a^2$		Hock, 1951[a]
Torpor conductance (C_t)	0.055	$mLO_2g^{-1}h^{-1}$	Hock, 1951
Specific heat capacity of tissue (S)	0.131	$mLO_2g^{-1}{}^{\circ}C^{-1}$	Thomas et al., 1990
Constant (k)	173	$h^{\circ}C^{-1}$	Calculated[b]
Maximum torpor duration ($t_{tor\text{-}max}$)	792	h	Brack and Twente, 1985
Arousal duration (t_{ar})	0.75	h	Thomas et al., 1990
Euthermic duration (t_{eu})	3	h	French, 1985
Winter duration (t_{winter})	4632	h	Fenton, 1970

[a]Polynomial equation permits Q_{10} to vary with T_a as reported by Hock (1951).
[b]Matches $t_{tor} = f(E_{tor})$ above and below $T_{tor\text{-}min}$.

WHAT THE MODEL TELLS US ABOUT HIBERNATING *M. LUCIFUGUS*

Temperature and the Energetics of Hibernation

Our model predicts that energy requirements for hibernation will be minimized at 2 °C and increase sharply with either an increase or decrease in hibernaculum temperature (figure 2.2A). This prediction is insensitive to

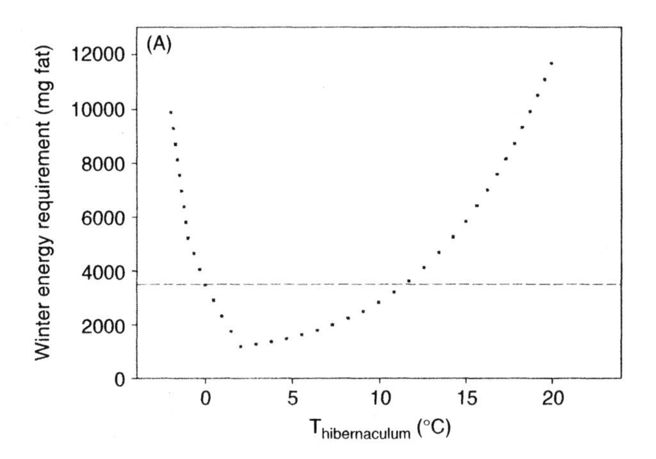

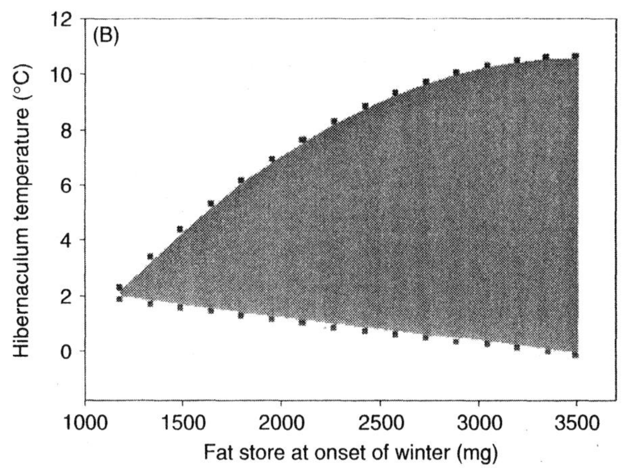

Figure 2.2 (A) The effect of ambient temperature on total winter energy requirements of *M. lucifugus*, based on a winter length of 193 days (Fenton, 1970). (B) The predicted range of suitable hibernaculum temperature is a function of the size of the fat store at the onset of winter, assuming a winter length of 193 days. Temperatures above or below the filled region result in metabolic rates that deplete fat stores prior to spring, resulting in overwinter mortality. Thus, an individual that initiates hibernation with a fat store of only 1200 mg can survive only if it finds a site offering an ambient temperature of 2 °C. An individual having a 3500 mg fat store is able to hibernate successfully at sites providing an ambient temperate anywhere within the range of 0.1 °C to 11 °C.

Table 2.2 Sensitivity of Model Predictions to Assumed Duration of *M. lucifugus* Euthermic Bouts

t_{eu} (h)	Optimal T_a (°C)	Min. E_{total}(g fat)[a]	Potential T_a Range (°C)
3	2.0	1.2	0.0–12.9
6	2.0	1.8	0.7–9.9
15	2.0	3.6	–

[a]Assuming winter duration of 193 days (Fenton, 1970).

large variations in the duration of euthermic intervals (table 2.2). That total energy requirements are minimized at $T_{tor\text{-}min}$, where torpor expenditure is minimized and torpor bout length is maximized, reflects the fact that bats spend the overwhelming proportion of winter time in torpor. This reflects the relative economy of their arousals at all temperatures (compared with other, larger bodied hibernators), and the tight temperature-dependency of torpor metabolism and bout length. Energy requirements should increase sharply below 2 °C because the energy costs of torpor, arousal, and euthermy all increase as temperature decreases. The increase is less marked above 2 °C because increases in torpor requirements are not as pronounced and are partially compensated by reductions in the cost of arousal and euthermy.

Given a typical winter length in the northern portion of this species' range (193 days; Fenton, 1970), the temperature-dependency of hibernation energetics should severely limit the potential range of temperatures where successful hibernation is possible and, within this range, have pronounced effects on the extent of fat store depletion. Hibernating at 2 °C for 193 days requires only 1.2 g of winter fat depletion, but temperatures only 3 °C lower or 11 °C higher increase energy requirements threefold, beyond the 3.5 g level, and therefore successful hibernation becomes unlikely (figure 2.2A). Thus, a bat entering hibernation with only 1.2 g of body fat requires a hibernaculum temperature between 1.7 and 2.9 °C to survive the winter, while an individual with 3.5 g of fat could successfully hibernate at any site between 0.1 and 11.0 °C (figure 2.2B). These predictions are, of course, highly sensitive to the actual duration of euthermic intervals T_{eu} (table 2.2).

Climate-based estimates for the onset, termination, and total duration of hibernation at different latitudes correspond well with dates and durations reported in the literature (table 2.3). The onset and termination of the hibernation period at a given cave involve a gradual, month-long, increase or decrease in the number of torpid bats occupying the cave, and are influenced by weather and individual body condition (Kunz et al., 1998). Thus, there is considerable variability in the duration of the hibernation period for different individuals, and likely in different years. However, our assumption that hibernation spans the period when average nightly minimum T_a is less that 0 °C appears to provide a reasonable approximation of how the minimum length of the hibernation period varies with latitude.

Table 2.3 Comparison of Observed Timing of *M. lucifugus* Hibernation with Dates Predicted from Occurrence of 0 °C Daily Minima

Parameter	**Latitude, Longitude**	**Predicted**	**Observed**	**Average Error (days)**	**References**
Immergence	43° N, 73° W	Oct. 8	Oct. 7	−1	Kunz et al., 1998
	52° N, 118° W	Sep. 23	Oct. 6	+13	Schowalter, 1980
Emergence	39° N, 87° W	Mar. 17	Mar. 16	+1	Whitaker and Rissler, 1992
t_{winter} (days)	40° N, 75° W	128	134	−6	Raesly and Gates, 1986
	45° N, 78° W	168	178	−10	Fenton, 1970
	47° N, 86° W	194	193	1	Fenton, 1970
Average				−1	

Longer winters narrow the range of ambient temperatures for which a given fat store will permit survival to spring. Again assuming winter survival becomes unlikely when hibernation requirements exceed 3.5 g of fat, we can predict the combinations of length of winter and hibernaculum temperature for which successful hibernation should and should not be possible (figure 2.3). Reported hibernaculum temperatures for *M. lucifugus* indicate that bats hibernate across a relatively wide temperature range within a geographic location, but this range does not extend to regions predicted unsuitable by the model (figure 2.3).

Surviving winter at high latitudes could require more or less energy than surviving winter at lower latitudes, depending on the relative energetic benefits of cool hibernacula and energetic costs of longer winters. The pronounced increase in energy requirements with small changes in T_a means that, generally, requirements will actually be less in colder localities. For example, the hibernation period in southern Ontario, Canada is 40% longer than in southern Missouri, USA, but total winter energy requirements are 26% *less* because hibernaculum temperatures are 6 °C cooler (table 2.4). It is, therefore, not surprising that some southern populations of *Myotis* spp. migrate north to hibernation sites (McNab, 1982). However, farther north, where annual average temperatures are below 2 °C ($T_{tor\text{-}min}$), predicted energy requirements rise sharply because winters are long, and costs of torpor, arousal, and euthermy all increase with declining temperature.

Temperature and Energetics Constrain Distribution

The pronounced increase in energy requirements at temperatures below the setpoint (figure 2.2A), combined with the high predictability of

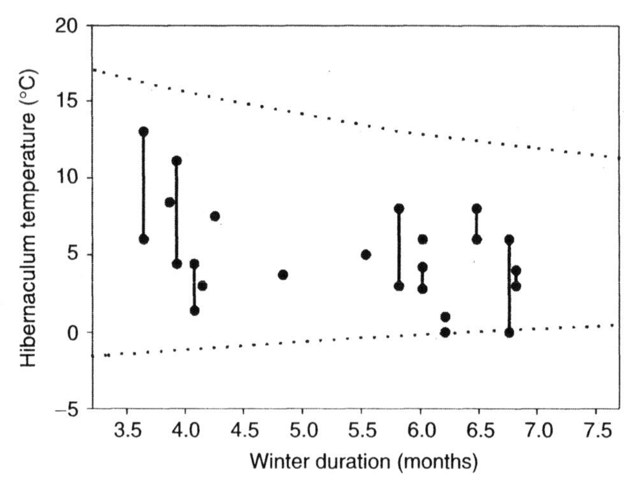

Figure 2.3 Comparison of predicted and observed hibernaculum temperatures of *M. lucifugus* hibernating at different latitudes. Combinations of winter length and hibernaculum temperatures outside the dotted line are predicted to require more than 3.5 g of fat to survive winter, which is the upper limit of fat storage in this species (Kunz et al., 1998). Winter duration for each location is estimated from the period when daily minima are below 0 °C. Filled circles represent observed hibernaculum temperatures of *M. lucifugus* at different geographical locations, and filled circles connected by a line represent a range of occupied temperatures reported from a single location (Twente, 1960; Menaker, 1962; Pearson, 1962; Davis and Hitchcock, 1964, 1965; Henshaw and Folk, 1966; Fenton, 1970; McManus, 1974; Nagorsen, 1980; Brack and Twente, 1985; Thomas, 1995).

Table 2.4 Total Energetic Cost of *M. lucifugus* Hibernation at Southern and Northern Latitudes

Location	Latitude	Average Annual T_a (°C)	$T_{hibernacula}$ (°C)	Winter Length (days)	Energy Requirement (g fat)
Missouri[a]	39° N	12.0	9.5	142	1.60
Ontario[b]	47° N	3.3	3.0	193	1.19

[a]Twente, 1955; [b]Fenton, 1970.

maximum cave temperatures in a given region (Dwyer, 1971; Parker et al., 1997), should allow for an accurate prediction of the northern limit for successful hibernation by *M. lucifugus*. Combining predicted maximum cave temperatures with estimated winter length and hibernation energy requirements, indicates more than 3.5 g of fat would be required to hibernate successfully throughout most of Alaska, Yukon, and NWT, and the northern portions of Manitoba, Ontario, and Quebec (figure 2.4). Given that this

Figure 2.4 Predicted and observed range distributions of *M. lucifugus* in Canada and Alaska. Dashed line represents the approximate distribution limits of summer activity, based primarily on museum specimen localities (Fenton and Barclay, 1980). Shaded line indicates the predicted distributional limit of hibernacula, based on combinations of winter length and cave temperatures where more than 3.5 g of fat would be required to survive hibernation. The width of the shading represents uncertainty based on a sensitivity analysis not presented in this chapter. Confirmed northern hibernacula locations (Fenton, 1970; Schowalter, 1980; Nagorsen et al., 1993; Parker et al., 1997) are indicated by stars.

level of fat accumulation is rarely exceeded by *M. lucifugus* (Kunz et al., 1998), successful hibernation seems extremely unlikely in these regions.

The predicted northern limit of suitable *M. lucifugus* hibernacula sites is consistently 250–350 km south of the northern limit of the species' summer distribution (figure 2.4). Given that *M. lucifugus* is known to migrate up to 300 km to hibernation sites, our model suggests that both the winter and summer distribution of this species may directly result from thermal effects on hibernation energetics. Although the northernmost confirmed hibernation sites in Canada and Alaska typically lie about 500 km south of their predicted northern limit (figure 2.4), knowledge of northern *M. lucifugus* hibernacula localities is very limited due to low human settlement and research effort in the northern portion of this species' distribution (Parker et al., 1997). Because of the 700 km separation between known summer distributions and hibernation sites, and the absence of winter sampling of caves in the intervening region, more northerly hibernation localities are believed to exist (Parker et al., 1997; D. Nagorsen, R. Barclay, B. Fenton, pers. comm.).

Our prediction for the northern limit also approximates the 0 °C (average annual temperature) isocline, which has been previously speculated as a northern barrier to *M. lucifugus* hibernation (Fenton, 1970). Hibernating *M. lucifugus* can tolerate temperatures well below 0 °C in the laboratory, but in natural hibernacula the ice build-up and absence of standing water associated with freezing cave temperatures could prevent successful hibernation (Thomas and Geiser, 1997). Present knowledge of northern hibernacula is too incomplete to discriminate between energetic or geophysical explanations for the northern range limit, but it seems likely that either of these thermal consequences, independently or in combination with each other, could limit the northern distribution.

Our model provides much less insight into the southern range limit of *M. lucifugus*. Although the model predicts substantial energetic costs associated with warm cave temperatures, it is difficult to predict the length of the hibernation season, available cave temperatures, and role of interspecific competition in southern regions. There is potential for insect feeding throughout winter in the region around the southern geographic limit of *M. lucifugus*, although individuals do not appear to take advantage of this opportunity (Whitaker and Rissler, 1993). Cooler-than-annual-average hibernaculum temperatures are readily available near cave entrances, but the instability of temperatures and low humidity in these areas may result in additional costs not accounted for by our model. Finally, the much greater diversity and abundance of aerial insectivores in southern regions (McNab, 1974; Bell et al., 1986) reduces the likelihood that distribution patterns can be accounted for by a simple, single-species bioenergetics model.

What Determines Hibernaculum Temperature Preference?

Given the pronounced increase in hibernation energy requirements above or below 2 °C, it is perhaps surprising that *M. lucifugus* is known to hibernate across a wide range of temperatures, even within a single cave (figure 2.3). Relatively few studies have quantified the microhabitat preferences of hibernating *M. lucifugus* in detail, but it is clear that many individuals hibernate in regions where our model predicts higher than necessary energy requirements. Evaporative water loss plays a major role in bat hibernation energetics (Thomas and Geiser, 1997), and thus humidity gradients are a likely second major axis of microhabitat preference. If warmer regions of caves generally have higher humidity, the reduced evaporative water loss experienced by bats in these regions may compensate for the predicted increase in thermal costs.

A second explanation for why some bats occupy warmer regions of caves may involve a general pattern of torpor minimization by endotherms. When provided with large energy reserves during hibernation, the eastern chipmunk, *Tamias striatus*, reduces both the depth and the duration of torpor bouts, indicating that there may be unidentified costs to the use of deep torpor. This response to variation in energy supplies is similar to that shown by species that use daily torpor, where the use of torpor is suppressed when energy is readily available. Although the benefits of euthermy or the costs of torpor during hibernation remain poorly resolved (French, 1992; Humphries et al., 2003), it may be that the selection of "warm" sites by hibernating bats is indeed shaped by the size of their energy reserves.

Hibernating bats face sufficiently severe energetic constraints so that nearly complete avoidance of torpor is impossible, to the extent observed in food-hoarding eastern chipmunks and daily torpor species (Humphries et al., 2002). This does not preclude the possibility that only those bats with dangerously small fat stores at the onset of hibernation seek microhabitats where energy requirements are minimized by expressing the deepest and most prolonged torpor bouts. Bats with larger fat stores may benefit from, or at least not suffer any significant negative consequences, occupying warmer regions and expressing shallower torpor. This explanation is consistent with observations of "programmed mass loss" in hibernating bats, and the general tendency of fatter bats to lose more mass during winter (Mrosovsky, 1976). Direct empirical support for this hypothesis requires demonstration of a positive relationship between the size of fat stores at the onset of winter and preferred hibernaculum temperature.

LITERATURE CITED

Bell, G.P., G.A. Bartholomew, and K.A. Nagy. 1986. The roles of energetics, water economy, foraging behavior, and geothermal refugia in the distribution of the bat, *Macrotus californicus*. Journal of Comparative Physiology B, 156: 441–450.

Blem, C.R. 1990. Avian energy storage. Current Ornithology, 7: 59–113.

Brack, V., and J.W. Twente, Jr. 1985. The duration of the period of hibernation of three species of vespertilionid bats. I. Field studies. Canadian Journal of Zoology, 63: 2952–2954.

Davis, W.H., and H.B. Hitchcock. 1964. Notes on sex ratios of hibernating bats. Journal of Mammalogy, 45: 475–476.

Davis, W.H., and H.B. Hitchcock. 1965. Biology and migration of the bat, *Myotis lucifugus*, in New England. Journal of Mammalogy, 46: 296–313.

Dwyer, P.D. 1971. Temperature regulation and cave-dwelling in bats: an evolutionary perspective. Mammalia, 35: 424–455.

Fenton, M.B. 1970. Population studies of *Myotis lucifugus* (Chiroptera: Vespertilionidae) in Ontario. Life Sciences Contributions of the Royal Ontario Museum, 77: 1–34.

Fenton, M.B., and R.M.R. Barclay. 1980. *Myotis lucifugus*. Mammalian Species, 142: 1–8.

French, A.R. 1985. Allometries of the durations of torpid and euthermic intervals during mammalian hibernation: a test of the theory of metabolic control of the timing of changes in body temperature. Journal of Comparative Physiology B, 156: 13–19.

French, A.R. 1992. Mammalian dormancy. Pp. 105–121. In: Mammalian Energetics (T.E. Tomasi and T.H. Horton, eds.). Comstock Publishing Associates, Ithaca, NY.

Geiser, F. 1988. Reduction of metabolism during hibernation and daily torpor in mammals and birds: temperature effect or physiological inhibition? Journal of Comparative Physiology B, 158: 25–38.

Geiser, F., and R.V. Baudinette. 1990. The relationship between body mass and rate of rewarming from hibernation and daily torpor in mammals. Journal of Experimental Biology, 151: 349–359.

Geiser, F., and G.J. Kenagy. 1988. Torpor duration in relation to temperature and metabolism in hibernating ground squirrels. Physiological Zoology, 61: 442–449.

Geiser, F., and T. Ruf. 1995. Hibernation versus daily torpor in mammals and birds: physiological variables and classification of torpor patterns. Physiological Zoology, 68: 935–966.

Gwinner, E. 1990. Bird Migration: Physiology and Ecophysiology. Springer, Berlin.

Heldmaier, G., and T. Ruf. 1992. Body temperature and metabolic rate during natural hypothermia in endotherms. Journal of Comparative Physiology B, 162: 696–706.

Henshaw, R.E., and G.E. Folk, Jr. 1966. Relation of thermoregulation to seasonally changing microclimate in two species of bats (*Myotis lucifugus* and *M. sodalis*). Physiological Zoology, 39: 223–236.

Hochachka, P.W., and M. Guppy. 1987. Metabolic Arrest and the Control of Biological Time. Harvard University Press, Cambridge, MA.

Hock, R.J. 1951. The metabolic rates and body temperatures of bats. Biological Bulletin, 101: 289–299.

Humphries, M.M., D.W. Thomas, C.L. Hall, J.R. Speakman, and D.L. Kramer. 2002. The energetics of autumn mast hoarding in eastern chipmunks. Oecologia, 133: 30–37.

Humphries, M.M., D.W. Thomas, and D.L. Kramer. 2003. The role of energy availability in mammalian hibernation: a cost benefit approach. Physiological and Biochemical Zoology, 76: 165–179.

King, J.R., and M.E. Murphy. 1985. Periods of nutritional stress in the annual cycle of endotherms: fact or fiction? American Zoologist 25: 955–964.

Kunz, T.H., J.A. Wrazen, and C.D. Burnett. 1998. Changes in body mass and fat reserves in pre-hibernating little brown bats (*Myotis lucifugus*). Ecoscience, 5: 8–17.

Lyman, C.P., J.S. Willis, A. Malan, and L.C.H. Wang. 1982. Hibernation and Torpor in Birds and Mammals. Academic Press, New York.

McManus, J.J. 1974. Activity and thermal preference of the little brown bat, *Myotis lucifugus*, during hibernation. Journal of Mammalogy, 55: 844–846.

McNab, B.K. 1974. The behavior of temperate cave bats in a subtropical environment. Ecology, 55: 943–958.

McNab, B.K. 1982. Evolutionary alternatives in the physiological ecology of bats. Pp. 151–200. In: Ecology of Bats (T.H. Kunz, ed.). Plenum Press, New York.

Menaker, M. 1962. Hibernation-hypothermia: an annual cycle of response to low temperature in the bat *Myotis lucifugus*. Journal of Cellular and Comparative Physiology, 59: 163–173.

Moen, R., J. Pastor, and Y. Cohen. 1997. A spatially explicit model of moose foraging and energetics. Ecology, 78: 505–521.

Mrosovsky, N. 1976. Lipid programmes and life strategies of hibernators. American Zoologist, 16: 685–697.

Nagorsen, D.W. 1980. Records of hibernating big brown bats (*Eptesicus fuscus*) and little brown bats (*Myotis lucifugus*) in Northwestern Ontario. Canadian Field Naturalist, 94: 83–85.

Nagorsen, D.W., A.A. Bryant, D. Kerridge, G. Roberts, A. Roberts, and M.J. Sarell. 1993. Winter bat records for British Columbia. Northwestern Naturalist, 74: 61–66.

Nedergaard, J., and B. Cannon. 1990. Mammalian hibernation. Philosophical Transactions of the Royal Society of London, Series B, 326: 669–686.

Nickerson, D.M., D.E. Facey, and G.D. Grossman. 1989. Estimating physiological thresholds with continuous two-phase regressions. Physiological Zoology, 62: 866–887.

Parker, D.I., B.E. Lawhead, and J.A. Cook. 1997. Distributional limits of bats in Alaska. Arctic, 50: 256–265.

Pearson, E.W. 1962. Bats hibernating in silica mines in southern Illinois. Journal of Mammalogy, 43: 27–33.

Raesly, R.L., and J.E. Gates. 1986. Winter habitat selection by north temperate cave bats. American Midland Naturalist, 118: 15–31.

Rainey, R.C. 1976. Flight behaviour and features of the atmospheric environment. Pp. 75–112. In: Insect Flight (R.C. Rainey, ed.). Blackwell Scientific, Oxford.

Richter, A.R., S.R. Humphrey, J.B. Cope, and V. Brack, Jr. 1993. Modified cave entrances: thermal effect on body mass and resulting decline of endangered Indiana bats (*Myotis sodalis*). Conservation Biology, 7: 407–415.

Scholander, P.F., R. Hock, V. Walters, F. Johnson, and L. Irving. 1950. Heat regulation in some arctic and tropical mammals and birds. Biological Bulletin, 99: 237–258.

Schowalter, D.B. 1980. Swarming, reproduction, and early hibernation patterns of *Myotis lucifugus* and *M. volans* in Alberta, Canada. Journal of Mammalogy, 61: 347–350.

Speakman, J.R. 2000. The cost of living: field metabolic rates of small mammals. Advances in Ecological Research, 30: 177–296.

Speakman, J.R., and D.W. Thomas. 2003. Physiological ecology and energetics of bats. Pp. 430–490. In: Bat Biology (T.H. Kunz and M.B. Fenton, eds.). University of Chicago Press, Chicago.

Stones, R.C., and J.E. Wiebers. 1967. Temperature regulation in the little brown bat, *Myotis lucifugus*. Pp. 97–109. In: Mammalian Hibernation III (K.C. Fisher, A.R. Dawe, C.P. Lyman, and F.E. South, eds.). American Elsevier, New York.

Thomas, D.W. 1995. The physiological ecology of hibernation in vespertilionid bats. Symposia of the Zoological Society of London, 67: 233–244.

Thomas, D.W., and F. Geiser. 1997. Periodic arousals in hibernating mammals: is evaporative water loss involved? Functional Ecology, 11: 585–591.

Thomas, D.W., M.B. Fenton, and R.M.R. Barclay. 1979. Social behavior of the little brown bat, *Myotis lucifugus*. I. Mating behavior. Behavioral Ecology and Sociobiology, 6: 129–136.

Thomas, D.W., M. Dorais, and J.-M. Bergeron. 1990. Winter energy budgets and cost of arousal for hibernating little brown bats, *Myotis lucifugus*. Journal of Mammalogy, 71: 475–479.

Twente, J.W., Jr. 1955. Some aspects of habitat selection and other behavior of cavern-dwelling bats. Ecology, 36: 706–732.

Twente, J.W., Jr. 1960. Environmental problems involving the hibernation of bats in Utah. Proceedings of the Utah Academy of Science and Arts Letters, 37: 67–71.

Twente, J.W., Jr., J. Twente, and V. Brack. 1985. The duration of the period of hibernation of three species of vespertilionid bats. II. Laboratory studies. Canadian Journal of Zoology, 63: 2955–2961.

Walsberg, G.E. 1985. Physiological consequences of micro-habitat selection. Pp. 389–413. In: Habitat Selection in Birds (M.L. Cody, ed.). Academic Press, New York.

Whitaker, J.O., Jr., and L.J. Rissler. 1992. Winter activity of bats at a mine entrance in Vermillion County, Indiana. American Midland Naturalist, 127: 52–59.

Whitaker, J.O., Jr., and L.J. Rissler. 1993. Do bats feed in winter? American Midland Naturalist, 129: 200–203.

3

Daily Heterothermy by Temperate Bats Using Natural Roosts

Craig K.R. Willis

Periods of heterothermy, or torpor, can save mammals and birds substantial amounts of energy. Temperate-zone insect-eating bats often use torpor to accommodate the thermoregulatory challenges associated with their large surface area to volume ratio and dependence on food that fluctuates within and between seasons. Torpor associated with hibernation has been well studied, but heterothermy during summer has received less attention, especially in free-ranging bats. Roosting ecology almost certainly influences patterns of daily heterothermy, but, to date, the majority of studies have focused only on individuals roosting in buildings, as opposed to those roosting in naturally occurring sites (e.g., tree cavities, rock crevices). In this chapter I review a number of predictions about torpor use and roost selection by bats as suggested by the costs and benefits of torpor in different circumstances. I evaluate these predictions with examples from the literature on bats in buildings and natural roosts, and with new data on bats roosting in natural sites. I also suggest avenues for future research addressing this important aspect of the physiological ecology of bats.

INTRODUCTION

Endothermic animals rely on behavioral, physiological, and morphological adaptations to reduce thermoregulatory costs (Pough et al., 1996). These adaptations are especially important for animals in temperate regions where food resources fluctuate within and between seasons (McNab, 1982). A number of birds and many mammals reduce thermoregulatory costs by using periods of heterothermy, or torpor (Geiser, 1996; Wang, 1989), defined as a state of inactivity and reduced responsiveness to stimuli during which metabolic rate (MR) and body temperature (T_b) fall below normothermic levels, and from which animals can actively arouse in the absence of an exogenous heat source (Barclay et al., 2001; Bligh and Johnson, 1973; Geiser, 1998).

Heterothermy in temperate insect-eating bats is common (Audet and Fenton, 1988; Brigham, 1987; Geiser and Ruf, 1995; Hamilton and

Barclay, 1994; Thomas et al., 1990) and the development of small (0.5–1.5 g) temperature-sensitive radiotransmitters (Barclay et al., 1996) has facilitated field research addressing their use in free-ranging bats. This chapter: (1) provides a brief introduction to daily heterothermy and its importance to insect-eating temperate-zone bats; (2) reviews predictions regarding patterns of heterothermy and roost selection in bats as they relate to the costs and benefits of torpor under different circumstances; (3) evaluates these predictions using examples from the literature and with new data; and (4) concludes with some suggestions for future research addressing the physiological ecology of daily heterothermy in bats.

DAILY HETEROTHERMY

Endothermy is energetically expensive for small animals, especially when the difference between the ambient temperature, T_a, and T_b is large, because a large body surface relative to a small volume results in the loss of metabolic heat to the environment (Schmidt-Nielsen, 1984). A number of mammals and birds, therefore, depend on periods of torpor to defray thermoregulatory costs, particularly during periods when foraging efficiency is too low to recover the energetic costs of thermoregulation. Torpor is characterized by short-term reductions in metabolic rate and body temperature (Bligh and Johnson, 1973; Geiser, 1998). Torpid animals realize significant energy savings at low T_a (up to 99% of daily energy requirements; Webb et al., 1993), relative to conspecifics that defend normothermic T_b (Geiser, 1993; Hosken and Withers, 1997).

Based on their patterns of torpor use, heterothermic species can be divided into two broad categories: daily torpor users and hibernators. Species that use daily torpor are best distinguished from those capable of hibernation on the basis of the duration of torpor bouts (Geiser and Ruf, 1995). Hibernating species often use long bouts of heterothermy (days or weeks), while species that use daily torpor are capable of only short bouts (hours; Geiser and Ruf, 1995). The depth of torpor, or magnitude of reduction in T_b and, especially, MR also differentiate daily torpor from hibernation. Hibernating species may allow T_b and MR to fall very low, while daily heterotherms are limited to shallower bouts (Geiser and Ruf, 1995). In this sense the torpor used by microchiropterans during summer is really hibernation rather than daily torpor because, if ambient conditions permit, bats are capable of remaining at very low T_b for long periods (Geiser and Brigham, 2000; Lausen and Barclay, 2003; Turbill et al., 2003a,b; this chapter). To avoid confusion, then, I refer to "daily heterothermy" instead of "daily torpor" in temperate bats, although the terms torpor and heterothermy are often used interchangeably.

Daily heterothermy in bats and birds may occur as a part of a daily thermoregulatory routine (e.g., Chruszcz, 1999; Chruszcz and Barclay, 2002; Hamilton and Barclay, 1994; Hickey and Fenton, 1996), or may be reserved for periods of energy emergency, such as unfavorable weather conditions that reduce foraging opportunities (e.g., Anthony et al., 1981; Kissner and Brigham, 1993; Kurta et al., 1987).

Daily Heterothermy in Bats

A number of factors likely compound the energetic challenges associated with a large surface to volume ratio for temperate insect-eating bats. A weather-dependent, variable food source (Rydell, 1989), an unusual reproductive strategy among small mammals (i.e., low annual fecundity and long periods of parental care: Racey, 1982), and an expensive method of locomotion, all increase the potential for bats to be exposed to large fluctuations in energy balance relative to other mammals of similar size (Hickey and Fenton, 1996; Kurta et al., 1989). In temperate areas, heterothermy appears essential to the survival of insect-eating bats.

Use of torpor in the context of overwinter hibernation has been relatively well studied in mammals generally (e.g., Boyer and Barnes, 1999; Daan 1973; Körtner and Geiser, 1998; McNab 1974), and in bats (e.g., Brigham, 1987; Park et al., 2000; Speakman and Racey, 1989; Thomas et al., 1990; Whitaker and Gummer, 1992). Daily heterothermy in bats, however, has received less attention, perhaps because large groups of hibernating bats are more easily studied than those using daily heterothermy, which may frequently switch roosts (e.g., Brigham et al., 1997; Kalcounis and Brigham, 1998; Vonhof and Barclay, 1996; Willis and Brigham, 2004).

PREDICTED PATTERNS OF HETEROTHERMY AND ROOST SELECTION

Reproductive Costs of Heterothermy

As with any behavioral or physiological strategy, the benefits of heterothermy must be balanced against potential costs. Increased exposure to predation during periods of inactivity may be one selective pressure favoring homeothermy. Predation risk may be mediated by morphological adaptations such as cryptic coloration in foliage-roosting bats (e.g., hoary bats *Lasiurus cinereus* or red bats *Lasiurus borealis*, or behavioral adaptations, such as cavity roosting (e.g., big brown bats *Eptesicus fuscus* or little brown myotis *Myotis lucifugus*; Brigham et al., 1998; Kalcounis and Brigham, 1998).

Delayed development of prenatal and neonatal young is the most often cited cost of torpor for bats (e.g., Audet and Fenton, 1988; Grinevitch et al., 1995; Hamilton and Barclay, 1994; Racey 1973; Tuttle and Stevenson, 1982). This cost is likely not bat-specific as a number of studies report an apparent reluctance by other mammals and birds to enter torpor during the breeding season (e.g., Brigham, 1992; Brigham et al., 2000; Csada and Brigham, 1994; Geiser, 1996). Torpor slows gestation and, thus, delays parturition in bats (Racey, 1973; Racey and Swift, 1981), and inhibits lactation (Racey and Swift, 1981; Tuttle, 1976; Wilde et al., 1995), both of which extend the juvenile growth phase. For bats that hibernate during winter, this may represent an especially acute selection pressure because juveniles must accumulate fat stores to survive hibernation (Kunz, 1987; Kunz et al., 1998) and overwinter starvation is an important cause of mortality (Kurta, 1986). Female bats are typically the sole providers of parental care (Racey, 1982), so the use of torpor while rearing young could reduce their reproductive success and overall fitness. These observations led to the prediction that reproductive female bats will avoid torpor relative to nonreproductive females and males (Barclay, 1991; Thomas, 1988).

The relationship between torpid metabolic rate and T_b is exponential, so that torpor is characterized by diminishing returns as T_b falls (Studier, 1981). For example, at a T_a of 10 °C, a reduction in T_b from 30 °C to 28 °C saves more energy than a reduction from 18 °C to 16 °C (Studier, 1981). For hibernating species, the energetic benefits of shallow torpor may be increased still further via metabolic inhibition during torpor entry (Geiser, 1988). During periods of low prey availability, then, females could employ shallow torpor to save some energy but still minimize the reproductive costs of low T_b. At the same T_a, male and nonreproductive female bats could maximize their energy savings by employing deeper torpor and permitting T_b to approach T_a (Grinevitch et al., 1995; Hamilton and Barclay, 1994). A second prediction, then, is that during periods of low temperature and reduced prey availability, reproductive females may employ shallow torpor and defend a relatively high torpid T_b while males and nonreproductive females may be thermoconforming and use deep torpor (Hamilton and Barclay, 1994).

Roost Selection and Torpor

Bats spend as much as half their lives roosting (Kunz, 1982), and thus selecting roosts with optimal characteristics will have selective implications. The hypothesis that microclimate is one such characteristic, especially for cavity-roosting bats, has been suggested by a number of studies (e.g., Kalcounis and Brigham, 1998; Kerth et al., 2001; Kunz and Lumsden, 2003; Sedgeley, 2001; Vaughan and O'Shea, 1976). Often, in studies where roost temperatures have not been measured directly, other roost characteristics (e.g., height of roost trees, tree diameter, roost orientation) for which bats show a significant preference are thought to influence microclimate (e.g., Betts, 1996; Vonhof and Barclay, 1996). In many colonial, cavity-roosting species the bats themselves modify roost microclimate (Burnett and August, 1981; Kunz, 1982; Racey, 1982; Willis, 2003). Low roost T_a (Racey, 1973; Racey and Swift, 1981) and parental torpor delays prenatal, neonatal, and juvenile development. Based on these observations, a third prediction is that to maximize offspring growth rates and help minimize thermoregulatory costs in the absence of torpor, reproductive female bats should select warm roosts (Kunz, 1987; Racey, 1982).

Male and nonreproductive female bats, in contrast to reproductive females, should select sites characterized by cool microclimates (Hamilton and Barclay, 1994). An additional consideration for these bats, however, is the energetic cost of rewarming from torpor. Shivering and nonshivering thermogenesis allow mammals to rewarm from low T_b (Wunder and Gettinger, 1996), but these processes are energetically expensive (Geiser et al., 1990; Wunder and Gettinger, 1996). A number of heterothermic endotherms rely on passive rewarming to arouse from torpor and, by doing so, save significant amounts of energy (Geiser and Drury, 2003; Geiser et al., 2004; Lovegrove et al., 1999). Thus, if possible, male bats should select roosts that allow them to rewarm passively to reduce the energetic costs of metabolic thermogenesis. A fourth prediction regarding the use of torpor and roost selection, then, is: male and nonreproductive female bats will select roosts with a variable microclimate, and sites that are cool much of the time but warm in the late afternoon permit passive rewarming (Hamilton and Barclay, 1994; Vaughan and O'Shea, 1976).

Why Differentiate between "Natural" and Human-Made Roosts?

Most field studies investigating daily heterothermy in bats have focused on large maternity colonies (i.e., tens to hundreds of individuals) in physically spacious, building roosts (Audet and Fenton, 1988; Grinevitch et al., 1995; Hamilton and Barclay, 1994), likely for the same reasons that hibernation has received more research attention than daily heterothermy. Lewis (1995) found that roost fidelity in bats is correlated with roost permanency. Bats show long-term fidelity to roosts in buildings, caves, and mines but tend to switch roosts frequently when using shorter-lived sites, such as tree hollows, rock crevices or open foliage (e.g., Brigham et al., 1997; Kalcounis and Brigham, 1998; Mager and Nelson, 2001; Vonhof and Barclay, 1996; Willis and Brigham, 2004). New data suggest that permanency is not the only factor influencing roost switching as, even in very stable tree cavities that may be used by populations of bats for up to 10 years (Willis et al., 2003), roost switching within years still occurs every few days (Willis and Brigham, in press). Nevertheless, in logistic terms this means that torpor use by bats is more easily studied in buildings than in natural sites. However, a higher level of fidelity to buildings and other more permanent roosts also suggests that these sites offer selective advantages. The availability of suitable microclimates could be one such advantage (Lewis, 1995).

Hamilton and Barclay (1994) noted that the building roost in their study likely offered bats a range of ambient temperatures at any given time. They suggested that bats in such a roost could switch positions throughout the day, following the heat when avoiding torpor, or moving between cool sites if they were using torpor. Vaughan and O'Shea's (1976) finding that pallid bats (*Antrozous pallidus*) often switch between shallow and deep positions in rock crevice roosts over the course of the day supports the idea of within-roost microclimate selection. The physical size of roost cavities could also have implications for clustering behavior because large roosts would allow bats to cluster with roost-mates when avoiding torpor, or roost solitarily when using torpor.

Spatial variation in microclimate may not be available to bats roosting in many natural situations. These individuals may spend energy actively thermoregulating when roost T_a extends outside the thermoneutral zone because of the small range of microclimates available. Recent evidence demonstrates that roost switching in some colonial, tree-cavity-living species results from the social structure of the colony (Kerth and König, 1999; Willis and Brigham, 2004). However, roost switching could also represent a search for optimal microclimates (Lewis, 1995). In both cases, naturally roosting bats must spend additional time and energy, otherwise available for foraging, in search of roost-mates and optimal roost microclimates as they switch roosts. Therefore, the energy balance of bats in natural roosts, and their use of torpor, could differ markedly from that of bats using buildings.

If the range of microclimates available to bats within a roost does influence torpor, then the behavior of bats roosting in different types of natural sites could also differ. Natural roosts characterized by microclimate choices at a given time (e.g., large tree cavities, deep rock crevices, caves) could facilitate torpor avoidance relative to roosts that offer little thermal choice (e.g., open foliage, tents, small tree cavities). Understanding how the spatial variability of microclimates within roosts influences torpor requires that different types of natural roosts be evaluated. Despite the logistic difficulties of studying naturally roosting bats, focusing only on bats in buildings may not provide an accurate picture of torpor patterns, energy balance, and roosting requirements.

EXAMPLES FROM THE FIELD

Big Brown Bats (*Eptesicus fuscus*) in Attic and Rock Crevice Roosts

Sex differences in use of torpor have been observed in *Eptesicus fuscus* roosting in buildings (Audet and Fenton, 1988; Grinevitch et al., 1995; Hamilton and Barclay, 1994). Using temperature telemetry, all three studies found that male *E. fuscus* used torpor more often than pregnant and lactating females. They also found that, when reproductive females did use torpor, they defended a higher T_b than did males. In combination with laboratory research, these studies are often cited as part of the basis for the first two predictions outlined above.

Interestingly, Park et al. (2000) reported no sex difference in the frequency or duration of hibernation torpor bouts outside the reproductive season in free-ranging *Rhinolophus ferrumequinum*. This finding is consistent with the hypothesis that sex differences in use of torpor during the active season reflect selection pressure against torpor by reproductive females. Grinevitch et al. (1995) suggested that torpor in reproductive female *E. fuscus* may be restricted to energy emergencies when foraging conditions are poor, while males employ torpor as part of their regular thermoregulatory strategy.

Despite some ambiguity between studies regarding differences in the use of shallow torpor between pregnancy and lactation, a number of field studies suggest that the use of deep torpor is especially limited during lactation (Audet and Fenton, 1988; Grinevitch et al., 1995; Hamilton and Barclay, 1994; Lausen and Barclay, 2003). Lausen and Barclay (2003) found that for big brown bats in rock crevices, the frequency of torpor use during lactation was actually higher than during pregnancy. However, their results are consistent with previous studies in that deep torpor, characterized by very low skin temperature (T_{sk}), was used more often during pregnancy (T_{sk} as low as *c.* 9 °C) than during lactation (T_{sk} as low as *c.* 16 °C). Moreover, they found that during pregnancy there was a significant effect of T_a on depth of torpor (i.e., colder T_a was correlated with lower daily minimum T_{sk}) but, during lactation, minimum T_{sk} was significantly warmer (2 °C on average), independent of T_a. This suggests that lactating bats defend a higher minimum T_b during lactation than during pregnancy and that avoidance of deep torpor during lactation reflects a greater cost (or smaller benefit) of deep torpor for lactating bats (Lausen and Barclay, 2003).

Bechstein's Bats (*Myotis bechsteinii*) in Tree Cavities and Bat Boxes

Complementing a number of convincing correlative studies (e.g., Kalcounis and Brigham, 1998; Sedgeley, 2001; Vaughan and O'Shea, 1976), Kerth et al. (2001) provided the first experimental evidence that bats select roosts based on microclimate. They generated a temperature gradient within a known population of available roosts by painting bat boxes black or white to affect their absorption of solar energy, and then hung pairs of black and white boxes on trees in shaded or sun-exposed locations. They monitored the roosts throughout the season and found a significantly higher number of bats in the warmest roosts during the lactation period, an observation consistent with the predictions outlined above that reproductive female bats will avoid use of torpor and select warm roosts. Lactating females also showed a significantly stronger preference for warm boxes than did nonreproductive bats (Kerth et al., 2001).

A surprising finding of the study by Kerth et al. (2001), however, was the significant preference for cool roosts by pregnant bats, given that torpor and low temperature delay the development of embryos. They suggested that this preference could reflect a trade-off for female bats between short-term survival and long-term fitness during a poor year. The spring of 1996 (the year of their experiment) was unusually cold in their study area and the percentage of females giving birth was lower than normal. The authors hypothesized that in a normal year pregnant females would have exploited the warmest roosts but, because the benefits of foraging in 1996 were low compared with the high costs of flight while pregnant, many bats sacrificed reproductive success in favor of their own survival by using torpor. During the lactation period, bats that were successful in bringing their offspring to parturition moved into the warmest boxes to avoid torpor and maximize postnatal growth. This experimental evidence demonstrates that cavity-roosting bats do select roosts on the basis of microclimate. It further suggests that their selection is based not only on the costs but also on the benefits of torpor.

Long-Eared Bats (*Myotis evotis*) in Rock Crevices

Chruszcz (1999) and Chruszcz and Barclay (2002) addressed the relationship between foraging behavior, roosting ecology, and thermoregulation in female long-eared bats (*Myotis evotis*) roosting in rock crevices in the badlands along the South Saskatchewan River Valley of southeastern Alberta, Canada. *Myotis evotis* is unique in the study area in that it often gleans prey from surfaces and, thus, may exploit terrestrial arthropods during cool weather (Brigham et al., 2000; Chruszcz, 1999). In keeping with this foraging flexibility, Chruszcz (1999) found that *M. evotis* foraged all night, every night, even when aerial insects were not abundant. Surprisingly, however, he also found that both pregnant and lactating bats used torpor each day.

Consistent with previous studies, Chruszcz (1999) and Chruszcz and Barclay (2002) reported that pregnant *M. evotis* entered deep torpor more frequently than during lactation. Chruszcz (1999) suggested that this difference could reflect increased wing loading, low temperatures, and low insect availability during pregnancy, all of which would reduce foraging efficiency and increase the benefits of torpor. Higher T_a and insect availability and lower wing loading during the lactation period could make torpor less beneficial. As for big brown bats, then, this difference in the use of deep torpor between pregnancy and lactation may not necessarily reflect a higher reproductive cost of torpor during lactation but could indicate that torpor is more beneficial during pregnancy.

Consistent with the findings of Kerth et al. (2001), Chruszcz and Barclay (2002) also showed differential roost selection between pregnant and lactating *M. evotis*. Pregnant bats roosted in horizontal cracks in large boulders, sites characterized by low minimum and high maximum T_as. Lactating females, on the other hand, roosted in vertical crevices characterized by lower maximum and higher minimum T_as. In other words, pregnant *M. evotis* selected roosts well suited to heterothermy, whereas lactating females selected sites suited to homeothermy. Once again, this likely reflects considerations of energy balance for pregnant and lactating bats. Roosts favoring heterothermy would facilitate deep torpor use during pregnancy when bats suffer high costs of flight and thermoregulation, and low foraging returns. Roosts suited to homeothermy would allow them to avoid deep torpor during lactation when foraging costs are low and returns high. These roosts would also reduce thermoregulatory costs for nonvolant young remaining in the roost when lactating females leave to forage (Chruszcz, 1999; Chruszcz and Barclay, 2002). Once again, roost selection seems closely linked with thermoregulation and bats roosting in natural sites have, in part, violated the predictions about use of torpor suggested in the literature.

Hoary Bats (*Lasiurus cinereus*) in Open Foliage Roosts

Bats that roost alone in the open foliage of trees face thermal conditions that natural-cavity and building roosting bats can avoid. The range of possible microclimates among which foliage-roosting bats may choose is small and these bats cannot modify the microclimate of their roost by clustering. They are also exposed to greater extremes of T_a than cavity-roosting species (Willis 2003). Thus, one might expect different torpor patterns in species with this roosting strategy. The hoary bat is an example of a foliage-roosting species that violates predictions about use of torpor. Free-ranging, reproductive females regularly use deep torpor (Hickey and Fenton, 1996; Koehler, 1991; this chapter). I have used temperature-sensitive radio-transmitters (Holohil Systems, Carp, Ontario, Canada) and datalogging radiotelemetry receivers (SRX-400, Lotek Wireless, Newmarket, Ontario, Canada) to assess use of torpor in different reproductive classes of hoary bats in the Cypress Hills of Saskatchewan, Canada. The Cypress Hills are an excellent place to study torpor in bats because they are characterized by dramatic intraseasonal variation in T_a and insect availability.

My preliminary findings are consistent with those of Hickey and Fenton (1996) and Koehler (1991) in that reproductive female *L. cinereus* in the Cypress Hills rely on torpor virtually every day and often use deep torpor (figures 3.1, 3.2). In fact, based on Geiser and Ruf's (1995) definition, the pregnant female represented in figure 3.1 is hibernating, despite the fact that this species migrates south in autumn and is thought to remain active during the winter (Nagorsen and Brigham, 1993). These data are consistent with the field studies outlined above which demonstrated greater

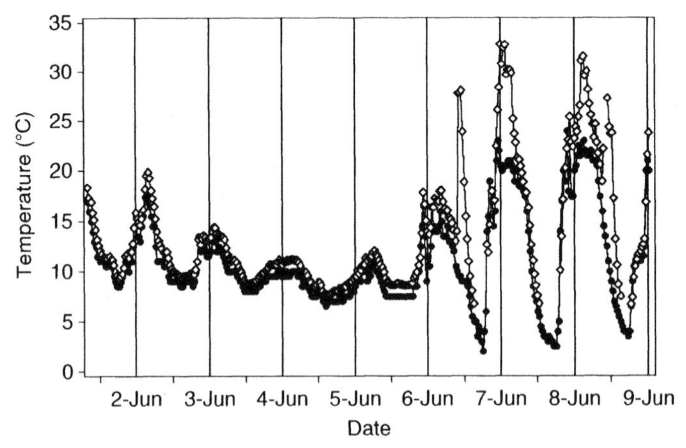

Figure 3.1 An example of hibernation (i.e., deep, prolonged torpor) in a free-ranging, reproductive female bat. Time courses of skin temperature (T_{sk}, open diamonds) and ambient temperature at the roost site (T_a, filled circles) obtained between June 1 and June 9, 2001 for a pregnant female *Lasiurus cinereus*. Vertical lines indicate 12:00 noon on each day. Gaps in the T_{sk} time course occurred on a few occasions when radio interference interrupted recording by the datalogging receiver.

use of deep torpor during pregnancy than lactation (e.g., Chruszcz, 1999; Hamilton and Barclay, 1994; Lausen and Barclay 2003). Lactating female *L. cinereus* are routinely captured in the Cypress Hills and figure 3.2 represents a typical time course of T_{sk} during lactation. To date, however, few pregnant females have been captured, so concluding that multi-day torpor is common in pregnant hoary bats is premature. Nonetheless, it seems clear that the use of torpor and deep torpor is important to reproductive female *L. cinereus*.

Interestingly, Cryan and Wolf (2003) found that pregnant female *L. cinereus* could rarely be induced to use torpor in the laboratory during spring migration and they never used deep torpor. This is an important finding because it suggests that the propensity for torpor use by reproductive females changes dramatically at some point during migration or just after individuals arrive at their summer range. It is possible that the difference reflects variation in the use of torpor between captive and free-ranging animals (Geiser et al., 2000), but this seems unlikely given the consistent responses of the large sample of animals studied by Cryan and Wolf (2003). While logistically difficult, further work addressing torpor patterns of free-ranging, pregnant female hoary bats at the onset of migration would help to confirm this finding.

The migratory overwintering strategy of *L. cinereus* appears to have exerted strong influence on the evolution of physiological traits in this

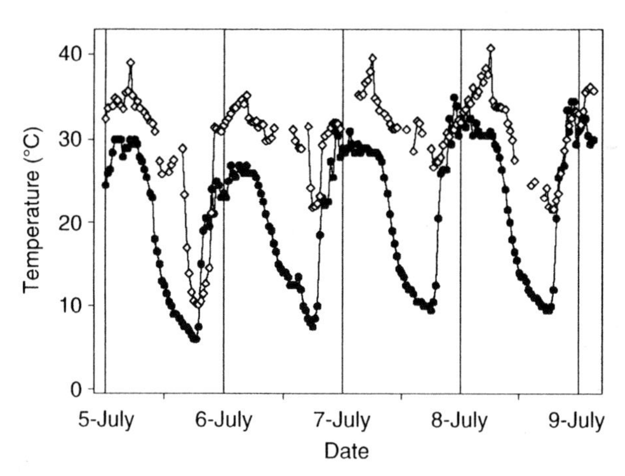

Figure 3.2 An example of daily heterothermy and deep torpor use in a free-ranging, reproductive female bat. Time courses of skin temperature (T_{sk}, open diamonds) and ambient temperature at the roost site (T_a, filled circles) obtained between July 5 and July 9, 2001 for a lactating female *L. cinereus*. Similar time courses have been recorded for other lactating *L. cinereus* captured in the study area ($n = 7$). Gaps in the T_{sk} time course indicate times when the bat was absent from the roost, presumably foraging, when the transmitter signal could not be detected by the datalogging receiver because of the position of the bat or due to radio interference that interrupted recording. Occasions when T_a appears to exceed T_{sk} reflect exposure of dataloggers to direct sunlight for brief periods when the bat was shaded.

species (Cryan and Wolf, 2003), and migration may also explain why reproductive females are able to use deep torpor so regularly. In species that hibernate, juveniles must accumulate fat stores to avoid starvation during hibernation (Kunz, 1987), a common cause of mortality in bats (Kurta, 1986). Migratory species may not face this challenge so juvenile growth rate need not be as rapid. Koehler and Barclay (2000) reported that growth rates of juvenile hoary bats are lower than those of most other temperate species, and they hypothesized that regular torpor use by mothers and juveniles could account for this finding. They also found that, in contrast to other temperate species (*M. lucifugus* and *E. fuscus*), adult females do not lose mass during lactation. A migratory wintering strategy and consequent reduction in reproductive costs of torpor, then, may have important selective implications for this species.

Despite the propensity of reproductive female hoary bats to use torpor, there is preliminary evidence that hoary bats select roost sites that are slightly warmer than conditions in randomly selected sites. During the Northern Hemisphere summer in 2000, I measured the compass orientations of roost sites, relative to the trunks of roost trees, for nine hoary bat roosts (4 lactating female, 5 male) in the open foliage of mature white spruce trees (*Picea glauca*) in the Cypress Hills. Roost sites fell within a narrow range of orientations (132° to 152°, i.e., southeast) and mean roost orientation (143.5° ± 6.3° SE) deviated significantly from random (V-test, $P < 0.001$). This southeast roost orientation could have important implications for roost microclimate because it would expose bats to morning sun following the cold early morning hours.

To test for a temperature difference between roost sites and random sites, I placed temperature dataloggers (iButton Thermocron, Dallas Semiconductor, Dallas, Texas, USA) as close to the roosting bats (<2 m) as possible and at randomly selected sites. Although roosts did reach their maximum temperature earlier in the day than random sites, there was not a significant difference in the temperature time courses of the two groups. Measurements of wind chill at roost and random sites, however, as opposed to simply temperature, may provide a measure of microclimate conditions more biologically relevant to foliage-roosting bats.

CONCLUSIONS

The patterns of torpor and roosting behavior outlined above are summarized in table 3.1. Geiser et al. (2000) found general differences in use of torpor between captive and free-ranging animals. The field data cited in this chapter support Geiser et al. (2000) because they show that bats often violate laboratory-based predictions about torpor patterns. It also seems that use of torpor by bats in buildings differs from that of bats roosting in natural sites. This could reflect the possibility that buildings offer a range of microclimates between which bats can switch throughout the day, thus permitting reproductive females to avoid torpor if necessary. If this is the case,

Table 3.1 Patterns of Torpor and Roost Selection in Free-Ranging Bats as Outlined in This Chapter

Species	Roost Type	Torpor Patterns	Roost Selection	References
Eptesicus fuscus	Large building maternity colonies	Males use torpor more than pregnant females	Males in the west of attics, females in the east	Audet and Fenton, 1988; Hamilton and Barclay, 1994; Grinevitch et al., 1995; Lausen and Barclay, 2003
Rhinolophus ferrumequinum	Cave hibernacula	Pregnant females use torpor more than lactating females	N/A	Park et al., 2000
Myotis bechsteinii	Bat boxes and tree cavities	No sex difference in torpor during hibernation N/A	Selected warm lactation roosts, and cool pregnancy roosts	Kerth et al., 2001
Myotis evotis	Rock crevices	Pregnant and lactating bats both use torpor every day. Deep torpor is more frequent while pregnant	Select stable T_a lactation roosts (vertical cracks) and variable T_a pregnancy roosts (horizontal cracks)	Chruszcz, 1999; Chruszcz and Barclay, 2002
Lasiurus cinereus	Open foliage of trees	Lactating bats use torpor often. Pregnant bats may use very deep torpor	Trend for warm sites, southeast side of trees	Koehler, 1991; Koehler and Barclay, 2000; this chapter

then depending on the range of microclimates available, patterns of torpor may differ among bats roosting in different types of natural roosts. Studies monitoring the range of microclimates within different roost sites, at the same time as monitoring use of torpor by different reproductive classes of bats, will help resolve this issue.

In keeping with predictions suggested by the reproductive costs of torpor, female *E. fuscus* in buildings avoid torpor during pregnancy and especially during lactation, relative to males. *M. evotis* in rock crevice roosts and *L. cinereus* in open foliage roosts, however, often depend on deep torpor while pregnant and lactating. This could reflect roosting behavior but other life-history traits may influence trade-offs between reproductive and energetic costs and benefits of torpor and activity. Migratory species, for example, may sacrifice rapid juvenile growth with relatively minor fitness consequences if their young need not fatten as much as those of hibernating species. Investigating intraspecific differences in torpor patterns in species such as *E. fuscus*, which are flexible in their roosting behavior, could help us tease apart the influences of wintering strategy versus roosting behavior on patterns of torpor.

Maternity-roost requirements of free-ranging bats are perhaps not as straightforward as has been assumed in the literature. Some authors have suggested that roost traits preferred by bats, such as distance to forest edge, or tree height, could result in warm roost temperatures that are the underlying reason for selection of those sites (e.g., Boonman, 2000; Vonhof and Barclay, 1996). The evidence reviewed here, however, suggests that in some circumstances reproductively active bats select cold roosts to help them enter torpor if the costs of normothermy and high wing loading during pregnancy cannot be recovered by foraging.

Few studies have addressed use of torpor in free-ranging bats, despite its importance to energy balance in many temperate species. Daily heterothermy, foraging behavior, reproduction, and roost selection are closely linked. Future studies, employing temperature telemetry in the field, could provide enormous insight into the physiological ecology of bats.

ACKNOWLEDGMENTS

I am grateful to Mark Brigham for his support and for helpful comments that improved the manuscript. Robert Barclay, Brock Fenton, Fritz Geiser, Lydia Hollis, and Cori Lausen all provided excellent suggestions on earlier drafts. This research was funded with contributions from Saskatchewan Environment and Resource Management, Mountain Equipment Co-op and the Natural Sciences and Engineering Research Council (Canada).

LITERATURE CITED

Anthony, E.L.P., M.H. Stack, and T.H. Kunz. 1981. Night roosting and the nocturnal time budget of the little brown bat, *Myotis lucifugus*: effects of reproductive status, prey density, and environmental conditions. Oecologia, 51: 151–156.

Audet, D., and M.B. Fenton. 1988. Heterothermy and the use of torpor by the bat *Eptesicus fuscus* (Chiroptera: Vespertilionidae): a field study. Physiological Zoology, 74: 1778–1781.

Barclay, R.M.R. 1991. Population structure of temperate zone insectivorous bats in relation to foraging behaviour and energy demand. Journal of Animal Ecology, 60: 165–178.

Barclay, R.M.R., M.C. Kalcounis, L.H. Crampton, C. Stefan, M. Vonhof, and R.M. Brigham. 1996. Can external radiotransmitters be used to assess body temperature and torpor in bats? Journal of Mammalogy, 77: 1102–1106.

Barclay, R.M.R., C.L. Lausen, and L. Hollis. 2001. What's hot and what's not? Defining torpor in free-ranging birds and mammals. Canadian Journal of Zoology, 79: 1885–1890.

Betts, B.J. 1996. Roosting behaviour of silver-haired bats (*Lasionycteris noctivigans*) and big brown bats (*Eptesicus fuscus*) in northeast Oregon. Pp. 55–61. In: Bats and Forests Symposium, October 19–21, 1995 (R.M.R. Barclay and R.M. Brigham, eds.). Research Branch, B.C. Ministry of Forests, Victoria, BC, Canada.

Bligh, J., and K.G. Johnson. 1973. Glossary of terms for thermal physiology. Journal of Applied Physiology, 35: 941–961.

Boonman, M. 2000. Roost selection by noctules (*Nyctalus noctula*) and Daubenton's bats (*Myotis daubentonii*). Journal of Zoology (London), 251: 385–389.

Boyer, B.B., and B.M. Barnes. 1999. Molecular and metabolic aspects of mammalian hibernation. Bioscience, 49: 713–724.

Brigham, R.M. 1987. The significance of winter activity by the big brown bat (*Eptesicus fuscus*): the influence of energy reserves. Canadian Journal of Zoology, 65: 1240–1242.

Brigham, R.M. 1992. Daily torpor in a free-ranging goatsucker, the common poorwill (*Phalaenoptilus nuttallii*). Physiological Zoology, 65: 457–472.

Brigham, R.M., M.J. Vonhof, R.M.R. Barclay, and J.C. Gwilliam. 1997. Roosting behavior and roost site preferences of forest-dwelling California bats (*Myotis californicus*). Journal of Mammalogy, 78: 1231–1239.

Brigham, R.M., S.J.S. Debus, and F. Geiser. 1998. Cavity selection for roosting, and roosting ecology of forest dwelling Australian Owlet Nightjars (*Aegotheles cristatus*). Australian Journal of Ecology, 23: 424–429.

Brigham, R.M., G. Körtner, T.A. Maddocks, and F. Geiser. 2000. Seasonal use of torpor by free-ranging Australian Owlet Nightjars (*Aegotheles cristatus*). Physiological and Biochemical Zoology, 73: 613–620.

Burnett, C.D., and P.V. August. 1981. Time and energy budgets for day roosting in a maternity colony of *Myotis lucifugus*. Journal of Mammalogy, 6: 211–218.

Chruszcz, B. 1999. Behavioral Ecology of *Myotis evotis*. M.Sc. Thesis, University of Calgary, Calgary, Alberta, Canada.

Chruszcz, B., and R.M.R. Barclay. 2002. Thermoregulatory ecology of a solitary bat, *Myotis evotis*, roosting in rock crevices. Functional Ecology, 16: 18–26.

Cryan, P.M., and B.O. Wolf. 2003. Sex differences in the thermoregulation and evaporative water loss of a heterothermic bat, *Lasiurus cinereus*, during its spring migration. Journal of Experimental Biology, 206: 3381–3390.

Csada, R.D., and R.M. Brigham. 1994. Reproduction constrains the use of daily torpor by free ranging common poorwills (*Phalaenoptilus nuttallii*) (Aves: Caprimulgidae). Journal of Zoology (London), 234: 209–216.

Daan, S. 1973. Activity during natural hibernation in three species of vespertilionid bats. Netherlands Journal of Zoology, 23: 1–71.

Geiser, F. 1988. Reduction of metabolism during hibernation and daily torpor in mammals and birds: temperature effect or physiological inhibition? Journal of Comparative Physiology B, 158: 25–37.

Geiser, F. 1993. Hibernation in the eastern pygmy possum, *Cercartetus nanus* (Marsupialia: Burramyidea). Australian Journal of Zoology, 41: 67–75.

Geiser, F. 1996. Torpor in reproductive endotherms. Pp. 81–86. In: Adaptations to the Cold: 10th International Hibernation Symposium (F. Geiser, A.J. Hulbert, and S.C. Nicol, eds.). University of New England Press, Armidale, Australia.

Geiser, F. 1998. Evolution of daily torpor and hibernation in birds and mammals: importance of body size. Clinical and Experimental Pharmacology and Physiology, 25: 736–740.

Geiser, F., S. Hiebert, and G.J. Kenagy. 1990. Torpor bout duration during the hibernating season of two sciurid rodents: interrelations with temperature and metabolism. Physiological Zoology, 63: 489–503.

Geiser F., and T. Ruf. 1995. Hibernation versus daily torpor in mammals and birds: physiological variables and classifications of torpor patterns. Physiological Zoology, 68: 935–966.

Geiser F., and R.M. Brigham. 2000. Torpor, thermal biology and energetics in Australian long-eared bats (*Nyctophilus*). Journal of Comparative Physiology B, 170: 153–162.

Geiser, F., J.C. Holloway, G. Körtner, T.A. Maddocks, C. Turbill, and R.M. Brigham. 2000. Do patterns of torpor differ between free-ranging and captive mammals and birds? Pp. 95–102. In: Life in the Cold: 11th International Hibernation Symposium (G. Heldmaier and M. Klingenspor, eds.). Springer, Berlin.

Geiser, F., and R.L. Drury. 2003. Radiant heat affects thermoregulation and energy expenditure during rewarming from torpor. Journal of Comparative Physiology B, 173: 55–60.

Geiser, F., R.L. Drury, G. Körtner, C. Turbill, C.R. Pavey, and R.M. Brigham. 2004. Passive rewarming from torpor in mammals and birds: Energetic, ecological and evolutionary implications. Pp. 51–62. In: Life in the Cold: Evolution, Adaptation, and Application (B.M. Barnes and H.V. Carey, eds.). Biological Papers of the University of Alaska, No. 27, Institute of Artic Biology, University of Alaska, Fairbam.

Grinevitch, L., S.L. Holroyd, and R.M.R. Barclay. 1995. Sex differences in the use of daily torpor and foraging time by big brown bats (*Eptesicus fuscus*) during the reproductive season. Journal of Zoology (London), 235: 301–309.

Hamilton, I.M., and R.M.R. Barclay. 1994. Patterns of daily torpor and day roost selection by male and female big brown bats (*Eptesicus fuscus*). Canadian Journal of Zoology, 72: 744–749.

Hickey, M.B.C., and M.B. Fenton. 1996. Behavioral and thermoregulatory responses of female hoary bats *Lasiurus cinereus* (Chiroptera: Vespertilionidae) to variations in prey availability. Ecoscience, 3: 414–422.

Hosken, D.J., and P.C. Withers. 1997. Temperature regulation and metabolism of an Australian bat, *Chalinolobus gouldii* (Chiroptera: Vespertilionidae) when euthermic and torpid. Journal of Comparative Physiology B, 167: 71–80.

Kalcounis, M.C., and R.M. Brigham. 1998. Secondary use of aspen cavities by tree roosting big brown bats. Journal of Wildlife Management, 62: 603–611.

Kerth G., and B. König. 1999. Fission, fusion and nonrandom associations in female Bechstein's bats (*Myotis bechsteinii*). Behavior, 136: 1187–1202.

Kerth, G., K. Weissmann, and B. König. 2001. Day roost selection in female Bechstein's bats (*Myotis bechsteinii*): a field experiment to determine the influence of roost temperature. Oecologia, 126: 1–9.

Kissner, K.J., and R.M. Brigham. 1993. Evidence for the use of torpor by incubating and brooding common poorwills, *Phalaenoptilus nuttali*. Ornis Scandinavica, 24: 333–334.

Koehler, C.E. 1991. The reproductive ecology of the hoary bat (*Lasiurus cinereus*) and its relation to litter size variation in vespertilionid bats. M.Sc. Thesis, University of Calgary, Calgary, Alberta, Canada.

Koehler, C.E., and R.M.R. Barclay. 2000. Post-natal growth and breeding biology of the hoary bat (*Lasiurus cinereus*). Journal of Mammalogy, 81: 234–244.

Körtner, G., and F. Geiser. 1998. Ecology of natural hibernation in the marsupial mountain pygmy-possum (*Burramys parvus*). Oecologia, 113: 170–178.

Kunz, T.H. 1982. Roosting ecology of bats. Pp. 1–55. In: Ecology of Bats (T.H. Kunz, ed.). Plenum Press, New York.

Kunz, T.H. 1987. Post-natal growth and energetics of suckling bats. Pp. 395–420. In: Recent Advances in the Study of Bats (M.B. Fenton, P. Racey, and J.V.M. Rayner, eds.). Cambridge University Press, Cambridge.

Kunz, T.H., and L.F. Lumsden. 2003. Ecology of cavity and foliage roosting bats. Pp. 3–89. In: Bat Ecology (T.H. Kunz and M.B. Fenton, eds.). University of Chicago Press, Chicago.

Kunz, T.H., J.A. Wrazen, and C.D. Burnett. 1998. Changes in body mass and body composition in pre-hibernating little brown bats (*Myotis lucifugus*). Ecoscience, 5: 8–17.

Kurta, A. 1986. Factors affecting the resting and postflight body temperature of little brown bats, *Myotis lucifugus*. Physiological Zoology, 59: 429–438.

Kurta, A., K.A. Johnson, and T.H. Kunz. 1987. Oxygen consumption and body temperature of female little brown bats (*Myotis lucifugus*) under simulated roost conditions. Physiological Zoology, 60: 386–397.

Kurta, A., G.P. Bell, K.A. Nagy, and T.H. Kunz. 1989. Energetics of pregnancy and lactation in free-ranging little brown bats (*Myotis lucifugus*). Physiological Zoology, 62: 804–818.

Lausen C.L., and R.M.R. Barclay. 2003. Thermoregulation and roost selection by reproductive female big brown bats (*Eptesicus fuscus*) roosting in rock crevices. Journal of Zoology (London) 260: 235–244.

Lewis, S.E. 1995. Roost fidelity of bats: a review. Journal of Mammalogy, 76: 481–496.

Lovegrove, B.G., G. Körtner, and F. Geiser. 1999. The energetic cost of arousal from torpor in the marsupial *Sminthopsis macroura*: benefits of summer ambient temperature cycles. Journal of Comparative Physiology B, 169: 11–18.

Mager, K.J., and T.A. Nelson. 2001. Roost-site selection by eastern red bats (*Lasiurus borealis*). American Midland Naturalist, 145: 120–126.

McNab, B.K. 1974. The behavior of temperate cave bats in a subtropical environment. Ecology, 55: 943–958.

McNab, B.K. 1982. Evolutionary alternatives in the physiological ecology of bats. Pp. 151–200. In: Ecology of Bats (T.H. Kunz, ed.). Plenum Press, New York.

Nagorsen, D.W., and R.M. Brigham. 1993. Bats of British Columbia. University of Vancouver Press, Vancouver, Canada.

Park, K.J., G. Jones, and R.D. Ransome. 2000. Torpor, arousal and activity of hibernating greater horseshoe bats (*Rinolophus ferrumequinum*). Functional Ecology, 14: 580–588.

Pough, F.H., J.B. Heiser, and W.N. McFarland. 1996. Vertebrate Life. Prentice-Hall, Englewood Cliffs, NJ.

Racey, P.A. 1973. Environmental factors affecting the length of gestation in heterothermic bats. Journal of Reproduction and Fertility, Supplement, 19: 175–189.

Racey, P.A. 1982. Ecology of bat reproduction. Pp. 57–95. In: Ecology of Bats (T.H. Kunz, ed.). Plenum Press, New York.

Racey, P.A., and S.M. Swift. 1981. Variation in gestation length in a colony of pipistrelle bats (*Pipistrellus pipistrellus*) from year to year. Journal of Reproduction and Fertility, 61: 123–129.

Rydell, J. 1989. Feeding activity of the northern bat, *Eptesicus nilsonii*, during pregnancy and lactation. Oecologia, 80: 562–565.

Schmidt-Nielson, K. 1984. Scaling: Why is Body Size so Important? Cambridge University Press, Melbourne, Australia.

Sedgeley, J.A. 2001. Quality of cavity microclimate as a factor influencing selection of maternity roosts by a tree dwelling bat, *Chalinolobus tuberculatus*, in New Zealand. Journal of Applied Ecology, 38: 425–438.

Speakman, J.R., and P.A. Racey. 1989. Hibernal ecology of the pipistrelle bat: energy expenditure, water requirement and mass loss, implications for survival and the function of winter emergence flights. Journal of Animal Ecology, 58: 797–813.

Studier, E.H. 1981. Energetic advances of slight drops in body temperature in little brown bats, *Myotis lucifugus*. Comparative Biochemistry and Physiology A, 70: 537–540.

Thomas, D.W. 1988. The distribution of bats in different ages of Douglas-fir forest. Journal of Wildlife Management, 52: 619–626.

Thomas, D.W., J. Dorais, and J.M. Bergeron. 1990. Winter energy budgets and costs of arousals for hibernating little brown bats, *Myotis lucifugus*. Journal of Mammalogy, 71: 475–479.

Turbill, C., G. Körtner, and F. Geiser. 2003a. Natural use of heterothermy by a small, tree-roosting bat during summer. Physiological and Biochemical Zoology, 76: 868–876.

Turbill, C., B. S. Law, and F. Geiser. 2003b. Summer torpor in a free-ranging bat from subtropical Australia. Journal of Thermal Biology, 28: 223–226.

Tuttle, M.D. 1976. Population ecology of the grey bat (*Myotis grisescens*): factors influencing growth and survival of newly volant young. Ecology, 57: 587–595.

Tuttle, M.D., and D. Stevenson. 1982. Growth and survival of bats. Pp. 105–150. In: Ecology of Bats (T.H. Kunz, ed.). Plenum Press, New York.

Vaughan, T.A., and T.J. O'Shea. 1976. Roosting ecology of the pallid bat (*Antrozous pallidus*). Journal of Mammalogy, 57: 19–42.

Vonhof, M.J., and R.M.R Barclay. 1996. Roost-site selection and roosting ecology of forest dwelling bats in southern British Columbia. Canadian Journal of Zoology, 74: 1797–1805.

Wang, L.C.H. 1989. Ecological, physiological and biochemical aspects of torpor in mammals and birds. Pp. 361–401. In: Advances in Comparative and Environmental Physiology (L.C.H. Wang, ed.). Springer, Berlin.

Webb, P.I., J.R. Speakman, and P.A. Racey. 1993. The implications of small reductions in body temperature for radiant and convective heat loss in resting endothermic brown long-eared bats (*Plecotus auritus*). Journal of Thermal Biology, 18: 131–135.

Whitaker, J.O., Jr., and S. Gummer. 1992. Hibernation of the big brown bat, *Eptesicus fuscus*, in buildings. Journal of Mammalogy, 73: 312–316.

Wilde, C.J., M.A. Kerr, C.H. Knight, and P.A. Racey. 1995. Lactation in vespertilionid bats. Symposia of the Zoological Society of London, No. 67: 139–149.

Willis, C.K.R. 2003. Physiological ecology of roost selection in female, forest-living big brown bats (*Eptesicus fuscus*) and hoary bats (*Lasiurus cinereus*). Ph.D. Thesis, University of Regina, Regina, Saskatchewan, Canada.

Willis, C.K.R., and R.M. Brigham. 2004. Roost switching, roost sharing and social cohesion: forest dwelling big brown bats (*Eptesicus fuscus*) conform to the fission–fusion model. Animal Behaviour, 68: 495–505.

Willis, C.K.R., K.A. Kolar, A.L. Karst, M.A. Kalcounis Rueppell, and R.M. Brigham. 2003. Medium- and long-term reuse of trembling aspen cavities as roosts by big brown bats (*Eptesicus fuscus*). Acta Chiropterologica, 5: 85–90.

Wunder, B.A., and R.D. Gettinger. 1996. Effects of body mass and temperature acclimation on the nonshivering thermogenic response of small mammals. Pp. 131–139. In: Adaptations to the Cold: 10th International Hibernation Symposium (F. Geiser, A.J. Hulbert, and S.C. Nicol, eds.). University of New England Press, Armidale, Australia.

4

Exploring the Evolution of the Basal Metabolic Rate in Bats

Ariovaldo P. Cruz-Neto & Kate E. Jones

Rates of energy expenditure are thought to play a crucial role in shaping the evolution of the behavior, ecology, and physiology of organisms. The most frequently measured rate of energy expenditure is the basal metabolic rate (BMR). Mammals show an enormous range in BMR, with the variability spanning five orders of magnitude. Here we review current hypotheses proposed to explain this variation and test these hypotheses using data from 95 species of bats from 10 families. Our analysis of the evolution of BMR in bats indicates that there is a significant phylogenetic component to BMR: closely related species have more similar rates of basal metabolism than more distantly related species. After controlling for this effect, the most important determinant of BMR in bats is body size, explaining 84% of the variation. We also found that several other life-history and ecological factors (independent of phylogeny and body size) played an important role in the evolution of BMR in bats, offering mixed support to current hypotheses. We also explore whether variation in BMR has allowed rates of diversification to change over evolutionary time but find no evidence to suggest that this has occurred.

INTRODUCTION

It is widely assumed that energy plays a pivotal role in shaping the behavior, ecology, and physiology of organisms. In fact, much of contemporary ecological theory attempts to understand the link between energetics and factors such as patterns of species richness, reproductive effort, distribution, activity patterns, and other life-history traits (e.g., Alexander, 1999; McNab, 1992a; Thompson, 1992). Among the several energetic parameters used to investigate such links, basal metabolic rate (BMR) stands as one of the most important. Originally defined as a way to index the minimum rate of energy necessary to maintain homeostasis, BMR is by far the most widely measured energetic parameter. In mammals, BMR accounts for more than 50% of the total free-ranging energy expenditure (Nagy et al., 1999; Speakman, 2000) and consequently may have overt ecological and evolutionary significance. Operationally, BMR is defined as occurring in postabsorptive animals within the zone of thermal neutrality in the absence of physical activity during the inactive phase of the normal circadian cycle (Kleiber, 1972; McNab, 1997). Although there is some dispute as to the extent to which these conditions can be completely fulfilled (McNab, 1997; Speakman et al., 1993), BMR provides a standardized measure of energy expenditure that can be used for intra- and interspecific comparisons. Mammals show an enormous variation in BMR from 78470 mL O_2/h in the western roe deer *Capreolus capreolus* to 4.81 mL O_2/h in the black myotis *Myotis nigricans* (Lovegrove, 2000). However, the proximate and, especially, the ultimate factors responsible for its magnitude and variability remain to be comprehensively explored (Lovegrove, 2000, 2001).

OBJECTIVES

We review the most common hypotheses proposed to explain variation in BMR and use an explicitly evolutionary approach to test these hypotheses in bats. Specifically, we examine whether variation in BMR has evolved in tandem with variation in morphological, behavioral, ecological, and physiological traits. We first analyze these variables separately and then statistically control for possible interactions between them using a multivariate approach to develop a model to explain the variation in BMR. We restrict our analysis to a particular group of mammals, the Chiroptera. Bats are among the most diverse groups of mammals (Neuweiler, 2000). Such diversity exists both in terms of numerical abundance and in terms of ecology, life history, and distribution (Kunz and Fenton, 2003). For example, bats span three orders of magnitude in body mass, exhibit one of the highest diversities of feeding habits among any mammalian group, occur in nearly all zoogeographical zones (and hence are exposed to a wide range of different climatic conditions), display an impressive array of behavioral, morphological, and physiological adaptations (partially associated with their unique form of locomotion and echolocation), and have evolved a complex pattern of life-history traits (Barclay and Harder, 2003; Jones and MacLarnon, 2001). Bats are therefore ideal models in which to investigate factors related to variation in rates of energy expenditure, including BMR. Although comprehensive reviews of the physiological ecology of bats exist (McNab, 1982 and, more recently, Speakman and Thomas, 2003), our study is unique in that it uses an explicit phylogenetic approach to evaluate a comprehensive set of hypotheses and variables (see below) and controls for the possible interactions between these variables.

HYPOTHESES, PREDICTIONS, AND DATABASES

Hypotheses and Predictions

We identified 10 recent hypotheses proposed to explain the variation in basal metabolic rate among mammals (see table 4.1 for a summary and specific

Table 4.1 Summary of the Hypotheses Proposed to Explain the Evolution of BMR in Mammals (see text for further details)

Hypotheses	Directional Predictions	References
1. Phylogenetic influence	Phylogenetic distance is positively correlated with similarity in BMR	Bennett and Harvey, 1987; Elgar and Harvey, 1987; Felsenstein, 1985; Harvey et al., 1990
2. Body mass	Body mass is positively correlated with BMR	Calder, 1996; Kleiber, 1932, 1972; West et al., 2001
3. Life histories	Mortality rates and age at maturity are negatively correlated with BMR, while annual fecundity is positively correlated with BMR	Hayssen et al., 1985; McNab, 1980; Symonds, 1999
4. Clade richness	Extinction risk and/or clade richness are positively/negatively correlated with BMR	Glazier, 1987; Marzluff and Dial, 1991; McNab, 1984; Purvis et al., 2003
5. Resource limitation and variability	Geographic range and island endemicity are positively and negatively correlated with BMR, respectively. Latitude is negatively correlated with BMR and geographic regions that are associated with high resource variability (Afrotropics, IndoMalaya, and Australasia) contain species with lower BMRs than more stable geographic regions (Nearctic and Palearctic)	Lovegrove, 2000; McNab, 1980, 1994a,b); Scholander et al., 1950

6. Resource quality	Diet quality (ranked from low to high, i.e., blood–insects–fruit–nectar–meat) is positively correlated with BMR	McNab, 1982, 1986, 1992a,b
7. Body temperature and regulation	Body temperature and degree of temperature regulation are positively correlated with BMR	Giloogy et al., 2001; McNab, 1992; Speakman et al., 1993; Speakman and Thomas, 2003
8. Roost temperature, type, and behavior	Roost temperature, roost size, and degree of openness of roost (foliage versus crevice-dwelling roosts) are positively correlated with BMR	McNab and Bonnaccorso, 2001; Speakman and Thomas, 2003
9. Organ mass	Body organ mass (relative to body mass) is positively correlated with BMR	Dann et al., 1990; Martin, 1981
10. Wing morphology	Aspect ratio is positively correlated with BMR	This study

directional predictions for their effects on BMR). Although these hypotheses are not exhaustive, they do effectively summarize the current state of research in this area. They are briefly described below.

Phylogenetic Influence Closely related species may be more similar to each other because of a shared ancestry rather than because of any adaptation to independent selective pressures (Harvey and Pagel, 1991). How much of the variation in BMR in mammals is due to ecological factors or to phylogenetic affiliation has been extensively debated (e.g., Elgar and Harvey, 1987; Hayssen and Lacy, 1985). This debate extends specifically to bats (e.g., Cruz-Neto et al., 2001; Speakman and Thomas, 2003) but has not been explored comprehensively. We explicitly test the influence that phylogenetic constraints have had on the evolution of BMR in bats. As we will see, how much of the variability in BMR in bats can be ascribed to phylogeny has methodological (cf. Garland et al., 1999) and theoretical (e.g., Elgar and Harvey, 1987; Hayssen and Lacy, 1985; McNab, 1992b) implications.

Body Mass One of the most influential discoveries in mammalian energetics was the dependence of rates of energy expenditure (including BMR) on body mass (Kleiber, 1932, 1972). Many studies have attempted to quantify the allometric relationship between these two variables to unravel the underlying functional explanation (reviewed in Calder, 1996). No one functional explanation for the allometric dependence of BMR has been widely accepted, although there have been significant recent advances in understanding the possible underlying processes (e.g., Bishop, 1999; Dodds et al., 2001; Gillooly et al., 2001; Riisgard, 1999; West et al., 1997, 2001). Despite the importance of body mass on the variation in BMR, it has long been recognized that differences in body mass alone are not the whole story (McNab, 1992a). For example, there is a tenfold difference in the residual variation in BMR observed between mammals of the same body mass (McNab, 1988, 1990). Additionally, the observation that there is also heterogeneity of slopes between different taxonomic levels and sizes has brought into question whether a single allometric exponent accurately describes the relationship between body mass and BMR (Dodds et al., 2001; Hayssen and Lacy, 1985; Lovegrove, 2000; but see Gillooly et al., 2001). Here we explore the extent to which BMR in bats is influenced by body mass and also assess the evidence for heterogeneity in slopes and the residual unexplained variation.

Life Histories The correlation between BMR and life history has been the subject of extensive speculation (e.g., Charnov, 1993, 2001; Harvey et al., 1991; Kozlowski and Weiner, 1997; McNab, 1980; Symonds, 1999; Thompson, 1992; West et al., 2001). It is predicted that the evolution of high BMR should be associated with the evolution of fast life-history traits (faster growth rates, earlier maturity, higher reproductive output), independent of body mass (McNab, 1980). However, evidence for a direct link between metabolic rate and life histories is not conclusive and plagued with methodological problems (Harvey et al., 1991; Reynolds and Lee, 1996).

We statistically investigate these patterns, and choose three life-history traits to represent the speed of an organism's life: age at sexual maturity, adult mortality, and annual fecundity. These variables were suggested as the main focus for selection pressures in an influential life history optimality model (Charnov, 1993, 2001): age at sexual maturity is optimized to maximize reproductive output in the face of environmental mortality.

Clade Richness Species richness of a clade is the sum of the rate of extinction and speciation, and here we examine whether changes in species richness of clades is correlated with changes in BMR. There are several directional predictions that could be made from theoretical considerations. For example, if BMR is directly related to life histories, then taxa with higher BMRs may have shorter generation times and are likely to be better able to track changing environments and recover more quickly from population crashes, both of which confer a reduced risk of extinction (Marzluff and Dial, 1991; Purvis et al., 2003). The associated rapid evolution and enhanced rate of adaptation to new niches (Van Valen, 1973) could increase the chance and rate of speciation. So either through decreased extinction or increased speciation, clades that contain species with a high BMR may be more speciose. Alternatively, in situations where energy availability is lower than individual energy demands, species with low BMRs would be favored as they can exploit resources more efficiently. In these environments, these species may have a lower risk of extinction, as they would be less affected by habitat resource limitation (McNab, 1994a,b). In this case, clades containing species with low BMR would be predicted to be more speciose (Glazier, 1987).

Resource Limitation and Variability In areas where food resources are thought to be limited or unpredictable, species with higher energetic demands (high BMR) may have an increased likelihood of extinction. Such conditions are more likely to occur for species that occupy smaller geographic ranges (exposed to a more limited range of resources and environmental conditions), occur on small islands, or inhabit geographic regions with unpredictable food resources due to climate variability. To adapt to these constraints, species may have evolved a lower energetic demand (Lovegrove, 2000; McNab 1994a,b). While quantifying geographic range or the degree of island endemicity of a species is relatively simple, defining resource variability is more complicated. Latitude has been widely used as a proxy for variability in many studies (Ashton et al., 2000; McNab, 1980; Speakman, 2000), since high latitudes have marked seasonal climatic variations. However, the use of this measure to quantify resource variability has been criticized and it has been suggested that using the geographic region in which species occur to measure variability may be more appropriate (Lovegrove, 2000). Moreover, latitude may also encompass variation in absolute temperatures as well as resource variability, further complicating the interpretation of this variable. We investigate the relationship of the variation in BMR to both possible measures of resource variability (latitude and geographic region) and range size (table 4.1).

Resource Quality Besides availability, the rate at which energy is acquired and used may have evolved in a correlated fashion with resource quality (McNab, 1980, 1986, 1992b; Thompson, 1992). For bats, an evolutionary shift in diet is typically associated with a concurrent shift in physiological, morphological, and behavioral features (Freeman, 2000; Schondube et al., 2001). Early studies (McNab, 1969, 1982) suggested that nectar-, meat-, and fruit-eating bats have a high residual BMR compared with species that feed on insects or blood. High BMR would be predicted to be associated with species that feed on easily digestible diets and/or food that contains fewer chemical toxins (McNab, 1986). However, despite its popularity in the literature, this hypothesis has not been rigorously examined in a taxonomically comprehensive sample of bats consuming different diets.

Body Temperature and Regulation Body temperature is an important determinant of metabolic rate in a wide range of organisms (Gillooly et al., 2001), including bats (McNab and Bonaccorso, 2001; Speakman and Thomas, 2003). These studies find that higher body temperatures (T_b) are correlated with higher rates of basal metabolism across species. However, this relationship has not been examined in an explicit phylogenetic or multivariate context. Additionally, there is considerable intraspecific variation in T_b. Although some mammals regulate T_b within narrow limits, others show appreciable variation (Lovegrove et al., 1991; McNab, 1969, 1982, 1988). Some species can down regulate their T_b and metabolism for certain periods, a process referred to as torpor (here we equate torpor with hibernation; for a complete description of these states see Geiser and Ruf, 1995). However, whether the propensity to enter torpor has allowed the evolution of lower BMRs remains unclear, as previous analyses have not considered other possible confounding characteristics. For example, some mammals that use torpor extensively also have lower BMRs (e.g., marsupials: Lovegrove, 1996; McNab, 1983; bats: McNab, 1969, 1982, 1983; McNab and Bonaccorso, 2001). However, this correlation might be due to the confounding effects of body mass. Additionally, torpor and a lower BMR might be simple parallel adaptations to a common problem: resource limitation, variability or quality. By using a multivariate approach, we address these questions more rigorously.

Roost Temperature, Type, and Behavior The roost types used by bats are amazingly diverse, and include caves, hollow trees, buildings, and foliage (for reviews see Kunz, 1982, and Kunz and Lumsden, 2003). The thermal characteristics of roosts are thought to be pivotal for the evolution of patterns of energy expenditure in bats (Bonaccorso et al., 1992; Rodriguez-Duran, 1995; Speakman and Thomas, 2003; chapter 3). Certain types of roosts may buffer variations in external conditions better than others. For example, caves may buffer external temperatures more than tree hollows and foliage. Thus, the energetic benefits associated with roost microclimate will likely vary with the roost temperature (Speakman and Thomas, 2003) and type of roost. Additionally, roost microclimate may be modified through the energetic benefits accrued from coloniality (McNab and Bonaccorso, 2001; Rodriguez-Duran, 1998). Thus, roost temperature, type, and behavior may have had an important influence on the evolution of metabolic energy requirements.

Organ Mass Different organs and tissues respire at different rates and may affect whole-organism metabolic rates (Singer et al., 1995). Organs that make energy available (liver, gastrointestinal tract), transport energy (heart), and control neuronal activities (brain) are found to have the highest metabolic rates even though they constitute only a small fraction of the total body mass (Hulbert and Else, 2000; Krebs, 1950; Martin, 1981). High maintenance costs of such metabolically expensive organs may contribute to the interspecific (Daan et al., 1990) and intraspecific (Konarzewski and Diamond, 1995) residual variation observed in BMR. We tested the organ-mass hypothesis by analyzing the effects of the gastrointestinal length, heart and brain mass on variation in BMR, controlling for the effects of body mass.

Wing Morphology Wings with high aspect ratio are associated with higher flight efficiencies and low flight costs (Norberg and Rayner, 1987). Even though aerodynamic models, which assume a constant efficiency of translating metabolic to mechanical power, do not predict a fixed ratio between BMR and flight cost (Speakman and Thomas, 2003), we assumed that flight cost would be positively correlated with BMR, irrespective of the precise value of the ratio. We based our argument on one of the assumptions of the aerobic model for evolution of endothermy, namely the existence of a functional link between BMR and rates of energy expenditure for activity (Bennett and Ruben, 1979; Hayes and Garland, 1995). Thus, we predict that the evolution of higher aspect ratio of wings would be correlated with the evolution of a low BMR.

Databases

Data were assembled from the literature, building on previous databases compiled by Jones (1998), Jones and MacLarnon (2001), and Jones et al. (2003). Variables for 95 bat species were gathered from 148 published primary and secondary literature sources. These data sources are highlighted in the Literature Cited, with a number in parenthesis given after the reference. The database is available from the authors.

Specifically, we assembled data for the following variables:

Basal Metabolic Rate (mL O_2 h^{-1}), according to the definition of McNab (1997);
Adult Body Mass (g);
Clade Richness: current extinction risk (from Hilton-Taylor, 2000), recoded after Purvis et al. (2000) as least concern = 0, near-threatened = 1, vulnerable = 2, endangered = 3, critically endangered = 4 (data-deficient and extinct species were excluded);
Life History: age at sexual maturity (months);

Adult Mortality Rate (years), following the definition in Purvis and Harvey (1995);
Annual Fecundity: number of female offspring per female per year;
Resource Limitation and Variability: current geographic range (km^2);
Island Endemicity: 0 = not endemic to islands, 1 = endemic;
Geographic Zone (following Lovegrove, 2000): 0 = Palearctic, 1 = Nearctic, 2 = Australasian, 3 = Afrotropical, 4 = Neotropical, 5 = Indomalayan;
Latitude (degrees);
Resource Quality: diet (following McNab, 1969): 1 = predominantly insectivorous, 2 = predominantly sangivorous, 3 = predominantly carnivorous, 4 = predominantly frugivorous, 5 = predominantly nectarivorous;
Body Temperature and Regulation: body temperature (°C); temperature regulation: 1 = deep torpor and hibernation, 2 = shallow torpor, mild heterothermy, 3 = homeotherm;
Roost Temperature, Type, and Behavior: roost temperature (°C); roost type: 1 = caves/mines/tree cavities/buildings only, 2 = foliage only, 3 = mixed roosting habits;
Coloniality: 1 = roost in groups from 1 to 10 individuals, 2 = roost in groups with more than 10 individuals;
Organ Mass: adult brain mass (mg), heart mass (g), stomach length (mm), intestine length (mm);
Wing Morphology: aspect ratio (wingspan^2 divided by wing area).

TESTING THE HYPOTHESES: DATA HANDLING AND ANALYSIS

If species have similar values of BMR because of a shared ancestry rather than because of independent adaptation, then species values do not represent statistically independent points for analysis. Treating these data as independent invalidates the assumptions of most statistical tests and leads to greater type I errors (rejecting a null hypothesis when in fact it is true) (Harvey, 2000). Because of the implications of this statistical problem, we tested for phylogenetic nonindependence in our data using the spatial autocorrelation statistic, Moran's *I* (Gittleman and Kot, 1990). This metric measures the similarity of trait values at different levels in the evolutionary history of the clade of interest. Here we investigated the amount of autocorrelation at four phylogenetic levels: species within genera, genera within families, families within suborders, and suborders within families (using the taxonomic designations in Koopman, 1993). BMR was significantly more similar at three out of the four levels in the phylogeny (figure 4.1). Because of this nonindependence, using individual species values would likely confound our analyses. We therefore employed a widely used phylogenetic comparative method, independent contrasts (Felsenstein, 1985), to generate values that attempt to control for similarity in traits due to a shared ancestry. These phylogenetically independent contrasts were generated using the

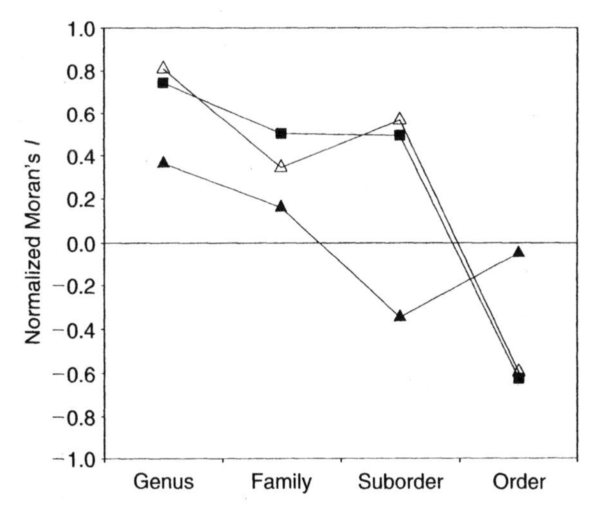

Figure 4.1 Phylogenetic correlation (Moran's I) of BMR (open triangles), body mass (squares), and residual BMR (filled triangles) at each taxonomic level (species within genera, genera within families, families within suborders, suborders within the order). Values of Moran's I vary from +1 to −1, where positive values indicate that a trait at a particular phylogenetic level is more similar than random and negative values indicate that traits are more different. $P < 0.05$ significance is reached at Moran's I values of $> +0.3$ and < -0.2. These results suggest that using species values for three traits as independent points is statistically invalid.

CRUNCH algorithm of the CAIC computer package (Purvis and Rambaut, 1995), based on algorithms in Felsenstein (1985) and Pagel (1992). This method requires that the relationships between all the species in the analysis are known. We used the bat phylogenetic supertree (Jones et al., 2002) to calculate the contrasts. This "supertree" (Sanderson et al., 1998) provides a complete estimate of the relationships among all bat species by formally summarizing all published bat phylogenetic information from a variety of methods and sources, such as informal analyses (i.e., taxonomies) and cladistic and distance-based analyses and using information from both molecular and morphological data.

Correlations between BMR and the trait of interest were investigated using least squares regression through the origin on the independent contrasts (Garland et al., 1992). First, we performed correlations using single model predictors (the trait of interest only) and two-predictor multiple regressions, where the trait of interest and body mass were entered as the predictors (i.e., residual BMR). All continuous variables were $\log_e$ transformed and all statistical tests were two-tailed and were performed using SPSS/PC (version 10.0). Because of inter-correlation among predictors, we also used multiple regression with model simplification in order to find minimum adequate models (MAMs) (Crawley, 1993) for BMR in bats, following the methodology used by Jones et al. (2003). We included all those variables in the MAM that were significant in the two-predictor regression model.

We investigated whether variation in BMR has had a significant influence on clade species richness in bats using the computer program MacroCAIC (Agapow and Isaac, 2002) to generate nested comparisons of sister clades using the bat supertree (Jones et al., 2002). Differences in the numerical diversity of sister clades were compared with differences in BMR at all branching points in the phylogeny except where both clades contained only one species. The richness contrasts used in these analyses are the logarithm of the ratio of species richness of the two clades, with the numerator being the number of species in the clade having the larger value of BMR (Purvis et al., 2003). Under the null hypothesis of no association and diversity, this quantity would have an expectation of zero. Correlations between these variables were tested for using least squares regression through the origin (Isaac et al., 2003). Additionally, we examined whether extinction was correlated with variation in BMR. We approximated bat extinction risk by using current extinction risk measured by IUCN Red List of threatened and endangered species (Hilton-Taylor, 2000) and calculated independent contrasts and assessed significant relationships as before.

RESULTS AND DISCUSSION

Is There a Phylogenetic Component in BMR?

Values of Moran's I vary from $+1$ to -1 where positive values indicate that BMR values at a particular phylogenetic level are more similar than random, whereas negative values indicate that they are more different (Gittleman and Kot, 1990). For species within genera, BMR had a Moran I value close to 0.8 (figure 4.1, open triangles), that is, BMR is significantly more similar among species than would be expected by chance. This conclusion also applies to genera within families and for families within suborders. Thus, our results reinforce the prediction that BMR has a strong phylogenetic component (Bennett and Harvey, 1987; Harvey et al., 1991). On the other hand, Moran I values for BMR for suborders within orders was significantly more different than would be predicted from random: Mega- and Microchiroptera are more different in their values of BMR than would be expected by chance ($P < 0.05$).

We also found a similar pattern of phylogenetic signal in body mass (figure 4.1, squares). Given the strong constraint that body mass may exert on BMR, the observed phylogenetic signal in BMR may be a consequence of the phylogenetic signal found in body mass. However, this is not the case. Residual BMR (obtained by calculating residuals from the independent contrast least squares regression slope of BMR against body size) was still significantly more similar, at least for species within genera. Interestingly, because only BMR of species within genera had a phylogenetic constraint, this may suggest that residual BMR may be more phylogenetically labile, perhaps undergoing many adaptive changes during evolution of bats across higher-level clades.

Is BMR Correlated with Clade Richness in Bats?

Variation in the absolute or residual BMR (relative to body mass) has not been responsible for changes in species richness within bats: regressions of ln clade richness ratio against BMR contrasts were not significantly different from zero ($t = 0.56$, $P = 0.36$, sister-clade contrasts = 42 and $t = 0.09$, $P = 0.90$, sister-clade contrasts = 42, for absolute and residual BMR, respectively). Additionally, we did not find any support for the suggestion that current extinction risk (as measured by the IUCN Red List categories; Hilton-Taylor, 2000) was significantly correlated with variation in BMR in bats ($t = -0.35$, $P = 0.73$, 69 contrasts, $t = -0.74$, $P = 0.46$, 69 contrasts for absolute and residual BMR, respectively). The proposed associations with species richness, that is, clades containing species with high BMRs are more (Marzluff and Dial, 1991) or less species rich (Glaizer, 1987), or at a higher current risk of extinction (McNab, 1994b), were not supported by our analysis. We conclude that BMR has not played a significant direct role in the speciation and/or extinction rate of bats.

It is possible, however, that the effects of BMR might be operating by a correlation with other important variables that themselves determine patterns of species richness, and not by a direct causal effect as we hypothesized. For example, geographic range size is the most important predictor of current extinction risk in bats (Jones et al., 2003) and species number in particular habitats (Ricklefs and Lovette, 1999). Increased extinction (and possibly, if speciation rates are unaffected, decreased clade richness) is correlated with smaller geographic ranges that contain species with lower BMRs (see later). Whatever the reason, we caution readers that only 10% of bat species have been measured for BMR and thus our results are dependent on more complete sampling of this clade.

What Are the Best Predictors for the Evolution of BMR in Bats?

The Importance of Body Size In line with previous analyses of bats (McNab, 1982; Speakman and Thomas, 2003) and mammals generally (Harvey et al., 1991; Hayssen and Lacy, 1985; Lovegrove, 2000), body mass is the most important predictor of BMR, explaining 84.4% (r^2) of the total variation (figure 4.2). Of the 10 families for which we have data, Pteropodidae, Megadermatidae, Rhinolophidae, and Phyllostomidae have the highest BMR, but also are the largest-bodied among these families. Controlling for body size, families Rhinolophidae, Emballonuridae, and Phyllostomidae have the highest BMR (table 4.2). These three families exhibit a major adaptive shift toward higher BMRs from their closest relatives, all of whom have lower BMRs than would be predicted from body size. Theories explaining the fundamental relationship between BMR and body mass crucially depend on the value of the allometric slope coefficient (Dodds et al., 2001; Giloogy et al., 2001; West et al., 2001). Our analyses of bats suggests that the scaling relationship follows the "three quarters" rule (independent contrasts slope coefficient $b = 0.76$, $SE = 0.04$, $t = 19.2$, $P \leq 0.001$, 69 contrasts, $r^2 = 0.84$).

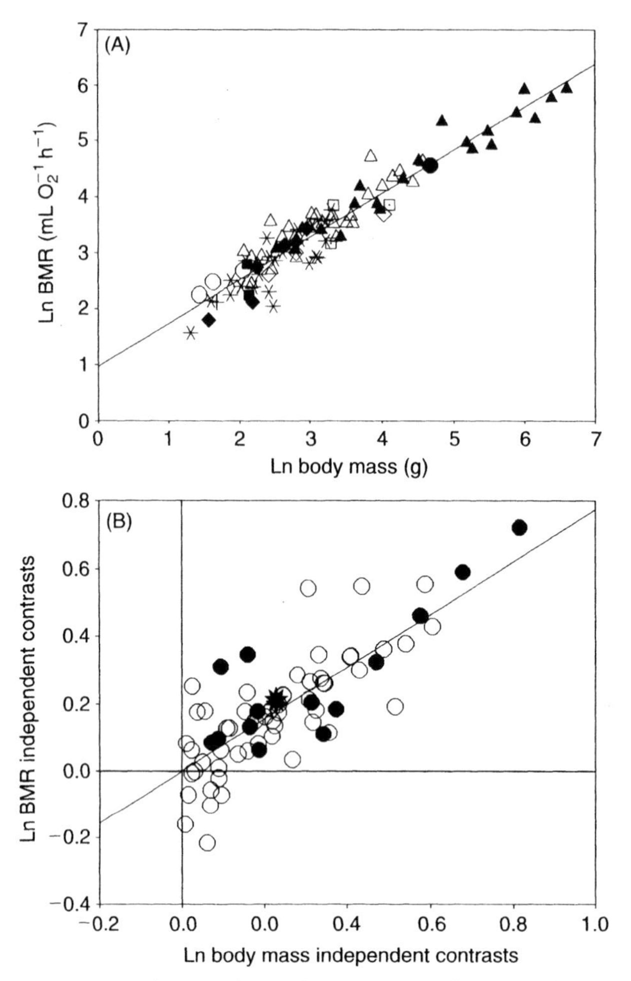

Figure 4.2 Scatter plot of ln basal metabolic rate and ln body mass for 95 species of bats in 10 families (Rhinolophidae is divided into subfamilies Rhinolophinae and Hipposiderinae). (A) Using species as independent data points (slope coefficient $b=0.77$, $t=30.7$, $P < 0.001$, $n=95$, $r^2=0.91$, constant$=0.97$), represents: ○, Emballonuridae; ■, Hipposiderinae; •, Megadermatidae; ◇, Molossidae; ◆, Mormoopidae; +, Natalidae; ⊡, Noctilionidae; △, Phyllostomidae; ▲, Pteropodidae; □, Rhinolophinae; $*$, Vespertilionidae. (B) Using independent contrasts (slope coefficient $b=0.76$, $t=19.2$, $P < 0.001$, $n=69$ contrasts, $r^2=0.84$) represent independent contrasts within microbats, ○ independent contrasts within megabats, •, and $*$ contrasts between megabats and microbats.

This slope value is unlikely to be affected by different reconstructions of the phylogenetic relationships, as the slope value obtained using only species values was very similar ($b=0.77$, SE$=0.03$, $t=30.68$, $P < 0.001$, $r^2=0.91$, constant$=0.97$, and that obtained using species values by Speakman and Thomas, 2003). Our result gives indirect support for a model developed by

Table 4.2 Evolution of Mean Absolute and Residual BMR in 10 Bat Families ($n = 95$ species) Traced across the Bat Supertree (Jones et al., 2002)

Phylogeny	Family	*n*	Mean Absolute BMR ± SE		Mean Residual BMR ± SE	
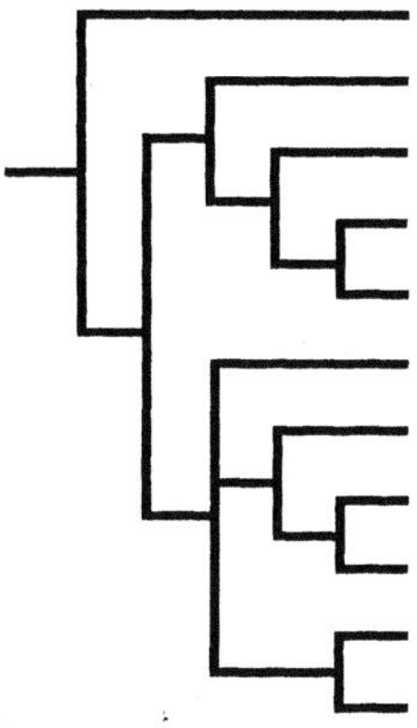	Pteropodidae	21	**134.10**	**(114.78)**	**0.02**	**(0.25)**
	Emballonuridae	3	11.93	(2.51)	**0.16**	**(0.06)**
	Megadermatidae	1	**94.34**		−0.01	
	Rhinolophinae	1	46.5		0.30	
	Hipposiderinae	2	12.78	(4.85)	−0.11	(0.40)
	Natalidae	1	8.32		−0.17	
	Noctilionidae	2	35.37	(16.41)	−0.10	(0.31)
	Mormoopidae	6	17.95	(9.73)	−0.31	(0.05)
	Phyllostomidae	30	40.78	(25.55)	**0.14**	**(0.26)**
	Molossidae	4	26.39	(11.84)	−0.17	(0.19)
	Vespertilionidae	24	17.82	(9.31)	−0.12	(0.31)

Rhinolophidae is divided into subfamilies Rhinolophinae and Hipposiderinae. Values in parentheses are standard errors of the mean. Residual BMR was obtained by calculating residuals from the independent contrast least squares regression slope of BMR against body mass (regression exponent, $b = 0.758$, $t = 19.2$, $P < 0.001$, $n = 69$ contrasts, $r^2 = 0.84$). Values in bold indicate that (for absolute BMR) families have a higher mean BMR than average (for residual BMR), or families have a higher BMR than would be predicted for their body mass.

West and colleagues (Gilloogy et al., 2001; West et al., 1997, 2001), which suggests that selection acting to optimize the efficiency of fractal resource delivery systems (such as lungs and blood vessels) produces a BMR scaling relationship of 0.75. However, other nonfractal models also predict this "three quarter" power rule and the ultimate functional relationship between these two variables remains an important and exciting area for further research.

Is there any heterogeneity in the relationship between BMR and body mass? Previous studies across mammals have suggested that there is no single allometric line that describes the relationship between BMR and body mass: either the relationship varies taxonomically (Hayssen and Lacey, 1985) or at small and large sizes (Lovegrove, 2000). However, within bats alone there is little evidence for different scaling relationships. For example, the fact that the Moran's *I* values for residual BMR comparing suborders within the order was not significantly different from random suggests that megabats and microbats do not scale differently. This is also supported by the independent contrasts plot. If there was a significant difference in the scaling of BMR and body mass within megabats and microbats, then the contrast comparing these two clades would be a significant outlier to the overall cloud of points (Purvis and Rambaut, 1995). This is clearly not the case; the contrast comparing this node (illustrated as an asterisk in figure 4.2B) is nested

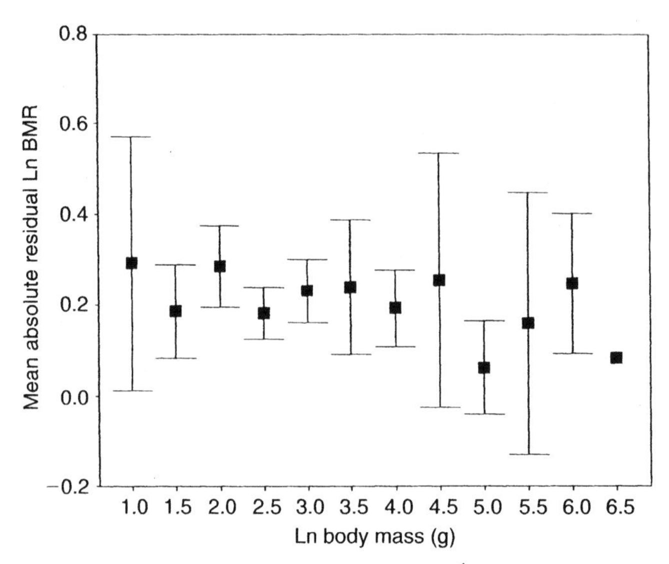

Figure 4.3 Mean absolute residual BMR (mL O_2 h^{-1}) calculated for 10 body size classes (bin size: ln = 0.5). The error bars are ± standard errors. There was no detectable increase in the size of residuals either above or below the constrained body size: 358 g (ln358 = 5.89, represented by the vertical line).

within the other contrasts. Lovegrove (2000) found a "bowtie" effect in the relationship between BMR and body mass across mammals as a whole. He suggested that there was greater variability in residual BMR for species with body masses that were lower or higher than the "constrained body mass" that he calculated to be 358 g. This he suggested caused heterogeneity in the scaling of BMR and further suggested different selective forces were acting above and below this size, invalidating many previous investigations into factors influencing the evolution of BMR. Is there any evidence for a similar "bowtie" effect in bats? We found no detectable "bowtie" trend in residual BMR values for bats either above or below the proposed constrained body mass of 358 g (figure 4.3), nor any evidence that there is a constrained body mass at any other size within the range of the 95 species for which we have data.

What Support Is There for the Other Hypotheses? Life Histories: Our results show that the evolution of lower BMRs is correlated with faster life histories (earlier ages at maturity, higher rates of mortality, and higher rates of annual reproduction, although the last was not significant when the confounding effects of body size are removed) (table 4.3). Our results suggest that BMR does play a role in shaping patterns of life-history evolution, contrary to previous findings across broad taxonomic levels of mammals and birds (Harvey et al., 1991; Read and Harvey, 1989). Symonds (1999) also found that variation in BMR was significantly correlated with life-history patterns in insectivores. He suggested that the failure of early studies to find a correlation between BMR and life-history traits was because such studies

Table 4.3 Results of Regressions of Independent Contrasts for Predicting Bat BMR for Each of the Hypotheses

Hypothesis	Variable	N_s	N_c	One-Predictor Model	Two-Predictor Model
Life histories	Age at sexual maturity	22	17	1.69	2.77*
	Adult mortality	12	9	−1.64	−4.77**
	Annual fecundity	64	47	−2.23*	0.89
Resource limitation and variability	Geographic range	91	67	2.24*	3.64**
	Island endemicity	95	69	−2.95**	−5.15***
	Zoogeographical zone	92	68	0.23	−0.06
	Latitude	95	69	1.02	1.02
Resource quality	Diet	95	69	3.21**	0.32
Body temperature and regulation	Body temperature	80	63	4.25***	5.44***
	Temperature regulation	85	65	6.66***	1.08
Roost temperature, type, and behavior	Roost temperature	49	43	−1.90	−1.45
	Roost type	95	69	−0.11	0.83
	Coloniality	81	62	−0.75	0.25
Organ mass	Brain mass	68	55	11.9***	0.12
	Heart mass	21	18	11.1***	1.30
	Stomach length	24	21	2.72*	−1.70
	Intestine length	27	23	6.53***	1.04
Wing morphology	Aspect ratio	62	48	0.40	−2.81**

N_s, number of species; N_c, number of contrasts.
t-statistics are presented for the one-predictor model (variable) and for two-predictor multiple regression, with the variable and body mass: *$P < 0.05$; **$P < 0.01$; ***$P < 0.001$.

looked across too wide a range of variation: pooling together large and small mammals may have obscured any pattern because BMR may constrain life-history traits differently in these groups (see also Lovegrove, 2000). Our results for bats are consistent with this hypothesis, potentially because they form a relatively heterogeneous small-bodied clade (compared with the range in size found within mammals generally; Jones and Purvis, 1997). However, the direction of the association between BMR and life-history patterns found in bats is opposite to that generally assumed (e.g., McNab, 1980) (table 4.3). Closer examination suggests that many of the life-history optimality models predict a relationship with increased production and faster life histories, rather than with higher metabolic rates per se (e.g., Charnov, 1993; 2001; Kozlowski and Weiner, 1997). This makes predicting the direction of any association between BMR and life histories difficult as the relationship between BMR and production is unclear. For example, production is a trade-off between assimilation and energy required for maintenance (Kozlowski and Weiner, 1997). Thus, although the energy required for maintenance increases with increasing BMR, it is not clear how this might affect assimilation and therefore overall lifetime production. In bats, if higher rates of metabolism decrease overall lifetime production (in effect, slowing life histories) then these results would be consistent with the current optimality models (Charnov, 2001; Kozlowski and Weiner, 1997). For example, a higher BMR may confer lower mortality through an increased capacity for efficient long-distance flight, thereby increasing an organism's ability to escape predators and exploit additional resources.

Resource Limitation and Variability: Several lines of evidence suggest that resource limitation and variability might influence the evolution of BMR (Lovegrove, 2000; McNab, 1994b). Small oceanic islands are associated with a limited resource base (McNab, 1994b) and, by the same reasoning, a similar relationship would be predicted between range size and resource availability. Controlling for phylogeny and body size, we found that bat species with smaller geographic ranges and/or restricted to small oceanic islands have evolved a lower BMR (table 4.3). This is consistent with McNab (1994b; see also McNab and Bonaccorso, 2001; McNab and Armstrong, 2001), who also reported these associations in a small number of pteropodid species (although not controlling for possible confounding variables).

Although we found support for the hypothesis that resource limitation influences BMR, we could find no evidence to suggest that resource variability had a similar effect. Lovegrove (2000) demonstrated that small species of mammals in geographical zones associated with high resource variability (Afrotropical, IndoMalaya, Australasia, and Neotropical) have lower BMR than species in more stable geographic regions (Nearctic and Palearctic). Our results suggest that this hypothesis does not hold for bats, either for absolute or for residual BMR (table 4.2). Lovegrove's (2000) analyses may not be comparable to ours as his were based on data across a wide range of mammalian orders (orders may respond differently) and did not control for the effects of body size or phylogeny (except within Rodentia). Similarly we found that latitude was unassociated with BMR in bats (supported by Speakman and Thomas, 2003). While the ability of latitude to reflect resource variability and how this result should be interpreted is unclear (Lovegrove, 2000), it seems unlikely from these analyses that the evolution of BMR is strongly influenced by resource variability. However, further analyses should employ more direct measures of variability such as predictability in annual rainfall or temperature to provide a more rigorous measure of variability.

Resource Quality: After controlling for the effects of a shared phylogenetic history, our results suggest that resource quality has led to an evolution of higher rates of BMR in bats (table 4.3), supporting the findings and predictions of McNab (1969, 1980, 1986, 1992b). However, fruit-, meat-, and nectar-feeding species generally tend to be large and therefore have a high BMR. Once the effect of body size is removed using residuals, we find no evidence to suggest that high resource quality is associated with high BMR. This conclusion is supported by an analysis of the association between feeding habits and BMR in the bats within the family Phyllostomidae (Cruz-Neto et al., 2001) (phyllostomids show the highest diversity of feeding habits within bats: Freeman, 2000), which detected no association of diet with residual BMR once the effects of phylogeny were controlled for. Overall, our results suggest that, as far as BMR is concerned, an evolutionary shift in diet is not correlated with a shift in the level of basal energy expenditure once the effect of body size is removed. However, a recent analysis suggests that the evolution of dietary preferences in some clades of bats may have been too recent to allow for morphological or physiological change (Lewis-Oritt et al., 2001). For example, Lewis-Oritt et al. (2001) suggest that the evolution of piscivory in *Noctilio* was recent enough to preclude adaptational divergence in other traits. The extent to which this type of evolutionary lag influences our results awaits further investigation.

Body Temperature and Regulation: Our results indicate that T_b is highly correlated with variation in BMR, even when the effects of phylogeny and body mass are controlled for (table 4.3). This offers further support for previous analyses of bats that also report species with high BMRs to have higher T_bs (McNab and Bonnacorso, 2001; Speakman and Thomas, 2003). Moreover, our result is consistent with a recent energetic model which suggests that most of the variation in organismal BMR is explained by T_b and body mass (Gilloogy et al., 2001). The amount of variation explained by these two variables is one of the highest of any of the two predictor multivariate models in our analysis (0.90). A more complicated picture of association was obtained when we investigated the correlation of BMR and propensity of species to enter torpor. Our analysis confirms the findings of previous nonphylogenetic studies that the capacity for temperature regulation is strongly correlated with BMR (McNab, 1969, 1982, 1983; McNab and Bonaccorso, 2001). Species with a lower body mass and, hence, a low BMR, also show a greater propensity to enter torpor (table 4.3). However, with the effects of body mass controlled for, the association disappears. This more robust analysis implies that once one controls for the fact that smaller species have a low BMR, the ability to enter torpor has not led to the evolution of lower BMRs.

Roost Temperature and Behavior: Our results indicate that neither roost temperature nor the type of behavior by bats within a roost has affected absolute or residual BMR (table 4.3). In contrast to a previous nonphylogenetic interspecfic study (Speakman and Thomas, 2003), we did not find a significant positive effect of roost temperature on BMR. Remembering that many of the same data are used in both studies, we suggest that Speakman and Thomas's result is an artifact of analysis that was uncontrolled for phylogenetic similarity. However, other intraspecific analyses have also suggested that the thermal profile of available roosts does affect energy expenditure of bats and that the distribution of suitable roosts may actually limit the distribution of some species (Arlettaz et al., 2000; Baudinette et al., 2000; Bell et al., 1986). This suggests that the proximate factors responsible for the maintenance of a given level of basal energy expenditure in bats might not be the same as those that promoted the evolution of this trait.

We also found no evidence that roost type influences BMR. It is possible that roost type is not an accurate reflection of the actual temperature to which given species are exposed in their roost, as individuals may alter their environment through behavioral changes. For example, clustering has been suggested to be an important behavioral mechanism by which bats modify the temperature of a roost and alter metabolic costs (Brown, 1999; Trune and Slobodchikoff, 1976; reviewed in Speakman and Thomas, 2003). However, we could find no evidence that our measure of clustering capacity within a roost (here measured as coloniality) was associated with an evolutionary increase in BMR. This conclusion is similar to a study of pteropodid bats that also failed to find an association (McNab and Bonaccorso, 2001). This does not necessarily invalidate clustering behavior as an important ultimate process influencing the evolution of BMR, because our measure of clustering may have been too crude to capture its variation. For example, the fact that bats are colonial does not necessarily mean that they cluster within a roost. Clearly we need better data on the behavior of bats within a roost to thoroughly test this hypothesis.

Organ Mass: We found no indication that increased organ size relative to body mass caused an increase in BMR in bats (table 4.3). How bat species afford relatively larger, more energetically expensive, tissues is not explained by increasing overall BMR. However, further analyses should explore the possibility of the occurrence of trade-offs between the sizes of energetically expensive tissues. For example, in primates it has been demonstrated that while overall BMR has remained the same, an increase in the relative size of one energetically expensive tissue (brain size) has led to a concomitant decrease in the size of others (intestine size) (Aiello and Wheeler, 1995; but see Jones and MacLarnon, 2004).

Wing Morphology: We found that one aspect of wing morphology (aspect ratio) has had a significant influence on the evolution of variation in BMR in bats (table 4.3). Once the effects of body mass have been removed, the evolution of longer and thinner wings has caused an associated decrease in rates of BMR. Wings with a high aspect ratio are associated with higher flight efficiencies and lower flight costs (Norberg and Rayner, 1987), and this increase in the efficiency of the use of energy required for flight may have allowed species to decrease their overall energy budget. However, this association may be caused by an interaction between BMR and other life-history or ecological traits that we are not detecting in this analysis. For example, aspect ratio is an important predictor of the variation in other traits such as colony size (Entwistle et al., 2000), current extinction risk (Jones et al., 2003), and smaller foraging ranges (Jones et al., 1995); thus further analyses of this intriguing association are required.

Minimum Adequate Model for BMR Much of our discussion is based on results from single- or two-predictor regression models and there may be interactions between other variables that would influence the associations found. For example, although we found that the variation in BMR can be explained by independent interactions between body mass and temperature, we do not know whether these relationships would still hold if we also considered the effect of life history in the same analysis. To address this problem, we used an approach that has been successful in a number of other similar studies exploring the determinants of extinction risk in mammals (Purvis et al., 2000; Jones et al., 2003). Specifically, we built a minimum adequate model to explain the variation in BMR. Our analyses (table 4.4) support the suggestion that large body mass (partial correlation, pc = 0.89; $P < 0.001$) and later ages at sexual maturity (pc = 0.60; $P < 0.05$) are independent predictors of BMR, explaining 94.5% of the total variance. Additionally, there was evidence that a decrease in body temperatures and lower wing aspect ratios were important independent predictors

Table 4.4 Minimum Adequate Model Predicting the Variation in BMR in All Bats

Variable #	***b***
N (contrasts)	14 (13)
r^2 (%)	94.5
Body mass	0.778***
Age at sexual maturity	0.544*
Body temperature	−3.976#
Aspect ratio	−1.288#
Geographical range	ns
Island status	ns

Values for traits are slope exponents, *b*. *N* indicates the number of species in the model with the number of contrasts in parentheses; ns indicates that the predictor was not significant in the model; # indicates that the predictor was significant at $P = 0.05$ with the other predictors remaining significant; $^*P < 0.05$, $^{***}P < 0.001$.

although not significant (pc = 0.36; $P = 0.051$ and pc = 0.35; $P = 0.054$ for temperature and aspect ratio, respectively). However, because of missing values for many of the predictors, the sample size of the minimum adequate model was low and consequently the power to detect significant relationships may also have been decreased. Although the usefulness of this multivariate approach is obvious, its practical application is limited because of these problems and further analysis is required with a more complete dataset to address them.

CONCLUSIONS

Evolutionary changes in the variation in BMR in bats are correlated with evolutionary changes in multiple variables. To a large extent, a species-specific BMR is determined by the species' phylogenetic history. If its ancestor had a high BMR then it is likely that it does too. Controlling for this phylogenetic similarity, the most important influence on the variation in BMR in bats is body mass. However, within this constraint, there have also been responses to different selection pressures. Higher rates of basal metabolism are associated with slower life histories and are also determined by the amount of resources available, since resource-limited areas are occupied by species with low BMRs. There is no evidence that species-specific BMRs reflect resource variability, although our measures may have been too crude to detect such a relationship. Despite the long reported association between diet and BMR, we found no evidence that BMR is associated with an adaptive shift toward higher quality diets (once the effect of body size has been controlled for). Body temperature does exert a significant influence on BMR, although the ability to enter torpor did not have an effect on BMR after controlling for body size. We could find no evidence that roost temperature, roost type, or behavior of individuals within a roost has played a major role in the evolution of BMR. Larger relative organ sizes have not been associated with a correlated increase in BMR. Wing morphology does seem to have played an important role in the evolution of BMR in bats, species with long narrow wings (high aspect ratios) having evolved low BMRs. These analyses suggest that these characters have acted independently on the evolution of BMR, but there are likely to be complex interactions between the variables that confound our interpretations. Controlling for the interrelationships between these variables, the minimum adequate model we generated supports only the body size and life-history effect. However, because of a low sample size, we lacked power to detect significant relationships and further analyses are required.

Finally, we also point out that the ultimate factors responsible for the evolution of BMR may not be the same as the proximate factors. Organisms show phenotypic plasticity in their basal rate of energy expenditure and may respond with adjustments of their BMR to proximate factors. For example, recent intraspecific studies show that species may adjust their BMR when faced with poor-quality resources (Koteja, 1996; Veloso and Bozinovic, 1993, 2000; but see Geluso and Hayes, 1999; Nussear et al., 1998). These results suggest that factors responsible for the maintenance of a given level of basal energy expenditure in bats might not be the same as those that promoted the evolution of this trait. Further analyses should consider how much of a role both proximate and ultimate factors have played in the evolution of this trait.

ACKNOWLEDGMENTS

We wish to thank J. Gittleman, M. Brigham, F. Bozinovic, N. Isaac, D. Thomas, C. Willis, A. Webster, and an anonymous referee for comments on previous versions of the manuscript and discussion of this and related work. Financial support was from NERC grant (UK) GR8/04371 and University of Virginia (K.E.J.) and FAPESP grant 0009968-8 (Brazil) (A.P.C.-N.).

LITERATURE CITED

Adams, J. 1989. *Pteronotus davyi*. Mammalian Species, 346: 1–5 (120).

Agapow, P.-M., and N.J.B. Isaac. 2002. MacroCAIC: revealing correlates of species richness by comparative analysis. Diversity and Distributions, 8: 41–43.

Aiello, L.C., and P. Wheeler. 1995. The expensive-tissue hypothesis. The brain and the digestive system in human and primate evolution. Current Anthropology, 36: 199–221.

Alexander, M.R. 1999. Energy for Animal Life. Oxford University Press, New York.

Arends, A., F.J. Bonaccorso, and M. Genoud. 1995. Basal rates of metabolism of nectarivorous bats (Phyllostomidae) from a semiarid thorn forest in Venezuela. Journal of Mammalogy, 76: 947–956 (11).

Arlettaz, R., C. Ruchet, J. Aeschimann, E. Brun, M. Genoud, and P. Vogel. 2000. Physiological traits affecting the distribution and wintering strategies of the bat *Tadarida teniotis*. Ecology, 81: 1004–1014.

Ashton, K.G., M.C. Tracy, and A. de Queiroz. 2000. Is Bergmann's rule valid for mammals? The American Naturalist, 156: 390–415.

Barbour, R.W., and W.H. Davis. 1969. Bats of America. University Press of Kentucky, Lexington, KY (146).

Barclay, R.M.R., and L.D. Harder. 2003. Life histories of bats: life in the slow lane. Pp. 209–256. In: Bat Ecology (T.H. Kunz and M.B. Fenton, eds.). University of Chicago Press, Chicago.

Baron, G., H. Stephan, and H.D. Frahm. 1996. Comparative Neurobiology in Chiroptera. Volume 1. Macromorphology, brain structures, tables and atlases. Birkhauser, Basel (53).

Bartels, W., B.S. Law, and F. Geiser. 1998. Daily torpor and energetics in a tropical mammal, the northern blossom bat *Macroglossus minimus* (Megachiroptera). Journal of Comparative Physiology B, 168: 233–239 (17).

Bartholomew, G.A., P. Leitner, and J.E. Nelson. 1964. Body temperature, oxygen consumption and heart rate in three species of Australian flying foxes. Physiological Zoology, 37: 179–198 (19).

Bates, P.J.J., and D.L. Harrison. 1997. Bats of the Indian Subcontinent. Harrison Zoological Museum, Sevenoaks, Kent, UK (72).

Baudinette, R.V., S.K. Churchill, K.A. Christian, J.E. Nelson, and P.J. Hudson. 2000. Energy, water balance and the roost microenvironment in three Australian cave-dwelling bats (Microchiroptera). Journal of Comparative Physiology B, 170: 439–466 (4).

Bell, G.P., G.A. Bartholomew, and K.A. Nagy. 1986. The roles of energetics, water economy, foraging behavior, and geothermal refugia in the distribution of the bat *Macrotus californicus*. Journal of Comparative Physiology B, 156: 441–450 (13).

Bennett, A.F., and J.A. Ruben. 1979. Endothermy and activity in vertebrates. Science, 206: 649–654.

Bennett, P.M., and P.H. Harvey. 1987. Active and resting metabolism in birds: allometry, phylogeny and ecology. Journal of Zoology (London), 213: 327–363.

Bergmans, W. 1990. Taxonomy and biogeography of African fruit bats (Mammalia, Megachiroptera) 3. The genera *Scotonycteris* Matschie 1894, *Casinycteris* Thomas 1910, *Pteropus* Brisson 1762 and *Eidolon* Rafinesque 1815. Beaufortia, 40: 111–177 (79).

Bergmans, W. 1994. Taxonomy and biogeography of African fruit bats (Mammalia, Megachiroptera). 4. The genus *Rousettus* Gray 1821. Beaufortia, 44: 79–126 (80).

Bergmans, W. 1997. Taxonomy and biogeography of African fruit bats (Mammalia, Megachiroptera). 5. The genera *Lissonycteris* Andersen 1912, *Myonycteris* Matschie, 1899 and *Megaloglossus* Pagenstecher, 1885; General remarks and conclusions; Annex: Key to all species. Beaufortia, 47: 11–90 (78).

Bernard, R.T.F., F.P.D. Cotterill, and R.A. Fergusson. 1996. On the occurrence of a short period of delayed implantation in Schreiber's long-fingered bat (*Miniopterus schreibersii*) from a tropical latitude in Zimbabwe. Journal of Zoology London, 238: 13–22 (102).

Bernard, R.T.F., J. Paton, and K. Sheppey. 1988. Relative brain size and morphology of some South African bats. South African Journal of Zoology, 23: 52–58 (58).

Bhatnagar, K.P., H.D. Frahm, and H. Stephan. 1990. The megachiropteran pineal organ: a comparative morphological and volumetric investigation with special emphasis on the remarkably large pineal of *Dobsonia praedatrix*. Journal of Anatomy, 168: 143–166 (57).

Bishop, C.M. 1999. The maximum oxygen consumption and aerobic scope of birds and mammals: getting to heart of the matter. Proceedings of the Royal Society of London, Series B, 266: 2275–2281.

Bonaccorso, F.J. 1998. Bats of Papua New Guinea. Conservation International, Washington, DC (75).

Bonaccorso, F.J., and B.K. McNab. 1997. Plasticity of energetics in blossom bats (Pteropodidae): impact on distribution. Journal of Mammalogy, 78: 1073–1088 (18).

Bonaccorso, F.J., A. Arends, M. Genoud, D. Cantoni, and T. Morton. 1992. Thermal ecology of moustached and ghost-faced bats (Mormoopidae) in Venezuela. Journal of Mammalogy, 73: 365–378 (8).

Boo Liat, L. 1970. Food habits and breeding cycle of the Malaysian fruit eating bat *Cynopterus brachyotis*. Journal of Mammalogy, 51: 174–177 (135).

Bradbury, J.W., and S.L. Vehrencamp. 1976. Social organization and foraging in emballonurid bats: I. Field studies. Behavioral Ecology and Sociobiology 1: 337–381 (115a).

Bradshaw, G.V.R. 1962. Reproductive cycle of the California leaf-nosed bat, *Macrotus californicus*. Science, 136: 645–646 (95).

Brown, C.R. 1999. Metabolism and thermoregulation of individual and clustered long-fingered bats, *Miniopterus schreibersi*, and the implications for roosting. South African Journal of Zoology, 34: 166–172.

Calder III, W.A. 1996. Size, Function, and Life History. Dover Publications, Mineola, NY.

Carpenter, R.E. 1968. Salt and water metabolism in the marine fish-eating bat *Pizonyx vivesi*. Comparative Biochemistry and Physiology, 24: 951–964 (28).

Carpenter, R.E. 1969. Structure and function of the kidney and the water balance of desert bats. Physiological Zoology, 42: 288–302 (37).

Carpenter, R.E., and J.B. Graham. 1967. Physiological responses to temperature in the long-nosed bat, *Leptonycteris sanborni*. Comparative Biochemistry and Physiology, 22: 709–722 (9).

Charnov, E.L. 1993. Life History Invariants. Some Explorations of Symmetry in Evolutionary Ecology. Oxford University Press, Oxford.

Charnov, E.L. 2001. Evolution of mammal life histories. Evolutionary Ecology Research, 3: 521–535.

Cheke, A.S., and J.F. Dahl. 1981. The status of bats on western Indian Ocean islands with special reference to *Pteropus*. Mammalia, 45: 205–238 (139).

Churchill, S.K. 1995. Reproductive ecology of the orange horseshoe bat, *Rhinonycteris aurantius* (Hipposideridae: Chiroptera), a tropical cave-dweller. Wildlife Research, 22: 687–698 (100).

Churchill, S.K. 1998. Australian Bats. Reed New Holland, Sydney (65).

Constantine, D.G. 1966. Ecological observations on lasiurine bats in Iowa. Journal of Mammalogy, 47: 34–41 (144).

Corbet, G.B. 1978. The Mammals of the Palaearctic Region: A Taxonomic Review. British Museum (Natural History). Cornell University Press, London (81).

Corbet, G.B., and J.E. Hill. 1992. The Mammals of the Indomalayan Region: A Systematic Review. Oxford University Press, Oxford (73).

Crawley, M.J. 1993. GLIM for Ecologists: Methods in Ecology. Blackwell Science, Oxford.

Cruz-Neto, A.P., and A.S. Abe. 1997. Taxa metabolica e termoregulacao no morcego nectarivoro *Glossophaga soricina* (Chiroptera, Phyllostomidae). Revista Brasileira de Biologia, 57:203–209 (12).

Cruz-Neto, A.P., T. Garland, Jr., and A.S. Abe. 2001. Diet, phylogeny, and basal metabolic rate in phyllostomid bats. Zoology, 104: 49–58 (10).

Cruz-Neto, A.P. 2002. Unpublished data (42).

Daan, S., D. Masman, and A. Groenewold. 1990. Avian basal metabolic rates: their association with body composition and energy expenditure in nature. American Journal of Physiology, 259: R333–R340.

Dickerman, R.W., K.F. Koopman, and C. Seymour. 1981. Notes on bats from the pacific lowlands of Guatemala. Journal of Mammalogy, 62: 406–411 (125).

Dodds, P.S., D.H. Rothman, J.S. Weitz. 2001. Re-examination of the "3/4-law" of metabolism. Journal of Theoretical Biology, 209: 9–27.

Douglas, A.M. 1967. The natural history of the ghost bat *Macroderma gigas* (Microchiroptera, Megadermatidae) in Western Australia. The Western Australian Naturalist, 10: 125–137 (116).

Dwyer, P.D. 1975. Notes on *Dobsonia moluccensis* (Chiroptera) in the New Guinea highlands. Mammalia, 39: 113–119 (136).

Eby, P. 1991. Seasonal movements of grey-headed flying foxes, *Pteropus poliocephalus* (Chiroptera: Pteropodidae) from two maternity camps in Northern New South Wales. Wildlife Research, 18: 547–559 (138).

Eisenberg, J.F. 1989. Mammals of the Neotropics: The Northern Neotropics. Volume 1. University of Chicago Press, Chicago (62).

Eisenberg, J.F., and K.H. Redford. 1992. Mammals of the Neotropics: The Southern Cone. Volume 2. University of Chicago Press, Chicago (64).

Eisenberg, J.F., and K.H. Redford. 1999. Mammals of the Neotropics: The Central Neotropics. Volume 3. University of Chicago Press, Chicago (61).

Eisentraut, M. 1950. Die Ernährung der Fledermäuse. Zool. Jahrbücher. Jena, 79: 114–177 (50).

Elgar, M.A., and P.H. Harvey. 1987. Basal metabolic rates in mammals: allometry, phylogeny and ecology. Functional Ecology, 1: 25–36.

England, A. 2002. Bat Conservation International. Personal communication (85).

Entwistle, A.C., P.A. Racey, and J.R. Speakman, 2000. Social and population structure of a gleaning bat, *Plecotus auritus*. Journal of Zoology (London), 252: 11–17.

Felsenstein, J. 1985. Phylogenies and the comparative method. The American Naturalist, 125:1–15.

Flannery, T.F. 1995. Mammals of New Guinea. Australian Museum/Reed Books, Chatswood, New South Wales, Australia (74).

Flannery, T.F. 1995. Mammals of the South-West Pacific and Moluccan Islands. Australian Museum/Reed Books, Chatswood, New South Wales, Australia (76).

Fleming, T.H. 1988. Short-tailed Fruit Bat. A study in Plant–Animal Interactions. University of Chicago Press, Chicago (107).

Fleming, T.H., D.E. Wilson, and E.T. Hooper. 1972. Three Central American bat communities: structure, reproductive cycles and movement patterns. Ecology, 53: 555–569 (93).

Forman, G.L. 1972. Comparative morphological and histochemical studies of stomachs of selected American bats. University of Kansas Science Bulletin, 49: 591–729 (43).

Freeman, P.W. 2000. Macroevolution in Microchiroptera: recoupling morphology and ecology with evolution. Evolutionary Ecology Research, 2: 317–335.

Gaisler, J. 1979. Ecology of bats. Pp 281–342. In: Ecology of Small Mammals (D.M. Stoddart, ed.). Chapman and Hall, London (113).

Garbutt, N. 1999. Mammals of Madagascar. Pica Press, Sussex, UK (68).

Garland, T., P.H. Harvey, and A.R. Ives. 1992. Procedures for the analysis of comparative data using phylogenetically independent contrasts. Systematic Biology, 41: 18–32.

Garland, T., Jr., P.E. Midford, and A.R. Ives. 1999. An introduction to phylogenetically based statistical methods, with a new method for confidence intervals on ancestral values. American Zoologist, 39: 374–388.

Geiser, F., and R.M. Brigham. 2000. Journal of Comparative Physiology B, 170:153–162 (35).

Geiser, F., and T. Ruf. 1995. Hibernation versus daily torpor in mammals and birds: physiological variables and classification of torpor patterns. Physiological Zoology, 68: 935–966.

Geiser, F., D.K. Coburn, G. Kortner, and B.S. Law. 1996. Thermoregulation, energy metabolism, and torpor in blossom bats, *Syconycteris australis* (Megachiroptera). Journal of Zoology (London), 239: 583–590 (21).

Geluso, K., and J.P. Hayes. 1999. Effects of diet quality on basal metabolic rate and internal morphology of European starlings (*Sturnus vulgaris*). Physiological and Biochemical Zoology, 72: 189–197.

Genoud M., and F.J. Bonaccorso. 1986. Temperature regulation, rate of metabolism and roost temperature in the greater white-lined bat *Saccopteryx bilineata* (Emballonuridae). Physiological Zoology, 59:49–54 (2).

Genoud, M. 1993. Temperature regulation in subtropical tree bats. Comparative Biochemistry and Physiology, 104: 321–332 (24).

Genoud, M., and F.J. Bonaccorso, and A. Arends. 1990. Rate of metabolism and temperature regulation in two small tropical insectivorous bats (*Peropteryx macrotis* and *Natalus tumidirostris*). Comparative Biochemistry and Physiology, 97: 229–234 (1).

Gerell-Lundberg, K., and R. Gerell. 1994. The mating behavior of the pipistrelle and the Nathusius pipistrelle (Chiroptera)—a comparison. Folia Zoologica, 43: 315–324 (106).

Gillooly, J.F., J.H. Brown, G.B. West, V.M. Savage, and E.L. Charnov. 2001. Effects of size and temperature on metabolic rate. Science, 293: 2248–2251.

Gittleman, J.L., and M. Kot. 1990. Adaptation: statistics and a null model for estimating phylogenetic effects. Systematic Zoology, 39:227–241.

Glazier, D.S. 1987. Energetics and taxonomic patterns of species diversity. Systematic Zoology, 36: 62–71.

Gopalakrishna, A., A.T. Varute, V.M. Sapkal, A.R. Unune, and G.C. Chari. 1985. Breeding habits and associated phenomena in some Indian bats: Part XI. *Miniopterus schreibersii fuliginosus* (Hodgson) Vespertilionidae. Journal of Bombay Natural History Society, 82: 594–601 (101).

Hall, E.R. 1981. The Mammals of North America, Volume 1. Wiley, New York (63).

Hamlett, G.W. 1947. Embryology of the molossid bat *Eumops*. Anatomical Record, 97: 340–341 (117).

Hanus, K. 1959. Body temperature and metabolism of bats at different environmental temperatures. Physiologia Bohemoslovaca, 8: 250–259 (27).

Happold, D.C.D., and M. Happold. 1989. Reproduction of Angola free-tailed bats (*Tadarida condylura*) and little free tailed bat (*Tadarida pumila*) in Malawi (central Africa) and elsewhere in Africa. Journal of Reproduction and Fertility, 85: 133–149 (91).

Harrison, D.L., and P.J.J. Bates. 1991. The Mammals of Arabia. Harrison Zoological Museum, Sevenoaks, Kent, UK (87).

Harvey, P.H. 2000. Why and how phylogenetic relationships should be incorporated into studies of scaling. Pp. 253–265. In: Scaling in Biology (J.H. Brown and G.B. West, eds.). Oxford University Press, Oxford.

Harvey, P.H., and M.D. Pagel. 1991. The Comparative Method in Evolutionary Biology. Oxford University Press, Oxford.

Harvey, P.H., M.D. Pagel, and J.A. Rees. 1991. Mammalian metabolism and life-history. The American Naturalist, 137: 556–566.

Hayes, J.P., and T. Garland, Jr. 1995. The evolution of endothermy: testing the aerobic capacity model. Evolution, 49: 836–847.

Hayssen, V., and R.C. Lacy. 1985. Basal metabolic rate in mammals: taxonomic differences in the allometry of BMR and body mass. Comparative Biochemistry and Physiology A, 81: 741–754.

Hayssen, V., A. van Tienhoven, and A. van Tienhoven. 1993. Asdell's Patterns of Mammalian Reproduction. Comstock Publishing Associates, Ithaca, NY (90).
Hayward, B.J., and E.L. Cockrum. 1971. The natural history of the western long-nosed bat *Leptonycteris sanborni*. Office of Research, Western New Mexico University, 1: 74–123 (130).
Heideman, P.D., and L.R. Heaney. 1989. Population biology and estimates of abundance of fruit bats (Pteropodidae) in Philippine submontane rainforest. Journal of Zoology (London), 218: 565–586 (109).
Herd, R.M. 1983. *Pteronotus parnelli*. Mammalian Species, 209:1–5 (121).
Hernandez, C.S., T.C. Osorio, and C.B.C. Tapia. 1986. Patron reproductivo de *Sturnira lilium parvidens* (Chiroptera: Phyllostomidae) en La Costa Central del Pacifico de Mexico. The Southwestern Naturalist, 31: 331–340 (131).
Hilton-Taylor, C. 2000. 2000 IUCN Red List of Threatened Species. IUCN, Gland, Switzerland (45).
Hosken, D.J. 1997. Thermal biology and metabolism of the greater long-eared bat, *Nyctophilus major* (Chiroptera: Vespertilionidae). Australian Journal of Zoology 45: 145–156 (31).
Hosken, D.J., and P.C. Withers. 1997. Temperature regulation and metabolism of an Australian bat, *Chalinolobus gouldii* (Chiroptera: Vespertilionidae) when euthermic and torpid. Journal of Comparative Physiology B, 167: 71–80 (23).
Hosken, D.J., and P.C. Withers. 1999. Metabolic physiology of euthermic and torpid lesser long-eared bats, *Nyctophilus geoffroyi* (Chiroptera: Vespertilionidae). Journal of Mammalogy, 80: 42–52 (30).
Hulbert, A.J., and P.L. Else. 2000. Mechanisms underlying the cost of living in animals. Annual Review of Physiology, 62: 207–235.
Humphrey, S.R., and J.B. Cope. 1970. Population samples of the evening bat, *Nycticeius humeralis*. Journal of Mammalogy, 51: 399–401 (115b).
Humphrey, S.R., and J.B. Cope. 1976. Population ecology of the little brown bat, *Myotis lucifugus*, in Indiana and North-Central Kentucky. Special Publication of the American Society of Mammalogists, 4: 1–81 (112).
Isaac, N.J.B., P.-M. Agapow, P. H. Harvey, and A. Purvis. 2003. Phylogenetically nested comparisons for testing correlates of species richness: a simulation study of continuous variables. Evolution, 57: 18–26.
Jones, G. 2002. Bristol University. Personal communication (60).
Jones, G., P.L. Duverge, and R.D. Ransome. 1995. Conservation biology of an endangered species: field studies of greater horseshoe bats. Symposia of the Zoological Society of London, 67: 309–324.
Jones, K.E. 1998. Evolution of Bat Life-Histories, Ph.D. Dissertation, University of Surrey Roehampton, London.
Jones, K.E., and A. MacLarnon. 2001. Bat life-histories: testing models of mammalian life-history evolution. Evolutionary Ecology Research, 3: 465–476.
Jones, K.E., and A. MacLarnon. 2004. Affording larger brains: testing hypotheses of mammalian brain evolution on bats. The American Naturalist, 164: 20–31.
Jones, K.E., and A. Purvis. 1997. An optimum body size for mammals? Comparative evidence from bats. Functional Ecology, 11: 751–756.
Jones, K.E., A. Purvis, A. MacLarnon, O.R.P. Bininda-Emmonds, and N.B. Simmons. 2002. A phylogenetic supertree of the bats (Mammalia: Chiroptera). Biological Reviews, 77: 223–259.
Jones, K.E., A. Purvis, and J.L. Gittleman. 2003. Bat extinction risk. The American Naturalist, 164: 601–614.

Jurgens, K.D., H. Bartels, and R. Bartels. 1981. Blood oxygen transport and organ weights of small bats and small non-flying mammals. Respiration Physiology, 45: 243–260 (39).

Jurgens, K.D., and J. Prothero. 1987. Scaling of maximal lifespan in bats. Comparative Biochemistry and Physiology, 88: 361–367 (56).

Kallen, F.C. 1977. The cardiovascular system of bats: structure and function. Pp. 290–484. In: Biology of Bats, Vol. 3 (W.A. Wimsatt, ed.). Academic Press, New York (41).

Kingdon, J. 1974. East African Mammals. Vol. 2, Part A. Academic Press, London (67).

Kitchener, D.J. 1975. Reproduction in female Gould's wattled bat *Chalinolobus gouldii* (Gray) (Vespertilionidae) in Western Australia. Australian Journal of Zoology, 23: 29–42 (143).

Kleiber, M. 1932. Body size and metabolism. Hilgardia, 6: 315–353.

Kleiber, M. 1972. Body size, conductance for animal heat flow and Newton's law of cooling. Journal of Theoretical Biology, 37: 139–150.

Konarzewski, M., and J. Diamond. 1995. Evolution of basal metabolic rate and organ masses in laboratory mice. Evolution, 49: 1239–1248.

Koopman, K.F. 1992. Biogeography of the bats of South America. Pp. 237–302. In: Mammalian Biology in South America (M. Mares and H. H.H. Genoways, eds.), Vol. 6. Special Publication Series, Pymatuning Laboratory of Ecology, University of Pittsburgh (69).

Koopman, K.F. 1993. Bats. Pp. 137–242. In Mammal Species of the World: A Taxonomic and Geographic Reference (D.E. Wilson and D. M. Reeder, eds.). Smithsonian Institution Press, Washington, DC (70).

Koteja, P. 1996. Limits to the energy budget in a rodent *Peromyscus maniculatus:* does gut capacity set the limit? Physiological Zoology, 69: 994–1020.

Kovtun, M.F., and N.F. Zhukova. 1994. Feeding and digestion intensity of chiropterans of different trophic groups. Folia Zoologica, 43: 377–386 (51).

Kozlowski, J., and J. Weiner. 1997. Interspecific allometries are by-products of body size optimization. The American Naturalist, 149: 352–380.

Krebs, H.A. 1950. Body size and tissue respiration. Biochimica Biophysica Acta, 4: 249–269.

Krutzsch, P.H., and E.G. Crichton. 1985. Observations on the reproductive cycle of female *Molossus fortis* (Chiroptera: Molossidae) in Puerto Rico. Journal of Zoology London, 207: 137–150 (118).

Kunz, T.H. 1982. Roosting ecology of bats. Pp. 1–55. In: Ecology of Bats (T.H. Kunz, ed.). Plenum Press, New York.

Kunz, T.H., and M.B Fenton (eds.). 2003. Bat Ecology. University of Chicago Press, Chicago.

Kunz, T.H., and L.F. Lumsden. 2003. Ecology of cavity and foliage roosting bats. Pp. 3–89. In: Bat Ecology (T.H. Kunz and M.B. Fenton, eds.). University of Chicago Press, Chicago.

Kunz, T.H., and A. Stern. 1995. Maternal investment and post-natal growth in bats. Symposia of the Zoological Society London, 67: 123–138 (36).

Kurta, A., and T.H. Kunz. 1988. Roosting metabolic rate and body temperature of male little brown bats (*Myotis lucifugus*) in summer. Journal of Mammalogy, 69: 645–651 (25).

Law, B.S. 1992. The adaptive evolution of the digestive system of the Chiroptera. Canadian Journal of Zoology, 69: 1853–1856 (49).

Lawrence, M.A. 1991. Biological observations on a collection of New Guinea *Syconycteris australis* (Chiroptera, Pteropodidae) in the American Museum of Natural History. American Museum Novitates, 3024 (126).

Leitner, P. 1966. Body temperature, oxygen consumption, heart rate and shivering in the California mastiff bat *Eumops perotis*. Comparative Biochemistry and Physiology, 19: 431–443 (5).

Lewis, S.E. 1992. Behavior of Peter's tent-making bat *Uroderma bilobatum* at maternity roosts in Costa Rica. Journal of Mammalogy, 73: 541–546 (133).

Lewis, S.E., and D.E. Wilson. 1987. *Vampyressa pusilla*. Mammalian Species, 292: 1–5 (134).

Lewis-Oritt, N., R.A. Van den Bussche, and R.J. Baker. 2001. Molecular evidence for evolution of piscivory in *Noctilio* (Chiroptera: Noctilionidae). Journal of Mammalogy, 82: 748–759.

Licht, P., and P. Leitner. 1967. Physiological responses to high environmental temperatures in three species of microchiropteran bats. Comparative Biochemistry and Physiology, 22: 371–387 (6).

Loughrey, W.J., and G.F. McCracken. 1991. Factors influencing female-pup scent recognition in Mexican free-tailed bats. Journal of Mammalogy, 72: 624–626 (119).

Lovegrove, B.G. 1996. The low basal metabolic rate of marsupials: the influence of torpor and zoogeography. Pp. 141–151. In: Adaptations to Cold: 10th International Hibernation Symposium (F. Geiser, A.J. Hulbert and S.C. Nicols, eds.). University of New England Press, Armidale, Australia.

Lovegrove, B.G. 2000. The zoogeography of mammalian basal metabolic rate. The American Naturalist, 156: 201–219.

Lovegrove, B.G. 2001. The evolution of body armor in mammals: plantigrade constraints of large body size. The American Naturalist, 55: 1464–1473.

Lovegrove, B.G., G. Heldmaier, and T. Ruf. 1991. Perspectives of endothermy revisited: the endothermic temperature range. Journal of Thermal Biology, 16: 185–197.

Madkour, G. 1977. Ear ossicles and tympanic bone of some Egyptian bats (Microchiroptera). The Annals of Zoology, 13: 63–81 (52).

Madkour, G.A., E.M. Hammouda, and J.G. Ibrahim. 1982. Histology of the alimentary tract of two common Egyptian bats. The Annals of Zoology, 19: 53–73 (48).

Maloney, S.K., G.N. Bronner, and R. Buffenstein. 1999. Thermoregulation in the Angolan free-tailed bat *Mops condylurus*: a small mammal that uses hot roosts. Physiological and Biochemical Zoology, 72: 385–396 (34).

Martin, R.D. 1981. Relative brain size and basal metabolic rate in terrestrial vertebrates. Nature, 293: 57–60.

Marzluff, J. M., and K.P. Dial. 1991. Life-history correlates of taxonomic diversity. Ecology, 72: 428–439.

McCracken, G.F., and J.W. Bradbury. 1981. Social organization and kinship in the polygynous bat *Phyllostomus hastatus*. Behavioral Ecology and Sociobiology, 8: 11–34 (108).

McLean, J.A., and J.R. Speakman. 1999. Energy budgets of lactating and non-reproductive brown long-eared bats (*Plecotus auritus*) suggest females use compensation in lactation. Functional Ecology, 13: 360–372 (33).

McNab B.K. 1989. Temperature regulation and rate of metabolism in three Bornean bats. Journal of Mammalogy, 70: 153–161 (3).

McNab, B.K. 1969. The economics of temperature regulation in neotropical bats. Comparative Biochemistry and Physiology, 31: 227–268 (14).

McNab, B.K. 1980. Food habits, energetics, and the population biology of mammals. The American Naturalist, 116: 106–124.

McNab, B.K. 1982. Evolutionary alternatives in the physiological ecology of bats. Pp 151–200. In: Ecology of Bats (T.H. Kunz. ed.). Plenum Press, New York.

McNab, B.K. 1983. Energetics, body size, and the limits to endothermy. Journal of Zoology (London), 199: 1–29.

McNab, B.K. 1986. The influence of food habits on the energetics of eutherian mammals. Ecological Monographs, 56: 1–19.

McNab, B.K. 1988. Complications inherent in scaling the basal rate of metabolism in mammals. Quarterly Review of Biology, 63: 25–54.

McNab, B.K. 1990. The physiological significance of body size. Pp. 11–23. In: Body Size in Mammalian Paleobiology: Estimation and Biological Implications (J. Damuth and B.J. MacFadden, eds.). Cambridge University Press, Cambridge.

McNab, B.K. 1992a. Energy expenditure: a short history. Pp. 1–15. In: Mammalian Energetics: Interdiscplinary Views of Metabolism and Reproduction (T.E. Tomasi and T.H. Horton, eds.). Cornell University Press, Ithaca. NY.

McNab, B.K. 1992b. A statistical analysis of mammalian rates of metabolism. Functional Ecology, 6: 672–679.

McNab, B.K. 1994a. Energy conservation and the evolution of flightlessness in birds. The American Naturalist, 144: 628–642.

McNab, B.K. 1994b. Resource use and the survival of land and freshwater vertebrates on oceanic islands. The American Naturalist, 144: 643–660.

McNab, B.K. 1997. On the utility of uniformity in the definition of basal rate of metabolism. Physiological Zoology, 70: 718–720.

McNab, B.K., and M.I. Armstrong. 2001. Sexual dimorphism and scaling of energetics in flying foxes of the genus *Pteropus*. Journal of Mammalogy, 82: 709–720.

McNab, B.K., and F.J. Bonaccorso. 1995. The energetics of pteropodid bats. Symposia of the Zoological Society London, 67: 111–122 (16).

McNab, B.K., and F.J. Bonaccorso. 2001. The metabolism of New Guinean pteropodid bats. Journal of Comparative Physiology B, 171: 201–214 (15).

Medellin, R.A., and H.T. Arita. 1989. *Tonatia evotis* and *Tonatia silvicola*. Mammalian Species, 334: 1–5 (132).

Mickleburgh, S.P., A.M. Hutson, and P.A. Racey. 1992. Old World Fruit Bats. An Action Plan for their Conservation. IUCN, Gland, Switzerland (96).

Mills, R.S., G.W. Barrett, and M.P. Farrell. 1975. Population dynamics of big brown bat (*Eptesicus fuscus*) in Southwestern Ohio. Journal of Mammalogy, 56: 591–604 (110).

Mitchell-Jones, A.J., G. Amori, W. Bogdanowicz, B. Krystufek, P.J.H. Reijnders, F. Spitzenberger, M. Stubbe, J.B.M. Thissen, V. Vohralik, and J. Zima. 1999. The Atlas of European Mammals. T&AD Poyser Natural History, Academic Press, London (82).

Morris, S., A.L. Curtin, and M.B. Thompson. 1994. Heterothermy, torpor, respiratory gas-exchange, water balance and the effect of feeding in Gould's long-eared bat *Nyctophilus gouldi*. Journal of Experimental Biology, 197: 309–335 (29).

Mutere, F.A. 1968. The breeding biology of the fruit bat *Rousettus aegyptiacus* E. Geoffroy living at 0,22'S. Acta Tropica, 25: 97–108. (98).

Myers, P. 1977. Patterns of reproduction of four species of vespertilionid bats in Paraguay. University of California Publications in Zoology, University of California Press, 107:1–38 (104).

Nagorsen, D.W., and R.M. Brigham. 1993. Bats of British Columbia. Vol. 1. The Mammals of British Columbia. University of British Columbia Press, Vancouver, Canada (89).

Nagy, K.A., I.A. Girard, and T.K. Brown. 1999. Energetics of free-ranging mammals, reptiles and birds. Annual Review of Nutrition, 19: 247–277.

Nelson, J.E. 1965. Movements of Australian flying foxes (Pteropodidae: Megachiroptera). Australian Journal of Zoology, 13: 53–73 (97).

Neuweiler, G. 2000. The Biology of Bats. Oxford University Press, New York.

Noll, U.G. 1979. Postnatal growth and development of thermogenesis in *Rousettus aegyptiacus*. Comparative Biochemistry and Physiology A., 63: 89–93 (20).

Norberg, U., and J.M.V. Rayner. 1987. Ecological morphology and flight in bats: wing adaptations, flight performance, foraging strategy and echolocation. Philosophical Transactions of the Royal Society of London, Series B, 316: 335–427 (59).

Nowak, R.M. 1994. Walker's Bats of the World. Johns Hopkins University Press, Baltimore (38).

Nussear, K.E., R.E. Espinoza, C.M. Gubins, K.J. Field, and J.P. Hayes. 1998. Diet quality does not affect resting metabolic rate or body temperature selected by an herbivorous lizard. Journal of Comparative Physiology B, 168: 183–189.

O'Brien, G.M., J.D. Curlewis, and L. Martin. 1993. Effect of photoperiod on the annual cycle of testis growth in a tropical mammal: the little red flying fox, *Pteropus scapulatus*. Journal of Reproduction and Fertility, 98: 121–127 (140).

O'Farrell, M.J., and E.H. Studier. 1970. Fall metabolism in relation to ambient temperatures in three species of *Myotis*. Comparative Biochemistry and Physiology, 35: 697–703 (26).

Okia, N.O. 1987. Reproductive cycles of East African bats. Journal of Mammalogy, 68: 138–141 (137).

Pagel, M.D. 1992. A method for the analysis of comparative data. Journal of Theoretical Biology, 156: 431–442.

Park, H., and E.R. Hall. 1951. The gross anatomy of the tongues and stomachs of eight new world bats. Transactions of the Kansas Academy of Science, 54: 64–72 (44).

Phillips, W.R., and S.J. Inwards. 1985. The annual activity and breeding cycles of Gould's long-eared bat *Nyctphilus gouldi* (Microchiroptera: Vespertilionidae). Australian Journal of Zoology, 33: 111–126 (105).

Pink, B., and O. von Helverson. 1995. Social behavior in a captive colony of *Glossophaga soricina* (Phyllostomidae). Bat Research News, 36: 101 (129).

Pirlot, P., and H. Stephan. 1970. Encephalization in Chiroptera. Canadian Journal of Zoology, 48: 433–444. (55).

Porter, F.L. 1979. Social behavior in the leaf-nosed bat *Carollia perspicillata*: I. Social organization. Zeitschrift für Tierpsychologie, 49: 406–417 (94).

Promislow, D.E.L., and P.H. Harvey. 1991. Mortality rates and the evolution of mammal life histories. Acta Oecologica, 12: 119–137.

Purvis, A., and P.H. Harvey. 1995. Mammalian evolution: a comparative test of Charnov's model. Journal of Zoology (London), 237: 259–283.

Purvis, A., and A. Rambaut. 1995. Comparative analysis by independent contrasts (CAIC): an Apple Macintosh application for analysing comparative data. Computer Applications to Biosciences, 11: 247–251.

Purvis, A., J.L. Gittleman, G. Cowlishaw, and G.M. Mace. 2000. Predicting extinction risk in declining species. Proceedings of the Royal Society of London, Series B, 267: 1947–1952.
Purvis, A., A.J. Webster, P.-M. Agapow, K.E. Jones, and N.J.B. Isaac. 2003. Primate life histories and phylogeny. Pp. 25–40. In: Primate Life History (P.M. Kappeler and M. Pereira, eds.). Cambridge University Press, Cambridge.
Racey, P.A. 1974. Ageing and assessment of reproductive status of Pipistrelle bats *Pipistrellus pipistrellus*. Journal of Zoology (London), 173: 264–291 (99).
Rachmatulina, I. 1992. Major demographic characteristics of populations of certain bats from Azerbaijan. Pp. 127–141. In: Prague Studies in Mammalogy (I. Horacek and V. Vohralik, eds.). Charles University Press, Prague (111).
Ransome, R. 1990. The Natural History of Hibernating Bats. Christopher Helm, London (83).
Read, A.F., and P.H. Harvey. 1989. Life-history differences among the eutherian radiation. Journal of Zoology (London), 219: 329–353.
Reeder, W.G., and K.S. Norris. 1954. Distribution type locality and habits of the fish-eating bat *Pizonyx vivesi*. Journal of Mammalogy, 35: 81–87 (142).
Reid, F.A. 1997. A Field Guide to the Mammals of Central America and Southeast Mexico. Oxford University Press, Oxford (88).
Reynolds, P.S., and R.M. Lee. 1996. Phylogenetic analysis of avian energetics: passerines and nonpasserines do not differ. The American Naturalist, 147: 735–759.
Ricklefs, R.E, and I.J. Lovette. 1999. The roles of island area per se and habitat diversity in the species-area relationships of four Lesser Antillean faunal groups. Journal of Animal Ecology, 68: 1142–1160.
Riisgard, H.U. 1999. No foundation of a "3/4 power scaling law" for respiration in biology. Ecological Letters, 1: 71–73.
Robin, H.A. 1881. Recherches anatomiques sur les Mammiferes de l'order des Chiropteres. Annales des Sciences Naturelles (Zoologie), 12: 1–180 (46).
Rodriguez-Durán, A. 1995. Metabolic rates and thermal conductance in four species of neotropical bats roosting in hot caves. Comparative Biochemistry and Physiology A, 110: 347–355 (7).
Rodriguez-Durán, A. 1998. Nonrandon aggregations and distribution of cave dwelling bats in Puerto Rico. Journal of Mammalogy, 79: 141–146.
Rodriguez-Duran, A., and T.H., Kunz. 1992. *Pteronotus quadridens*. Mammalian Species, 395:1–4 (122).
Sanderson, M.J., A. Purvis, and C. Henze. 1998. Phylogenetic supertrees: assembling the trees of life. Trends in Ecology and Evolution, 13: 105–109.
Scholander, P.F., R. Hock, V. Walters, and L. Irving. 1950. Adaptation to cold in arctic and tropical mammals and birds in relation to body temperature, insulation, and basal metabolic rate. Biological Bulletin, 99: 259–271.
Schondube, J.E., L. Gerardo Herrera-M., and C.M. del Rio. 2001. Diet and the evolution of digestion and renal function in phyllostomid bats. Zoology, 104: 59–73.
Sherman, H.B. 1945. The Florida yellow bat *Dasypterus floridanus*. Proceedings of the Florida Academy of Science, 7: 193–197 (145).
Singer, D., O. Schunck, F. Bach, and H.-J. Kuhn. 1995. Size effects on metabolic rate in cell, tissue, and body calorimetry. Thermochimica Acta, 251: 227–240.
Sluiter, J.W., and M. Bouman. 1951. Sexual maturity in bats of the genus *Myotis*: 1. Size and histology of the reproductive organs during hibernation in connection

with age and wear of the teeth in female *Myotis myotis* and *Myotis emarginatus*. Proceedings of the Koninklijke Nederlandse Akademie van Wetenschappen, Series C, Biological and Medical Sciences, 54: 594–601 (103).

Snyder, G.K. 1976. Respiratory characteristics of whole-blood and selected aspects of circulatory physiology in common short-nosed fruit bat *Cynopterus brachyotis*. Respiration Physiology, 28: 239–247 (40).

Speakman, J.R. 2000. The cost of living: field metabolic rates of small mammals. Advances in Ecological Research, 30: 178–297.

Speakman, J.R., and D.W. Thomas. 2003. Physiological ecology and energetics of bats. Pp. 430–492. In: Bat Ecology (T.H. Kunz and M.B. Fenton, eds.). University of Chicago Press, Chicago (22).

Speakman, J.R., R.M. McDevitt, and K.R. Cole. 1993. Measurement of basal metabolic rate: don't lose sight of reality in the quest for comparability. Physiological Zoology, 66: 1045–1049.

Stephan, H., J.E. Nelson, and H.D. Frahm. 1981. Brain size comparison in Chiroptera. Zoologische Systematik und Evolution-Forschung, 19: 195–222 (54).

Strahan, R. 1995. Mammals of Australia. Australian Museum/Reed Books, Chatswood, New South Wales, Australia (66).

Symonds, M.R.E. 1999. Life histories of the Insectivora: the role of phylogeny, metabolism and sex differences. Journal of Zoology (London), 249: 315–337.

Taddei, V.A. 1976. The reproduction of some Phyllostomidae (Chiroptera) from the Northwestern region of the State of São Paulo. Bolm. Zoologia, Universidade São Paulo, São Paulo, Brazil, 1: 313–330 (127).

Tedman, R.A., and Hall, L.S. (1985) The morphology of the gastrointestinal tract and food transit times in the fruit bats *Pteropus alecto* and *P. poliocephalus*. Australian Journal of Zoology, 33: 625–640 (47).

Thompson, M.J.A. 1987. Longevity and survival of female pipistrelle bats (*Pipistrellus pipistrellus*) on the Vale of York, England. Journal of Zoology (London), 211: 209–214 (114).

Thompson, S.D. 1992. Gestation and lactation in small mammals: basal metabolic rate and the limits of energy use. Pp. 213–259. In: Mammalian Energetics: Interdiscplinary Views of Metabolism and Reproduction (T.E. Tomasi and T.H. Horton, eds.). Cornell University Press, Ithaca, NY.

Trune, D.R., and C.N. Slobodchikoff. 1976. Social effects of roosting on metabolism of the pallid bat (*Antrozous pallidus*). Journal of Mammalogy, 57: 656–663.

van der Merwe, M. 1987. Adaptive breeding strategies in some South African bats between 22 degrees S and 28 degrees S. South African Journal of Science, 83: 607–609 (147).

Van Valen, L. 1973. Body size and numbers of plants and animals. Evolution, 27: 27–35.

Van Zyll de Jong, C.G. 1985. Handbook of Canadian Mammals: 2. Bats. National Museums of Canada, Ottawa, Canada (86).

Veloso, C., and F. Bozinovic. 1993. Dietary and digestive constraints on basal metabolic rate in a small herbivorous rodent. Ecology, 74: 2003–2010.

Veloso, C., and F. Bozinovic. 2000. Interplay between acclimation time and diet quality on basal metabolic rate in females of degus *Octodon degus* (Rodentia: Octodontidae). Journal of Zoology (London), 252: 531–533.

Webster, W.D., and C.O. Handley. 1986. Systematics of Millar's long-tongued bat *Glossophaga longirostris* with a description of two new subspecies. Occasional Papers, The Museum, Texas Tech University, 100: 1–22 (128).

West, G.B., J.H. Brown, and B.J. Enquist. 1997. A general model for the origin of allometric scaling laws in biology. Science, 276: 122–126.
West, G.B., J.H. Brown, and B.J. Enquist. 2001. A general model for ontogenetic growth. Nature, 413: 628–631.
Wilkins, K.T. 1989. *Tadarida brasiliensis.* Mammalian Species, 331: 1–10 (92).
Willig, M.R. 1985. Reproductive patterns of bats from Caatingas and Cerrado biomes in Northeast Brazil. Journal of Mammalogy, 66: 668–681 (123).
Wilson, D.E. 1971. Ecology of *Myotis nigricans* (Mammalia: Chiroptera) on Barro Colorado Island, Panama Canal Zone. Journal of Zoology (London), 163: 1–13 (148).
Wilson, D.E. 1984. Population biology of *Artibeus jamaicensis* on Barro Colorado Island. Bat Research News, 25 3–4: 50–51 (124).
Wilson, D.E., and S. Ruff. 1999. The Smithsonian Book of North American Mammals. Smithsonian Institution Press, Washington D.C. (71).
Wolton, R.J., P.A. Arak, C.J. Godfray, and R.P. Wilson. 1982. Ecological and behavioural studies of the megachiropteran of Mount Nimba, Liberia and notes on microchiropterans. Mammalia, 46: 419–448 (141).
Yongzu, Z. 1997. Distribution of Mammalian Species in China. China Forestry Publishing House, Beijing, China (84).
Yoshiyuki, M. 1989. A Systematic Study of the Japanese Chiroptera. National Science Museum, Tokyo, Japan (77).

II

Functional Morphology

Elizabeth R. Dumont & William A. Schutt, Jr.

Functional morphologists are united by the basic question: What is the relationship between structural design and how organisms interact with the world around them? Researchers work toward answering this question using tools and approaches from disciplines including, and often integrating, engineering and evolutionary biology. Although the quantity of functional studies in mammals lags far behind those for other vertebrates, analyses of bats continue to generate innovative approaches and techniques. These chapters were selected to provide a snapshot of the approaches to functional morphology used by bat biologists today and are roughly divided between analyses of cranial and postcranial morphology.

Three fundamentally distinct approaches and sets of techniques for investigating functional morphology are exemplified by this series of chapters. First, the chapter by Evans on molar tooth structure and function uses confocal microscopy and computer modeling which represent an engineering-based approach to functional morphology. By analyzing molar teeth from the perspective of tool design, Evans defines variables with specific functional meaning. He further identifies specific functional features of molar teeth that are preserved throughout the wear process and suggests that these elements are subject to strong selective pressures.

The next two chapters take an explicitly experimental approach to linking form and function through assessments of performance. Swartz et al. present a biomechanical analysis of bat flight that breaks new ground with its suggestion that traditional models of bat flight are in need of significant revision. Their combination of computer models, strain analysis, and three-dimensional kinematic data clearly demonstrate that we are only beginning to understand the physical principles that have guided the evolution of flight in bats. The goal of Nicolay and Winter's experimental study is to integrate data from cranial morphology, feeding behavior, and nectar-feeding performance to evaluate nectar feeding in an ecological context. While the authors demonstrate a link between cranial morphology and the bats' ability to extend their tongues, the strength of this experimental approach to evaluating function lies in the finding that both behavioral plasticity and as yet untested ecological factors appear to play a more significant role in mediating nectar-feeding performance.

Whether implicit or explicit, evolution is an underlying theme in most functional studies and strong functional signal is frequently interpreted as indirect evidence of adaptation. Two final chapters explicitly address evolutionary issues by including strong phylogenetic components in their analyses. By illustrating the analogous, independent evolution of quadrupedalism in three distinct clades of bats, Schutt and Simmons are able to suggest a range of selective pressures that may have guided the evolution of this unique form of locomotion. Similarly, Dumont uses modern comparative techniques (independent contrasts and squared change parsimony) to investigate the evolutionary interface between structural and behavioral variation in plant-visiting phyllostomids. Her demonstration of correlated evolution in cranial morphological and feeding behavior suggests a tight association between form, function, and behavior in the evolution of feeding strategies in this clade.

5

Quantifying Relationships between Form and Function and the Geometry of the Wear Process in Bat Molars

Alistair R. Evans

Despite many important predictive features of bat dentition having been described in the literature, a number of these have not been given sufficient attention or measured on actual teeth. I aim to address this by establishing criteria for the choice of useful functional parameters of chiropteran dentition. These are: (1) they apply to tooth components rather than an entire tooth, where the three main tooth components are cusps, crests, and basins; and (2) they have a predictive element, so that changes in the parameter are readily interpretable in terms of changes in function, such as increased force or energy required for the component to function. Parameters that influence the function of cusps and crests, in forced crack propagation of tough foods are described in detail: tip and cusp sharpness, cusp occlusion relief, rake angle, crest relief, approach angle, food capture, edge sharpness, and fragment clearance. All the parameters were measured for the upper molars of the microchiropteran *Chalinolobus gouldii*. Many of the parameters did not change significantly with heavy wear, pointing to geometrical and design characteristics for the maintenance of shape with wear in the dilambdodont tooth form.

INTRODUCTION

Molar function as a means of understanding animal diets has received considerable attention in recent years (e.g., Fortelius, 1985; Kay, 1984), and such work aids in the interpretation of teeth in both extant and fossil animals. This view is supported by demonstrations of the importance of the dentition to the nutritional ecology of an animal (Lanyon and Sanson, 1986; McArthur and Sanson, 1988; Pérez-Barberia and Gordon, 1998).

Examining teeth as tools for the breakdown of food aids in analyzing tooth function (Lucas, 1979). As for any tool, the shape of a tooth is a significant determinant of its function, and its examination should be revealing. The goal of functional studies of dentition should be to understand the interaction between tooth shape and function, allowing inferences of dental function and predictive measures of tooth function from morphology.

Previous analyses, however, have not quantified aspects that relate shape to function, perhaps because of the difficulty of doing so with traditional methods. The aim of this chapter is to closely examine tooth shape and function using engineering principles and computer modeling. My particular focus is dilambdodont molars, possessed by most microchiropterans and some insectivores, but the principles and findings can equally be applied to similar tooth forms (e.g., zalambdodont, tribosphenic).

Variation in dilambdodont-like (in particular, primate) tooth shape has been quantified in previous studies by measuring changes in the relative or absolute lengths of the crests (Kay, 1975; Kay et al., 1978), or the sum of all main crest lengths on a tooth standardized for tooth length or area (Strait, 1993). Another approach, by Freeman (1988), was to analyze tooth structure by comparing the relative proportions of the palate or tooth row dedicated to molars, the dilambdodont ectoloph, premolars, canines or incisors. However, none of the characters in these studies adequately describe the relationship between the complex tooth morphology and the way the tooth works.

Consideration of dental function is complicated by tooth wear. Change in the shape of teeth is likely to have an effect on their function. The common assumption is that the wear on a mammal's tooth results in an alteration of shape, and therefore function, of the tooth. However, wear may or may not change the shape of teeth. The teeth of most herbivores, such as the selenodont molars of bovids, are essentially nonfunctional when they first erupt. Moderate wear by tooth–tooth or tooth–food contact is required to transform the tooth shape into its functional configuration (Janis and Fortelius, 1988; Luke and Lucas, 1983). Beyond this point, shape, and therefore most probably function, is static despite the large amount of wear that occurs on teeth during most of the remainder of the animal's life. Eventually, these high-crowned teeth stop growing and the functional form rapidly degrades.

For dilambdodont-like teeth, the relationship between wear and shape has been largely neglected. The preformed occlusal morphology is presumed to be fully functional, requiring no wear for the tooth to become operational (Luke and Lucas, 1983). It is implicitly assumed that wear adversely affects the function of the teeth; however, there is little substantive data to support this proposition.

Functional occlusion of dilambdodont molars is complex, because precise alignment of the many interlocking crests and cusps is required. The possibility of malocclusion due to wear-induced changes has not been addressed in studies of this tooth form.

There may be design features of dilambdodont teeth that minimize changes in tooth shape during the wear process, thereby maintaining tooth shape and function as well as proper occlusion. Such features are common in herbivore teeth (such as the vertical enamel pillars of the molars: Janis and Fortelius, 1988) but have not been adequately described in dilambdodont teeth. Most functional studies have examined only unworn or slightly worn molars, as the measures of function used are inadequate to describe the changes in function with wear.

In one of the few studies on the effect of wear on tooth shape, Ungar and Williamson (2000) used surface reconstruction and topographic software to investigate gorilla teeth. Their study mostly relates to changes in the tooth shape as a whole (e.g., slope, surface area of the whole tooth, and topographic aspect), and thus are difficult to explicitly relate to any specific change in function. The teeth also differ from those under consideration here in that precise alignment may not be as important.

FUNCTIONAL PARAMETERS

Criteria for Functional Parameters

Two different types of dental measurements have been used as surrogates of function. The first may relate to the quantity of food processed. For example, shear length measurements (Kay, 1975; Kay et al., 1978; Strait, 1993) and relative areas of talonid/trigonid basins (Kay, 1975) are of this type. These variables approximate the amount of food processed or the relative allocation of the dentition to distinct types of food processing, such as "shearing" or "grinding." They also largely reflect the size of the features rather than their shape.

The second type of measure specifically relates the shape of tooth components to their function. These are interpretable in terms of differences in function, such as the amount of force or energy required, and describe the shape of the features rather than size. In lineages showing a decrease in the emphasis on shearing, as shown by decreases in size of crests or the trigonid basin, there may be concurrent modifications in the shape of the crest. The latter will have an equal if not greater effect on the crest's function.

Not all measures of shape will be useful in determining function. First, the parameter should relate to specific components of the teeth rather than the shape of the whole tooth surface at once. Variables relating to the shape of the entire tooth are difficult, if not impossible, to interpret. Dilambdodont teeth can be seen as combinations of three different types of tools, or tooth components: cusps, crests, and basins. These tools may not be independent of one another. For example, cusps often occur at the ends of crests. Analysis of tooth function is greatly simplified, and arguably more powerful, if the tooth is considered a conglomerate of tool components, each of which can be more successfully analyzed separately.

Second, the variables must be predictive, so that changes are clearly testable in terms of changes in function. Most often, this should be in terms of the force and/or energy that will be required for the component to function. The comparison may be quantitative or qualitative, but the direction of functional change should be clear from the change in the functional variable.

Variables that fulfill these aspects will be useful in comparing unworn molars between individuals of the same species, individuals of species with different diets, and unworn and worn molars of the same species.

Functional Variables

Several tooth variables relating to shape, with a predictive value, can be found in the engineering literature (e.g., Nee, 1998; Ostwald and Muñoz, 1997) and theoretical discussions of tooth shape and function (e.g., Frazzetta, 1988; Lucas, 1979, 1982; Lucas and Luke, 1984; Osborn and Lumsden, 1978). The function of the cusps and crests of dilambdodont teeth can be analyzed with reference to at least nine functional variables: tip sharpness, cusp sharpness, cusp occlusion relief, edge sharpness, rake angle, crest relief, approach angle, capture area, and fragment clearance. These appear to cover all major aspects of shape as they relate to the function of cusps and crests.

Most of these variables have not been measured in teeth, in part because of technical difficulties. Recently developed techniques, however, allow the reconstruction and visualization of occluding three-dimensional tooth surfaces (Evans et al., 2001; Jernvall and Selänne, 1999; Ungar and Williamson, 2000). These techniques permit the three-dimensional measurement of important shape parameters that were previously difficult to evaluate (e.g., curvature of tooth surface) with accuracy and relative ease, irrespective of wear state.

Basins represent a third major topographic feature of teeth. Quantification and interpretation of the function of basins is difficult. An increase in the size of a basin will most likely lead to a greater amount of food processed, but the force and energy required for such processing is likely to increase. A greater understanding of the processes of food division carried out within a basin is needed, although the functional role of basins will not be considered here.

Cusps Cusps are important for initial penetration of food, either alone (in puncture-crushing) or followed by crest-against-crest contact in occlusion. I will examine the effect of cusp and crest shape on the breakdown of tough foods, which are those that resist crack propagation and do not undergo brittle fracture (Strait, 1993).

Tip Sharpness: The stress required for a cusp to initiate a crack in food will depend on the surface area of contact between the cusp and food. Tip sharpness of a cusp is measured as the radius of curvature at its tip, so that a cusp with higher tip sharpness has a smaller radius of curvature (Evans and Sanson, 1998; Freeman and Weins, 1997). A smaller radius of curvature will give a smaller area of contact (for a given elastic modulus of the food), and thus produce a higher stress in the food (Lucas, 1982) and allow crack initiation with lower force. Tip sharpness has been measured in different ways for human cusps (Lucas, 1982), bat canines (Freeman and Weins, 1997), and lemur molars (Yamashita, 1998). Empirical tests on several foods by Freeman and Weins (1997) and Evans and Sanson (1998) demonstrated that higher tip sharpness decreases the force necessary to penetrate food.

Cusp Sharpness: Once a cusp has initiated a crack in a tough food, it must be continually driven into the food to sustain propagation of the crack. The force and energy required will in part depend on the volume of the tool and the amount of food displaced. This can be quantified as "cusp sharpness," the volume of the cusp at increasing distances from the tip (Evans and Sanson, 1998). A cusp with higher cusp sharpness has a smaller cusp volume for a given distance from the tip. For a cusp with high cusp sharpness (smaller volume), fewer bonds in the material will be broken or strained when it is driven through a tough food, and so a lower force and less energy will be required. This variable has only been measured for artificial cusps used in force testing (Evans and Sanson, 1998), where cusps with higher cusp sharpness required lower force and energy to penetrate food.

Cusp Occlusion Relief: In occlusion, cusps often move into a valley between two cusps or crests. Friction between the adjacent tooth surfaces and food caught between the surfaces that push apart occluding teeth will increase the force required to maintain tooth contact (also see below under "Relief of Crests"). Relief behind the point at which the cusp occludes (cusp occlusion relief) can reduce the friction and the tendency for occluding teeth to be pushed apart.

Crests Crests are important in dividing food through "cutting," which can be defined as dividing a tough material using a blade or crest, or "forced crack propagation using a blade." Crests can function either singly, such as on the side of a tooth when driving through food (e.g., some bat canines: Freeman, 1992), or in occluding pairs in an analogous manner to a guillotine, the latter particularly during chewing.

Edge Sharpness: The edge sharpness of a crest is the radius of curvature of the crest edge. A crest with a smaller radius of curvature will have higher edge sharpness, except that the maximum curvature is extended in one dimension. Higher edge sharpness will decrease the area of contact and thus increase stress on the food. Edge sharpness has been most extensively discussed by Popowics and Fortelius (1997), where its functional implications were described in some detail and measured for a range of mammal species. It was also measured for bat canines by Freeman (1992).

Rake of Crests: The angle between the leading surface of a tool and a line perpendicular to the direction of tool movement is termed the rake angle (figure 5.1a). A crest with a positive rake angle has its leading face angled away from the material to be divided (figure 5.1b). Compared with a crest with negative rake, less force is required to fracture food for a number of reasons. First, a negative rake crest must displace more material than one with positive rake, and thus requires greater force. Second, the surface area of contact between the food and a crest with negative rake will be greater, and so a positive rake crest will achieve higher stress for a given force. Third, the rake angle affects the force required to keep occluding crests together. When a rake surface contacts food, the crest tends to be forced in the direction perpendicular to the surface. Thus, crests with a negative rake angle will be

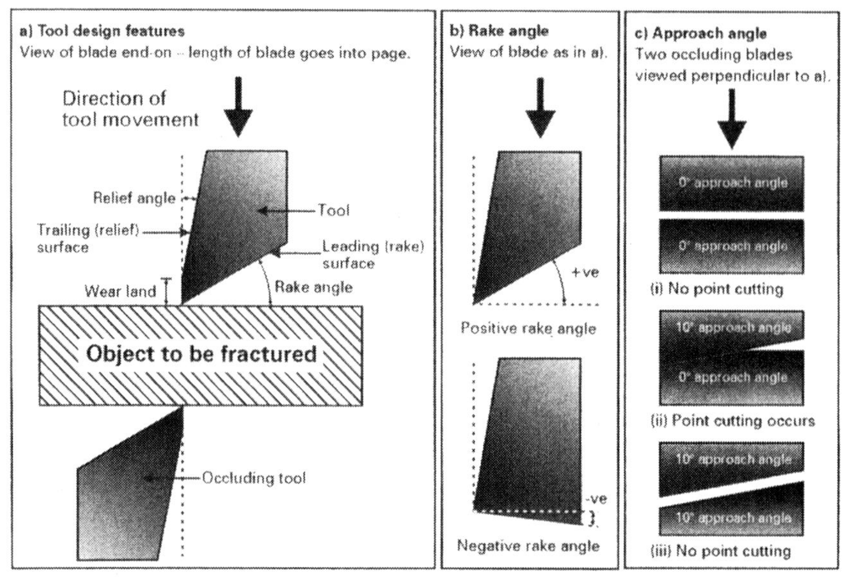

Figure 5.1 Tool design features. (a) Basic tool design features: leading surface, trailing surface, rake angle, relief angle, wear land; (b) positive and negative rake angles; (c) different approach angles of two occluding blades and the effect on point cutting.

forced apart when food is trapped between them, whereas a positive rake angle helps to direct a crest toward its opposing crest.

The importance of rake angle is widely recognized in engineering (e.g., Ostwald and Muñoz, 1997) and has been discussed with regard to teeth (Osborn and Lumsden, 1978). However, rake angle has not been measured for the crests of the multitude of tooth forms in which it is important.

Relief of Crests: The space between the direction of tooth movement and the trailing surface of the crest is referred to as relief (figure 5.1a). Relief behind crests (also called "clearance" in some engineering texts) reduces the effect of crests being forced apart by material caught between them, which would require greater force to maintain the proximity of the crests. It also reduces crest-against-crest friction and friction attributable to material caught between the crests. Once sufficient relief is attained, no further benefit is achieved by increasing the angle (e.g., between 4° and 7° is best for sharp metal or glass sectioning blades shearing plant material for histological sections: Atkins and Vincent, 1984). Relief is important in the design of machine cutting tools (e.g., Nee, 1998; Ostwald and Muñoz, 1997), and also has been noted by some workers in dental morphology (Osborn and Lumsden, 1978). Notwithstanding, measurements of relief of crests have not been reported for actual teeth.

A number of different aspects relate to the amount of relief for a crest. The first is that many teeth and tools, either by design or following wear, have an area of no relief immediately behind the cutting edge (figure 5.1a). This is known as the "wear land" (Nee, 1998; Ostwald and Muñoz, 1997). In teeth, it is the attrition facet—a planar surface where tooth–tooth contact occurs. The size of the wear land in a histological section is the width of the attrition facet on the crests. Second is the volume of space behind a cusp, which represents the amount of space into which food could be directed, and is estimated by the height of the crest above the adjacent valley. Third is the relief angle behind the wear land, which is the angle between the trailing surface and the direction of tooth movement (figure 5.1a).

Approach Angle: The approach angle of a crest is the angle between the long axis of the crest and a line perpendicular to the direction of movement (figure 5.1c). The mechanical advantage (MA) of a crest will depend on the approach angle (α) of the crest, where $MA = 1/\cos(\alpha)$ (Abler, 1992; Evans and Sanson, 1998), so that a larger approach angle will have a greater mechanical advantage. The approach angle of crests will affect the "point cutting" of the system, which occurs when the long axes of two occluding crests are not parallel. As a result, only one point (or two points if at least one crest is concave) meets at a time rather than the entire length of the crest (figure 5.1c). Point cutting will decrease the amount of crest surface area in contact at any one time (if there is a wear land or no relief), which will increase pressure and decrease friction between the crests. Point cutting does not imply that the crest only fractures food at one point at a time, as fracture of food will occur along the entire length of the crest in contact with food (Abler, 1992). This is like a pair of scissors cutting thick rubber: the majority of the material is cut by the crests before the cutting points come into close proximity.

Approach angle has been discussed by a few authors with reference to teeth (Abler, 1992; Evans and Sanson, 1998), and point cutting by others (e.g., Crompton and Sita-Lumsden, 1970; Seligsohn, 1977). Approach angle has not been measured on actual teeth but the effect of approach angle on the required occlusal force has been measured for sharp blades (Abler, 1992) and facsimile tooth crests (Worley and Sanson, 2000).

Food Capture: It is advantageous to "trap" the food between crest edges, preventing it from escaping off the ends of the crests and therefore being incompletely divided. Crests can be concave so that the ends meet first, enclosing the food before it is divided. This is advantageous for foods with a high Poisson's ratio (e.g., over 1.0; Lucas and Luke, 1984), which may be particularly hard to trap between crests. A crest may be notched or curved (a sharp or rounded concavity respectively). The amount of food trapped can be estimated by the area enclosed by one crest. The importance of food capture has been recognized for many teeth (such as carnassial and dilambdodont: Abler, 1992; Freeman, 1979; Savage, 1977).

Fragment Clearance: The function of crests will improve if the fractured material is directed away from the crests and off the rake surfaces. If material is trapped on rake surfaces, it may prevent fracture of any remaining food between the crests. Where there is insufficient space into which fractured material can flow, food may be compressed between opposing tooth structures and prevent the occlusion of crests. It is advantageous to create flow channels and exit structures through which food can flow away from the crests.

The movement and clearance of food has been given some attention in the dental literature: it was incorporated into Rensberger's (1973) models of herbivore teeth, discussed in terms of "food escapement" by Seligsohn (1977), as "sluiceways" by Sanson (1980), and as clearance provided by "gullets" between successive teeth by Frazzetta (1988).

METHODS

Molars of 20 specimens of an insectivorous microbat, Gould's wattled bat *Chalinolobus gouldii*, were examined in the present study. This species is common in Australia, is approximately 10–18 g in body mass, and its upper second molar ranges from 1.2 to 1.4 mm in length. Methods for digitization of tooth surfaces follow Evans et al. (2001); briefly, casts of the upper second molar were stained with eosin and imaged using fluorescence confocal microscopy. Geographic information systems (GIS) software (Surfer v6.04, Golden Software Inc.) and algorithms written by the author were used to calculate the variables. The main two cusps (paracone and metacone) and the four ectoloph crests (pre- and postparacrista and pre- and postmetacrista) of these teeth were measured for the nine functional variables for cusps and crests, respectively.

Specimens of three wear states (10 unworn or lightly worn, five moderately worn and five heavily worn) were measured. Wear states were determined as follows (figure 5.2): unworn or lightly worn—only a small amount of wear had occurred along the crests, usually apparent as attrition facets on the relief surfaces; moderate wear—approximately one quarter to a one half of the rake surface is exposed dentine; heavy wear—the majority of the rake surface is exposed dentine, and the height of the cusps has been substantially reduced.

Occlusion of the upper and lower molars was simulated using Virtual Reality Modeling Language (VRML) reconstructions of the molar tooth row with the VRML browser CosmoPlayer 2.1 (Computer Associates, Inc.), allowing the occlusion of unworn and worn molars to be compared (Evans et al., 2001). Examination of the occlusal relations of teeth is only an exploratory study.

RESULTS

The results of measurements of the nine functional parameters for one cusp, the metacone, and one crest, the postmetacrista, for low and high wear states are given in table 5.1. In terms of capture area and relief angle, there was no significant difference between the unworn and highly worn molars. Approach angle was the only variable to exhibit a large difference that would decrease the force and/or energy required for components to function. Moderately increased force and/or energy for function would be required with increasing

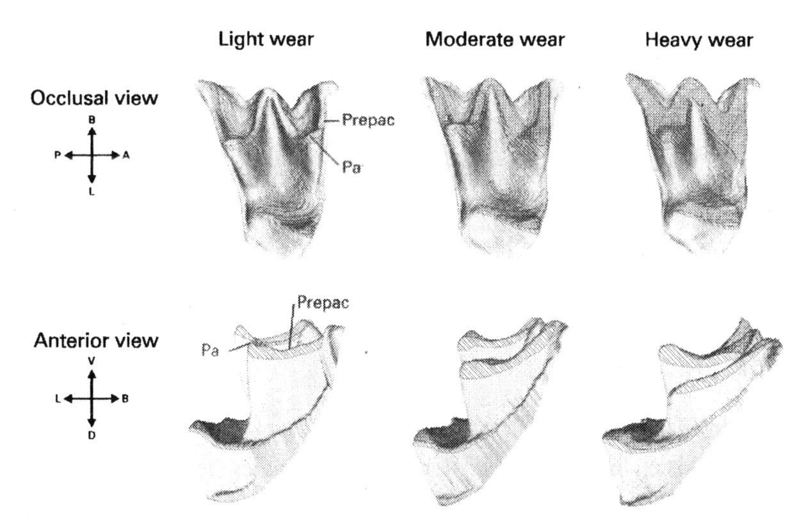

Figure 5.2 Occlusal and anterior views of the right upper second molar of *Chalinolobus gouldii*. Three wear states are shown: light, moderate and heavy wear. Wear on the rake surface is shown by cross-hatching; wear on the relief surface by diagonal lines. Illustrations are generated from VRML reconstructions of confocal scans. A, anterior; B, buccal; D, dorsal; L, lingual; P, posterior; V, ventral; Pa, ⌧; prepac.

Table 5.1 Measurements for the Nine Functional Parameters for Cusp and Crest Function Considered in the Current Chapter for the Upper Second Molar of *Chalinolobus gouldii*

Component	**Parameter**	**Low Wear**	**High Wear**
Metacone	Tip sharpness (μm)	25.6 ± 2.6	50.3 ± 8.0
	Cusp sharpness to 100 μm (10^3 μm^3)	1084.6 ± 118.0	1670.6 ± 280.7
	Cusp occlusion relief of trigon groove (μm)	129.7 ± 11.5	33.2 ± 6.0
Postmetacrista	Edge sharpness (μm)	14.4 ± 2.1	26.8 ± 1.8
	Rake angle (deg)	14.6 ± 5.4	−40.8 ± 6.0
	Relief:		
	—Wear land (qual.)	Small	Moderate
	—Volume (qual.)	Large	Moderate
	—Angle (qual.)	Small	Small
	Approach angle (deg)	37.8 ± 0.8	50.0 ± 1.5
	Food capture (10^3 μm^2)	63.5 ± 3.1	78.9 ± 13.5
	Fragment clearance (qual.)	High	Moderate

Low Wear/High Wear are the mean and SE (or median for qualitative data; qual.) for the metacone and postmetacrista at low and high wear states.

wear for cusp sharpness, edge sharpness, and wear land, and large increases force and/or energy for tip sharpness, cusp occlusion relief, rake angle, relief volume, and fragment clearance.

Computer reconstructions of the upper and lower molars in occlusion for several wear states showed that occlusion is maintained between crests despite heavy wear. Where crest shape and arrangement were altered due to wear, corresponding changes in opposing crests insured that proper alignment of crests was retained.

DISCUSSION

Functional Variables

Wear-related changes in nine functional variables of crests and cusps measured in this study have a predictable influence on the function of the tooth components. Only one variable (approach angle) showed significant improvement with wear, where the increased angle improved the mechanical advantage of the crest. The majority of variables showed no significant change after heavy wear. However, dramatic change was evident in some. A change of 55° in rake angle is very likely to have a large effect on the force required for the crest to function (where a significant difference in forces required was found between blades with rake angles of 0° and 30°; N. Aranwela, pers. comm.). It would be expected that the overall effectiveness of these teeth would decrease with wear, requiring more force and energy to divide tough food. This is the first demonstration of changes in measurable tooth features that predict an increase in the force or energy required for a tooth to function (i.e., decreased tooth "efficiency" with wear). Previously, this had been assumed.

There are independent lines of evidence that support this prediction. Carraway et al. (1996) and Verts et al. (1999) showed that insectivorous shrews increase the efficiency of their jaw mechanics with age, increasing the available bite force. It was only assumed in these studies that worn teeth were less effective at dividing food.

Wear scratches on the enamel relief surface of crests appear to be deeper or wider at higher wear states in *C. gouldii*, or are at least more visible (A.R. Evans, pers. obs.). Deeper or wider scratches would likely require a greater force to produce, supporting the hypothesis that greater force is necessary for worn teeth to function (Teaford, 1988; Ungar and Spencer, 1999). Alternative explanations for such a difference, such as change in enamel structure with increased depth, may also account for this observation.

There are some important implications of decreased tooth efficacy with increased wear. Greater energy and possibly time (in number of chews) must be expended in dividing food, reducing the amount available for other important biological processes, including food searching and gathering, and social interactions. The assumption of Carraway et al. (1996) was that older animals would need to switch to a diet of softer food. Their search for such a dietary shift assumes that the worn teeth have retained the ability to divide "soft" food; however, worn teeth may be equally ineffective in dividing "soft" and "hard" foods. Increases in efficiency of jaw mechanics would then be more important, as occurs in shrews (Verts et al., 1999).

In principle, all of the nine variables may vary independently, although there is a high dependence among some of these variables. For example, allowing food capture requires a change in approach angle of the crest to make it concave. It cannot be assumed that a single quantitative measure, taken either individually from teeth or as a complex combination of any of the functional variables discussed here, can be used to predict or represent the function of the entire tooth. This is compounded by the presence of multiple components on teeth, where their number, arrangement, and shape will vary. This should not be seen as a failure of functional morphology, rather as an advance in our ability to comprehend the full complexity of tooth shape and function. However, if it could be shown that the majority of the variables alter in synchrony in actual teeth, an estimate of the change in effectiveness of the teeth using one or several parameters as surrogates for a given tooth form may be possible.

Maintenance of Shape and Function during Wear

Despite the noticeable change in tooth shape that occurs due to wear, many of the functional features retain a reasonably advantageous state with wear. The design of certain features of these teeth means that they retain many aspects of good functional shape longer than expected according to the traditional view of dilambdodont teeth as ineffective in coping with wear. This may occur either through particular design characteristics of teeth or through the geometry of the wear process.

Food Capture: A concave crest is able to maintain its shape to some degree merely through its use. If the entire crest cavity is filled with food and all of the food is divided, then the middle of a concave crest will divide more material than the ends, and may be under greater pressure (or certainly under pressure for a longer time) from the food. Wear on the rake surface of a crest caused by food abrasion will be at least partly determined by the amount of material the crest divides and by the pressure exerted on the surface. Thus, greater wear will occur in the middle of a crest compared with the ends. Relative wear at positions along a crest will also be influenced by the thickness of the enamel along a crest. Thinner enamel on the middle, with crests compared with the ends, will cause the middle to wear more rapidly. The enamel appears thicker toward the lingual end of each ectoloph crest in the specimens of *C. gouldii* examined (A.R. Evans, pers. obs.). The greater amount of dentine that must be worn at the buccal end of the crests would have the same effect as the thicker enamel on the other end, reducing the rate at which the height of the buccal end of the crest is decreased. Both of these variables will preserve or possibly even increase the concavity of the crest following wear.

Cusp Sharpness: Wear in the center of the crests on two-crested cusps (such as the paracone and metacone) maintains higher cusp sharpness to some extent. Thicker enamel on the rounded, lingual faces of the cusps will preserve the height of the cusp, whereas wear on the rake surface of the adjoining crests will reduce the volume of the cusp, maintaining high cusp sharpness. To a limited extent, this occurs on the upper molar cusps, where in some instances a highly worn cusp has higher cusp sharpness than moderately worn cusps.

Relief: The incidence and proportion of wear on the rake and relief surfaces will influence the amount of relief of a crest (figure 5.3a). Wear on the relief surface only would cause the majority of the relief to be removed, producing a wide wear land. If the rake surface is also worn, then this wear land will be removed and the relief will be retained. If sufficient wear occurs on both the rake and the relief surfaces, then relief will be maintained. *Chalinolobus gouldii* molars have high degrees of wear on the rake surface relative to wear on the relief surface, reducing the wear land on the relief surface and maintaining relief (figure 5.2).

A non-zero relief angle can be maintained during wear when the relief surface is straight, and will even increase, given a convex curvature of the relief surface (figure 5.3b). Relief surfaces, particularly those on the lingual side of upper molar cusps, are either straight or slightly convex, so that the relief angle is maintained as tooth wear progresses.

Edge Sharpness: The edge sharpness of crests is also maintained to some extent during wear. This is likely to be attributable to both the thickness of the enamel and its microstructure. A thin enamel edge will wear more slowly than the surrounding dentine, resulting in a ridge of enamel higher than the dentine (as occurs in many herbivore teeth). Enamel prisms in the relief surface of crests are arranged parallel to the rake surface, so that when prisms are removed by wear a sharp edge is retained (Stern et al., 1989).

Capture area and relief can be maintained to some extent purely by geometrical relations of wear, a point that has never been noted in considerations of the effect of wear on teeth. Maintenance of these features can be achieved or aided by specific design considerations, such as enamel thickness and prism structure. In addition, the maintenance of these features points to controlled wear. Wear cannot be avoided, but teeth can be shaped to dictate which regions of the tooth wear more than others. The utilization of cusps and crests, and distribution and structure of enamel, can encourage wear of particular areas, allowing function to be maintained.

Molar Occlusion: For proper occlusion to be maintained, the shape of the crest edges, when viewed along the vector of tooth movement, must be the same for the upper and lower molars (figure 5.4). The shape of the crest profile can be reduced to three features that must correspond between the upper and lower molars: the projected two-dimensional lengths of crests; the angle of the crests; and curvature at the junction between crests. In principle, each of these could be altered to different extents and at various rates during tooth wear. Thus, it is not guaranteed that worn upper and lower

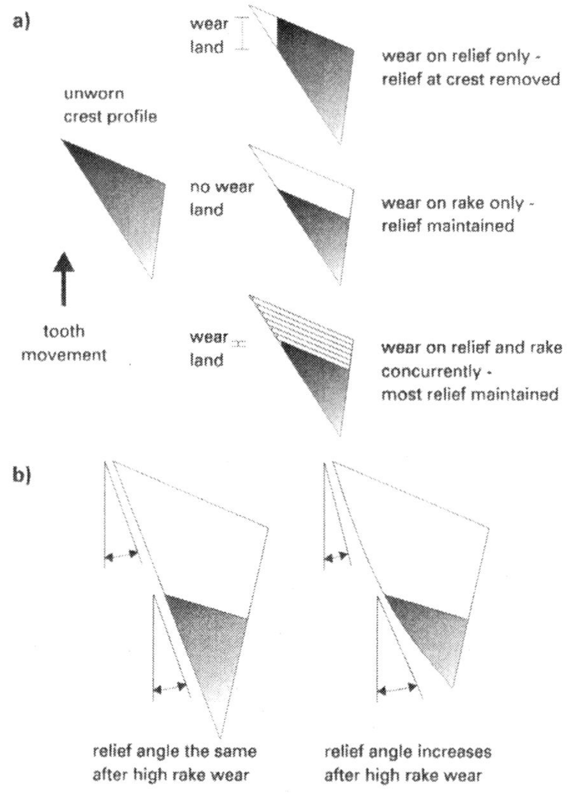

Figure 5.3 (a) Effect of relative wear on rake and relief surfaces on the relief behind a crest. If wear occurs only on the relief surface, then relief behind the crest is removed (as represented by a wear land). If wear occurs only on the rake surface, relief is maintained. For the more realistic situation, where wear occurs on both the rake and relief surfaces concurrently, substantial relief (indicated by a small wear land) is maintained. (b) Relief angle is maintained after wear on the rake surface for a linear relief surface but increases if the relief surface is convexly curved. Tooth profile is represented by shaded areas; unshaded areas represent tooth removed by wear.

molars maintain occlusion. In occlusal view, the tips of cusps and the junctions of crests of unworn teeth are relatively pointed, approximating a "W" shape. With wear, the upper molar crests are now closer to the base of the paracone and metacone, where it is slightly wider than in the unworn state. Also, the two-dimensional crest length is reduced (figure 5.4). Concurrent changes in the occluding features must also occur in the lower molar. If this were not the case, the cusp would fail to fit into its embrasure, or the crests on either side of a junction would fail to meet.

CONCLUSIONS

Functional characteristics elaborated and quantified here makes it possible for the first time to determine changes in function with change in shape

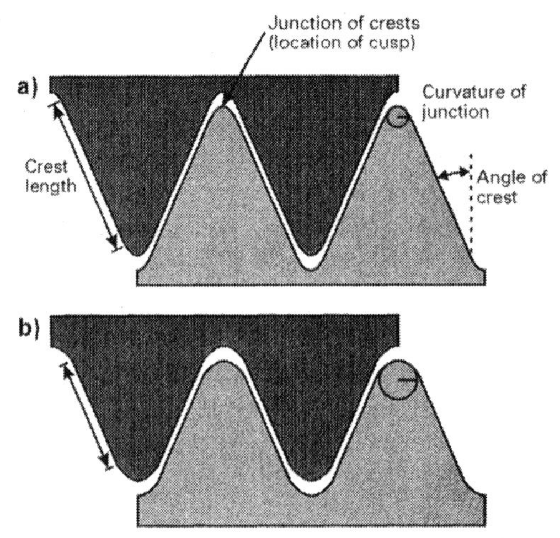

Figure 5.4 Changes in the shape of molars during wear must be concurrent in upper and lower molars in order for occlusion to be maintained. This shows diagrammatic profiles of the crests of upper (dark) and lower (light) molars as viewed along the direction of tooth movement. Three main elements must correspond between each pair of upper and lower occluding crests: two-dimensional length of the crest, angle of the crest, and curvature at cusp points. Two of the three (two Θ dimensional crest length and curvature) are shown to alter between a light wear state (a) and a heavy wear state (b).

resulting from wear in dilambdodont or tribosphenic teeth. Wear is a significant process that must be considered when examining tooth form and function. It must have been a highly influential selective force in shaping teeth during evolution. Wear is an additional constraint on tooth shape that may not be inherent in unworn tooth function, and should be considered when evaluating the apparent function of unworn teeth. Although not as adapted as some other tooth forms to cope with high wear and maintain function, there are significant functional features of dilambdodont and tribosphenic teeth that are retained with wear.

Despite their high complexity in shape and relations between features of opposing teeth, dilambdodont teeth are able to maintain functional occlusion through many wear states. New approaches and technology now make it possible to investigate the design of teeth in greater detail and, despite wear, how the design ensures the shapes of the upper and lower crests remain in tandem.

ACKNOWLEDGMENTS

I am most grateful to Gordon Sanson for discussions on the topics presented here and for advice on drafts of this chapter. I also thank Gudrun Arnold, Deb Archer, Nuvan Aranwela and Elizabeth Dumont for comments on earlier drafts, and to Trish Freeman and an anonymous reviewer for many helpful suggestions to improve the chapter. Thanks to Lina Frigo of Melbourne Museum for loan of specimens.

LITERATURE CITED

Abler, W.L. 1992. The serrated teeth of tyrannosaurid dinosaurs and biting structures in other animals. Paleobiology, 18: 161–183.

Atkins, A.G., and J.F.V. Vincent. 1984. An instrumented microtome for improved histological sections and the measurement of fracture toughness. Journal of Materials Science Letters, 3: 310–312.

Carraway, L.N., B.J. Verts, M.L. Jones, and J.O. Whitaker, Jr. 1996. A search for age-related changes in bite force and diet in shrews. American Midland Naturalist, 135: 231–240.

Crompton, A.W., and K. Hiiemae. 1970. Molar occlusion and mandibular movements during occlusion in the American opossum, *Didelphis marsupialis* L. Zoological Journal of the Linnean Society, 49: 21–47.

Crompton, A.W., and A. Sita-Lumsden. 1970. Functional significance of the therian molar pattern. Nature, 227: 197–199.

Evans, A.R., and G.D. Sanson. 1998. The effect of tooth shape on the breakdown of insects. Journal of Zoology (London), 246: 391–400.

Evans, A.R., I.S. Harper, and G.D. Sanson. 2001. Confocal imaging, visualization and 3-D surface measurement of small mammalian teeth. Journal of Microscopy, 204: 108–119.

Fortelius, M. 1985. Ungulate cheek teeth: developmental, functional, and evolutionary interrelations. Acta Zoologica Fennica, 180: 1–76.

Frazzetta, T.H. 1988. The mechanics of cutting and the form of shark teeth (Chondrichthyes, Elasmobranchii). Zoomorphology, 108: 93–107.

Freeman, P.W. 1979. Specialized insectivory: beetle-eating and moth-eating molossid bats. Journal of Mammalogy, 60: 467–479.

Freeman, P.W. 1988. Frugivorous and animalivorous bats (Microchiroptera): dental and cranial adaptations. Biological Journal of the Linnean Society, 33: 249–272.

Freeman, P.W. 1992. Canine teeth of bats (Microchiroptera): size, shape and role in crack propagation. Biological Journal of the Linnean Society, 45: 97–115.

Freeman, P.W., and W.N. Weins. 1997. Puncturing ability of bat canine teeth: the tip. Pp. 225–232. In: Life among the Muses: Papers in Honor of James S. Findley (T.L. Yates, W.L. Gannon, and D.E. Wilson, eds.). The Museum of Southwestern Biology, Albuquerque, NM.

Janis, C.M., and M. Fortelius. 1988. The means whereby mammals achieve increased functional durability of their dentitions, with special reference to limiting factors. Biological Review, 63: 197–230.

Jernvall, J., and L. Selänne. 1999. Laser confocal microscopy and geographic information systems in the study of dental morphology. Palaeontologia Electronica, 1: 12 pp. http://palaeo-electronica.org/1999_1/confocal/issue1_99.htm

Kay, R.F. 1975. The functional adaptations of primate molar teeth. American Journal of Physical Anthropology, 43: 195–216.

Kay, R.F. 1984. On the use of anatomical features to infer foraging behavior in extinct primates. Pp. 21–53. In: Adaptations for Foraging in Non-human Primates: Contributions to an Organismal Biology of Prosimians, Monkeys and Apes (P.S. Rodman and J.G.H. Cant, eds.). Columbia University Press, New York.

Kay, R.F., R.W. Sussman, and I. Tattersall. 1978. Dietary and dental variations in the Genus *Lemur*, with comments concerning dietary–dental correlations among Malagasy primates. American Journal of Physical Anthropology, 49: 119–128.

Lanyon, J.M., and G.D. Sanson. 1986. Koala (*Phascolarctos cinereus*) dentition and nutrition. II. Implications of tooth wear in nutrition. Journal of Zoology (London), 209: 169–181.
Lucas, P.W. 1979. The dental–dietary adaptations of mammals. Neues Jahrbuch für Geologie und Paleontologie Monatshefte, 8: 486–512.
Lucas, P.W. 1982. Basic principles of tooth design. Pp. 154–162. In: Teeth: Function, Form and Evolution (B. Kurtén, ed.). Columbia University Press, New York.
Lucas, P.W., and D.A. Luke. 1984. Chewing it over: basic principles of food breakdown. Pp. 283–301. In: Food Acquisition and Processing in Primates (D.J. Chivers, B.A. Wood and A. Bilsborough, eds.). Plenum Press, New York.
Luke, D.A., and P.W. Lucas. 1983. The significance of cusps. Journal of Oral Rehabilitation, 10: 197–206.
McArthur, C., and G.D. Sanson. 1988. Tooth wear in eastern grey kangaroos (*Macropus giganteus*) and western grey kangaroos (*Macropus fuliginosus*), and its potential influence on diet selection, digestion and population parameters. Journal of Zoology (London), 215: 491–504.
Nee, J.G. 1998. Fundamentals of Tool Design. Society of Manufacturing Engineers, Dearborn, MI.
Osborn, J.W., and A.G.S. Lumsden. 1978. An alternative to "thegosis" and a re-examination of the ways in which mammalian molars work. Neues Jahrbuch für Geologie und Paleontologie Abhandlung. 156: 371–392.
Ostwald, P.F., and J. Muñoz. 1997. Manufacturing Processes and Systems. Wiley, New York.
Pérez-Barberia, F.J., and I.J. Gordon. 1998. The influence of molar occlusal surface area on the voluntary intake, digestion, chewing behaviour and diet selection of red deer (*Cervus elaphus*). Journal of Zoology (London), 245: 307–316.
Popowics, T.E., and M. Fortelius. 1997. On the cutting edge: tooth blade sharpness in herbivorous and faunivorous mammals. Annales Zoologici Fennici, 34: 73–88.
Rensberger, J.M. 1973. An occlusal model for mastication and dental wear in herbivorous mammals. Journal of Paleontology, 47: 515–528.
Sanson, G.D. 1980. The morphology and occlusion of the molariform cheek teeth in some Macropodinae (Marsupialia: Macropodidae). Australian Journal of Zoology, 28: 341–365.
Savage, R.J.G. 1977. Evolution in carnivorous mammals. Palaeontology, 20: 237–271.
Seligsohn, D. 1977. Analysis of species-specific molar adaptations in strepsirhine primates. Pp. 1–116. In: Contributions to Primatology, Vol. 11 (F.S. Szalay, ed.). S. Karger, Basel, Switzerland.
Stern, D., A.W. Crompton, and Z. Skobe. 1989. Enamel ultrastructure and masticatory function in molars of the American opossum, *Didelphis virginiana*. Zoological Journal of the Linnean Society, 95: 311–334.
Strait, S.G. 1993. Molar morphology and food texture among small-bodied insectivorous mammals. Journal of Mammalogy, 74: 391–402.
Teaford, M.F. 1988. A review of dental microwear and diet in modern mammals. Scanning Microscopy, 2: 1149–1166.
Ungar, P., and M. Williamson. 2000. Exploring the effects of tooth wear on functional morphology: a preliminary study using dental topographic analysis. Palaeontologia Electronica, 3: 18 pp. http://palaeo-electronica.org/2000_1/gorilla/issue1_00.htm
Ungar, P.S., and M.A. Spencer. 1999. Incisor microwear, diet, and tooth use in three Amerindian populations. American Journal of Physical Anthropology, 109: 387–396.

Verts, B.J., L.N. Carraway, and R.A. Benedict. 1999. Body size- and age-related masticatory relationships in two species of *Blarina*. Prairie Naturalist, 31: 43–52.

Worley, M., and G. Sanson. 2000. The effect of tooth wear on grass fracture in grazing kangaroos. Pp. 583–589. In: Proceedings of the Plant Biomechanics Meeting (H.-C. Spatz and T. Speck, eds.). Georg Thieme, Freiburg-Badenweiler, Germany.

Yamashita, N. 1998. Functional dental correlates of food properties in five Malagasy lemur species. American Journal of Physical Anthropology, 106: 169–188.

6

Dynamic Complexity of Wing Form in Bats: Implications for Flight Performance

Sharon M. Swartz, Kristin Bishop, & Maryem-Fama Ismael Aguirre

For many decades, students of animal flight have taken advantage of the body of engineering theory used to design fixed-wing aircraft. Here, we review important ways in which bat wings and bat flight violate the assumptions of this body of theory, and we identify features of bat wing morphology and kinematics that may be of particular importance for flight performance. In particular, the complexity and heterogeneity of the mechanical properties of wing membrane skin and the low material and structural stiffness of wing bones result in wings that deform extensively under aerodynamic loading in ways not seen in aircraft engineered by humans.

INTRODUCTION

Biologists have long used aircraft aerodynamics to estimate lift and drag forces encountered by wings, aspects of flight energetics, and limits of flight performance, particularly regarding speed and maneuverability. Here, we examine how well the wings of bats meet the assumptions of conventional aerodynamic theory, and try to identify those areas where such theory can be effectively employed to understand bat flight, as well as those for which conventional aerodynamic theory poorly models bat flight. Aerodynamic principles suited to large fixed-wing aircraft have been applied to bat flight largely because of the lack of more appropriate theory; the mechanics and aerodynamics of small, flexible, flapping wings has traditionally been beyond the scope of fluid mechanics. Recent technological, computational, and theoretical advances, however, have dramatically changed the field of fluid dynamics. As a consequence, biologists can gain considerable new insight into bat flight and its morphological basis. We may be able to better identify the most functionally significant features of wing morphology and flight kinematics, and more accurately estimate the mechanics and energetics of flight. This will enable us to draw more realistic and informative comparisons among taxa that differ in wing shape, body size, etc. One important first step in adopting these methods is to distinguish between those aspects of flight for which relatively simple theoretical approaches are sufficient and those that require more sophisticated but complex and labor-intensive analysis. We hope to demonstrate that this approach holds great promise for enhancing our understanding of the function and evolution of the distinctive features of the flight apparatus of bats.

We focus on three aspects of the structural design and functional performance of bat wings: (1) the mechanical properties of wing tissues, (2) the three-dimensional complexity of wing shape, and (3) the dynamic changes in wing form during the wingbeat cycle. In doing so, we rely on information obtained from approaches in materials science to understand biological tissues, detailed computer modeling of flight mechanics and aerodynamics, and three-dimensional visualization of natural wing motions in a diversity of living bats. For each of these subjects, we consider how closely the biological reality matches the assumptions of aerodynamic theory, and identify areas where future work holds particular promise.

WING STRUCTURE

Mechanical Properties of Wing Tissues: Wing Membrane Skin

In most large, human-engineered aircraft, wings are stiff enough to deflect only insignificantly during normal flight; lightweight gliders are one obvious exception beyond the scope of the present discussion. Moreover, an aircraft wing structurally is relatively homogeneous, with all parts of the wing, from its attachment at the body or fuselage to the wingtip, made of the same constituent materials. Many of the materials traditionally used to manufacture aircraft are also nearly isotropic, that is their stiffness or elastic modulus and strength do not vary with the direction in which they are loaded. The degree of anisotropy and homogeneity both strongly influence the behavior of a mechanical structure subjected to aerodynamic loading. Because bat wings differ from engineered airfoils in this regard, their performance may differ significantly from that of comparably shaped airfoils at similar velocities or Reynolds numbers (where comparable Reynolds numbers indicate dynamically similar movements with comparable ratio of inertial to viscous forces; see also general texts on biological fluid mechanics, such as Denny, 1993; Vogel, 1994).

Extreme Nonlinearity of Stress–Strain Relationships in Wing Membrane Skin

Metals and many other materials widely used in engineering fabrication display a characteristic relationship between applied loads and resulting deformations. When loads are exerted on structures made of these so-called linearly elastic materials, deformation increases in direct proportion to load up to a point beyond which the structure begins to deform disproportionately, or undergo plastic deformation (for more on this general subject, see introductory texts in engineering mechanics or Gordon, 1978; or Wainwright

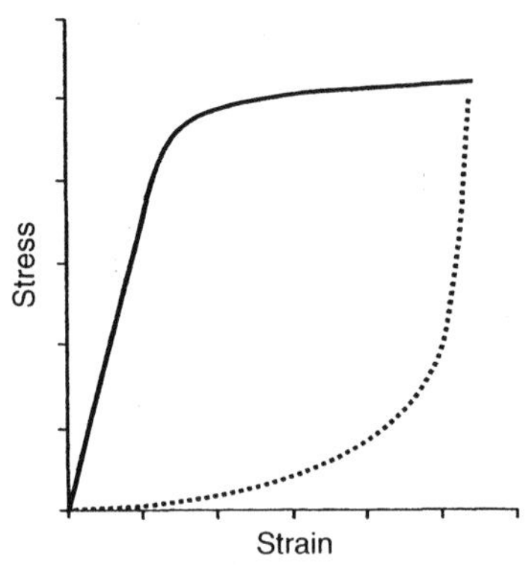

Figure 6.1 The relationship between stress and strain differs between relatively stiff, linearly elastic materials such as bone (continuous line) and soft, highly flexible tissues such as wing membrane skin (dotted line). In linearly elastic materials, the relationship between stress and strain (elastic modulus or stiffness) is linear up to a yield point beyond which the material deforms tremendously with only small increments of stress. For materials with characteristically J-shaped stress–strain curves, stiffness changes continuously until an inflection point beyond which the material is relatively stiff and linearly elastic. Most biological materials with this kind of stress–strain relationship are only stressed to their linearly elastic region during unusually large accidental loads.

et al., 1976). The ratio of the applied stress to the resulting relative deformation is defined as its Young's or elastic modulus, and corresponds closely to our colloquial sense of stiffness (figure 6.1).

In contrast, most biological materials, including ligament, tendon, and skin, show a complex, nonlinear relationship between load and deformation or stress and strain. In structures made of these materials, low loads typically induce large deformations in comparison with the loads that deform stiff skeletal materials such as bone; that is, at low stresses, these materials have very low stiffness (Wainwright et al., 1976). As such, structures reach higher stress levels, their stiffness begins to increase, eventually reaching values tens or hundreds of times greater than their stiffness at low stresses, resulting in a characteristically J-shaped stress–strain curve (Gordon, 1978). The range of material properties found in the diverse biological materials that are not linearly elastic is enormous; for example, tendon is characterized by stiffnesses several orders of magnitude higher than that of skin.

Mechanically, mammalian skin shares many features with other vertebrate soft tissues such as tendon and ligament, particularly in the nonlinearity of its stress–strain relationships (Gibson, 1977; Nordin et al., 2001). Wing membrane skin is somewhat similar to the skin of other mammals, displaying characteristic J-shaped stress–strain curves (Swartz et al., 1996; figure 6.1). However, the elastic modulus of a single skin sample can vary more than 1000-fold between the lowest measurable values to highest values achieved before mechanical failure (Swartz et al., 1996). The "toe-region," where the modulus changes rapidly from its lowest to higher values, can be extremely steep, occupying as little as 10% of the total range of skin strain. This combination of nonlinearity and rapid change in modulus raises particularly difficult challenges for understanding the function of the membrane during flight. The stiffness of the skin and its resultant deformations under a given aerodynamic load determine the three-dimensional conformation of the wing by determining how much the membrane billows. It is virtually impossible, however, to predict the degree of billowing without precise knowledge of the aerodynamic forces exerted on the wing when the load–deformation relationship is complex and load-dependent.

Anisotropy of Wing Membrane Skin Much mammalian skin is nearly isotropic: it stretches the same amount under load regardless of the load's orientation within the skin plane (Gibson et al., 1969; Gibson, 1977; Shadwick et al., 1992). If one could account for the nonlinearity in stress-strain relationships in skin, modeling how the three–dimensional shape of a region of wing membrane would respond when a uniform pressure is applied to the wing would be relatively straightforward if bat skin were isotropic. However, bat wing membrane is many times stiffer in the chordwise than the spanwise direction (figure 6.2) (Swartz et al., 1996; Skene 2000). The lift acting on the wing surface, therefore, will cause a greater spanwise than chordwise

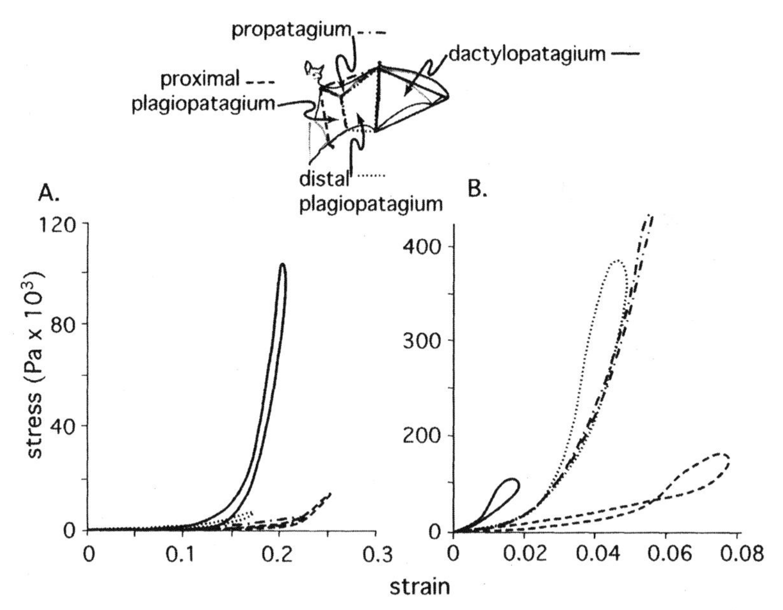

Figure 6.2 Bat wing membrane skin behaves in mechanically complex ways. The mechanical behavior of skin from different regions of the wing membrane (dactylopatagium, continuous line; proximal plagiopatagium, dashed line; distal plagiopatagium, dotted line; propatagium, dashed-dotted line) is highly heterogeneous. There are also great differences between the stiffness characteristics of skin loaded spanwise (A) versus chordwise (B); note the different scales on both axes of the two graphs.

deformation, resulting in differential billowing in these two directions. This effect greatly complicates making accurate theoretical estimates of wing form during flight from anatomical data alone.

Anatomical and Taxonomic Variation in Mechanical Properties of Wing Membrane Skin Our understanding of the functional complexity of wing membranes is also limited because of the variation in skin mechanical properties among wing regions and among taxa (Swartz et al., 1996). The dactylopatagium, distal plagiopatagium, propatagium, and proximal plagiopatagium show a gradient of decreasing stiffness, and similar patterns of variation in strength, and other characteristics (figure 6.2). This variation can be many-fold, and will strongly influence the three-dimensional shape of the wing when a uniform pressure differential is applied across it. The less stiff regions, such as the proximal plagiopatagium, will billow substantially more than the stiffer ones, particularly the propatagium, the region of the wing's leading edge. Studies have also shown considerable variation in all mechanical properties of wing skin in comparisons of diverse microchiropteran taxa (figure 6.3) (Swartz et al., 1996). Functional interpretation of this variation remains uncertain because of our rudimentary understanding of how skin properties affect wing shape and thus aerodynamic characteristics and because the limited comparisons carried out to date cannot reliably distinguish among effects of body size, feeding mode, flight style, and phylogenetic affinity. More study of the mechanics of wing skin is clearly needed to answer broad questions concerning the role of skin and its three-dimensional conformation in chiropteran aerodynamics, and also to successfully understand the meaning of the within- and among-species variation in wing properties. This is likely to suggest new research questions for field biologists as well as bioengineers.

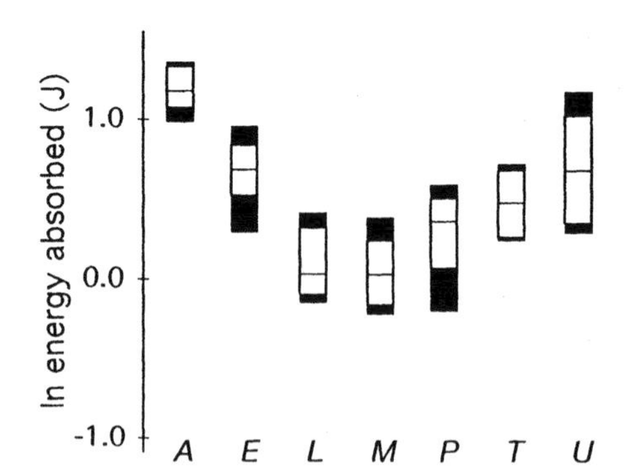

Figure 6.3 There is considerable interspecific variation in mechanical properties of skin, including, in this example, the amount of energy the skin can absorb before rupture. Black line, mean; white box, two standard deviations; black box, range. A, *Artibeus jamaicensis*; E, *Eptesicus fuscus*; L, *Lasiurus cinereus*; M, *Myotis lucifugus*; P, *Pteronotus parnellii*; T, *Tadarida brasiliensis*; U, *Uroderma bilobatum*.

Mechanical Properties of Wing Tissues: Wing Bones

The morphology and composition of the bones of the wing, the structural supports for the elastic wing membrane, are key determinants of the shape and function of the wings as airfoils. Aircraft wings are relatively rigid, and we conventionally consider bone as a material that is functionally rigid in vertebrate limbs. However, how a bone will respond to the mechanical loads imposed by locomotion depends on both the geometry and the mineralization of the bone, and our experience with medium to large terrestrial vertebrates may not be an adequate guide to the nature of the bat wing skeleton, or, indeed, the skeletons of any small mammals. In particular, three aspects of the design of the handwing bones interact to produce structural support elements that can deform a great deal in flight.

First, the mineral content of the bones of the bat wing decreases significantly in a proximal to distal gradient; the distalmost elements are often completely lacking in mineral except immediately adjacent to the articular cartilage through which joint forces are transmitted (Papadimitriou et al., 1996; Huber, 2000). In these respects, bats are unique among vertebrates (Swartz, 1997). Because the mineralization of a bone is one of the critical determinants of its stiffness, strength, and ability to absorb energy (Currey, 1979), the low mineral content of the handwing bones, particularly the phalanges, predisposes them to large deformations under relatively small loads. The stiffness of a structure's constituent material is not, however, the only determinant of its mechanical behavior. Beams that are short relative to their cross-sectional dimensions (low aspect ratio), and beams whose cross-sectional geometry maximizes the second moment of area, or moment of inertia, will deflect relatively little for a given load (Gordon, 1978; Wainwright et al., 1976). The bones of bat wings are prone to large deformations on both counts: they are greatly elongated and their cross-sectional shapes (relatively small outer diameters with small to nonexistent medullary cavities) tend to minimize the second moment of area (Swartz, 1997).

It is possible to measure directly the absolute amount of bone deflection that may arise from the interactions of aerodynamic forces and the wing skeleton. We have used in vivo strain gauge analysis to determine the surface strains in a number of wing bones, including metacarpals and phalanges, of the grey-headed flying fox, *Pteropus poliocephalus* (Swartz et al., 1992, 1993). Building on these data, it is possible to instrument a dissected metacarpal or phalanx in a manner similar to that used to record bone loading in flying animals, place the bone specimen in a mechanical testing machine with one end stably fixed, and apply a controlled load to the bone until the strains recorded from the bone surface match those of the flying animals. When we carry out these physical simulations, we find that the free end of a 75 mm bone is deflected as much as 20 mm with respect to its fixed end. This suggests that deformations of the wing skeleton may be important determinants of dynamically changing wing form during flight.

Mechanical Properties of Wing Tissues: Significance

How important is the variability in mechanical properties of the skin and bones of the bat wing? Are morphological and material characteristics important for wing strength, flight energetics, and/or maneuverability? These questions are exceptionally difficult to answer with direct or experimental methods. It is not possible to experimentally modulate tissue stiffness, bone geometry, etc., to test their effects on the flight parameters that interest us. These questions call for alternative approaches, particularly for computer modeling and simulation. With an effective and realistic computer simulation, it is possible to vary elements of a model that cannot be altered in nature, and to tease apart individual factors that may be intrinsically coupled in living organisms. Computer modeling poses a number of challenges of its own, however, and it is important to examine these to ascertain the value and limits of computer-based approaches.

Computer Modeling of Bat Flight

A computer model of bat flight may achieve several objectives. First, it is simpler to construct a realistic model of bat flight than to model the flight of pterosaurs, or insects. We obviously have far more detailed information about the wing tissues and flight kinematics of bats than of pterosaurs. Although insect wings are in some ways even simpler than those of vertebrates, with neither joints nor musculature within the wing itself, it is now clear that the aerodynamics of insect flight are extremely complex and dominated by a variety of unsteady effects (e.g., Birch and Dickinson, 2001; Dickinson et al., 1999; Dudley, 2000; Liu et al., 1998) that may not apply to animals as large and fast-flying as bats. In addition to the mechanical properties and data discussed above, there are some kinematic and bone-loading data that can serve as model inputs and/or sources for model validation.

Validation is a particularly important issue; models of complex phenomena that are difficult to study directly are of little value if there is no way to determine whether the model has captured important aspects of the real situation. But, if it is possible to construct a model for bat flight using only some of the available inputs, and to use the model to calculate values against which checks can be made, we can rigorously assess the accuracy and precision of the model.

Our recent computational model is based on three types of information: kinematic data (flight speed, wingbeat frequency, and three-dimensional movement patterns of anatomical landmarks), morphological data (body mass, bone diameter and cortical thickness, wing shape, and wing mass distribution), and mechanical properties of wing bones and skin (Watts et al., 2001). In the computer model, the body and wings of the bat moving through the air are partitioned into a number of anatomical subdivisions, each of which are represented by point mass at the three-dimensional location of the centroid of the segment (figure 6.4). Within the program, each point mass retains certain key characteristics—its mass, surface area, tissue mechanical

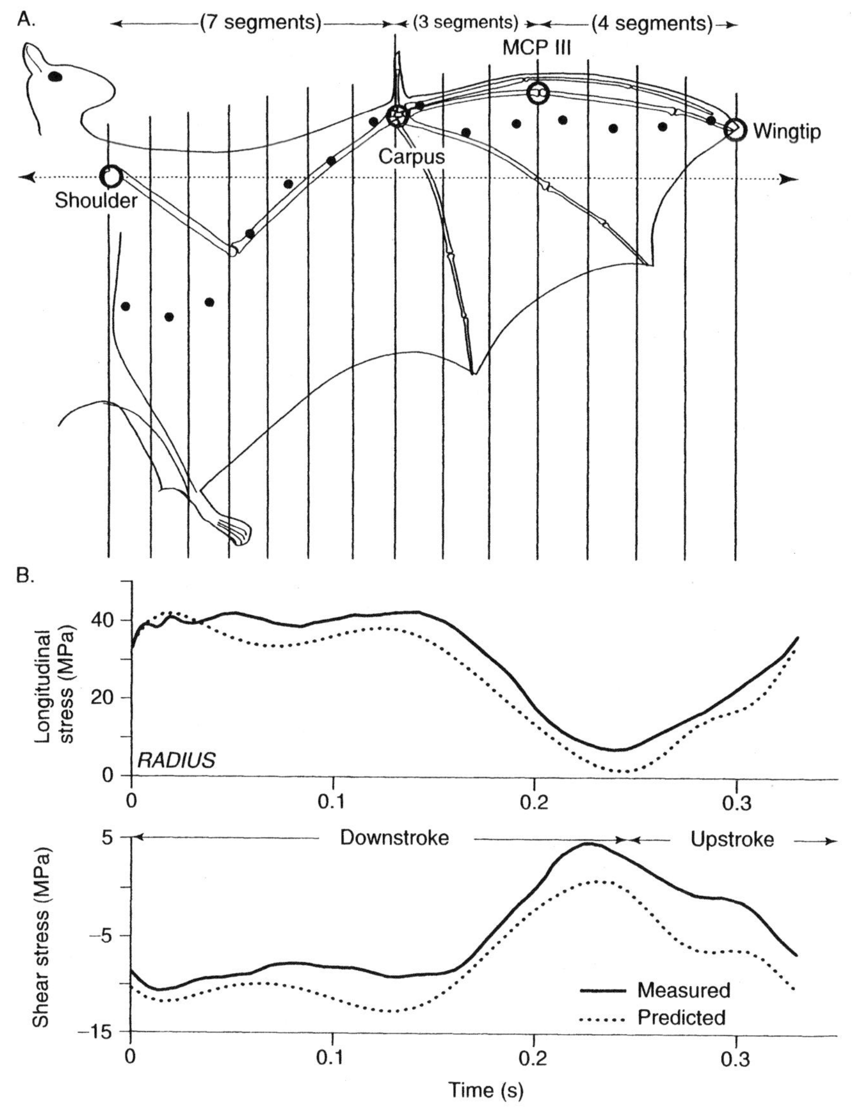

Figure 6.4 (A) Plan-view of the ventral surface of the wing of a *Pteropus poliocephalus* held in a horizontal plane illustrating the idealization of anatomy employed in computer modeling (Watts et al., 2001). The wing is subdivided into 14 chordwise strips; there are seven segments between the shoulder and the carpus, three between the carpus and the metacarpophalangeal joint of the third digit (MCP III), and four between the MCP III joint and the wingtip. The locations of the centers of mass of these strips relative to a reference line through the shoulder joint are indicated by the filled circles. The large labeled circles (shoulder, carpus, MCP III, wingtip) were used as digitizing markers for collecting kinematic input data. (B) Validation of the computer model by comparison of midshaft bone stresses during a complete wing-beat cycle measured in vivo (continuous lines) with those computed by the model (dotted lines). The horizontal axis gives time in seconds.

properties, bone structure, and spatial position—throughout the wingbeat. We determined values for these parameters from dissections, analysis of high-speed films of *P. poliocephalus* in a wind tunnel (Bartholomew and Carpenter, 1973), and mechanical testing of wing bones and skin (Papadimitriou et al., 1996; Swartz et al., 1996). For 40 time increments within a wingbeat, the model computes, using principles of mechanics and aerodynamics, how each segment experiences the following forces: gravity, inertia, added mass force, aerodynamic force, drag, force due to tension in the plagiopatagium (segments 1–7 only). From the balance of forces on each segment, it is then possible to compute the internal force carried by the wing structures within each segment. From these forces we can calculate joint forces and moments, bone stresses and strains, and skin stresses and strains, and use these results to estimate derived parameters including mechanical power and energetic cost associated with each segment and time increment, propensity for buckling, and maneuverability (Watts et al., 2001).

The degree of confidence one places in computer models of biological phenomena must ultimately come from an assessment of the model's realism. In this case, we have compared particular model results directly with empirically measured values of the same parameters. Earlier studies have measured the strains developed in the humerus and radius during level, steady flight in *P. poliocephalus* (Swartz et al., 1992). We tested the reliability of the model by comparing its predictions with previously acquired empirical results, asking the following questions: (1) Does the model predict the appropriate type of skeletal loading (tension versus compression)? (2) Does the model appropriately predict how skeletal stresses change in relation to the overall kinematics of the wingbeat (downstroke, upstroke, etc.)? (3) Are the stresses of appropriate absolute magnitude? and (4) Can the model accurately predict the details of changes in bone stresses during wing movement? Based on these increasingly stringent criteria, our model appears to capture much of the important mechanics and aerodynamics of level flight in *P. poliocephalus* (figure 6.4) (Watts et al., 2001).

A number of outputs from this model point to previously undescribed features of the mechanics of bat flight. For example, within the plane of the wing, bone stresses are extremely sensitive to the stiffness of the wing membrane skin (figure 6.5A, B). Peak bone stresses increase almost linearly with skin modulus. Given typical failure stresses for mammalian compact cortical bone of approximately 150 MPa (Currey, 1984), this result suggests that the wing membrane skin is not likely to be stretched to its stiffest values (see figure 6.2) during flight due to risk of fracture in the skeleton. These model results reinforce the need to better understand the relationship between skin mechanics and aerodynamics and mechanics of flight.

This computer model also allows us to modulate wing morphology to examine aspects of flight performance for wing designs not observed in any living bats. For example, the mass of the handwing is typically less than 5% of the total mass of a bat (Swartz, 1997; Watts et al., 2001). If that small mass is doubled, the peak stresses in the skeleton can increase significantly

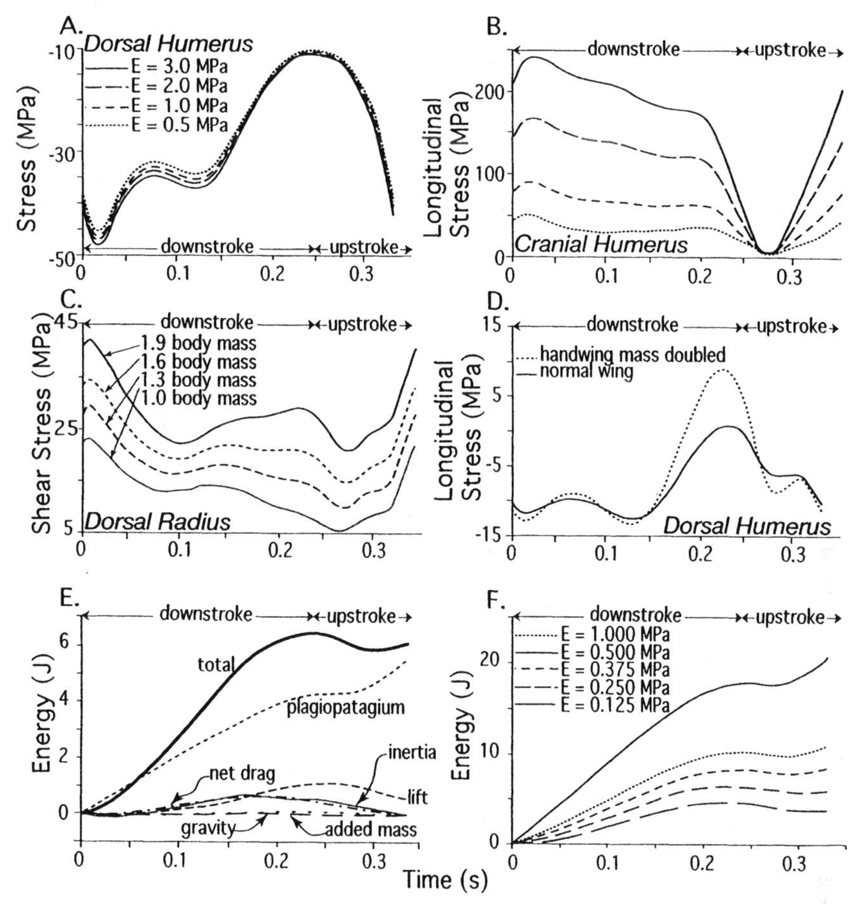

Figure 6.5 (A), (B) Longitudinal stresses computed by the model at the midshaft of the humerus throughout the wingbeat cycle as a function of spanwise elastic modulus of the skin, E_{xx}. On the dorsal surface of the bone, stresses are insensitive to skin stiffness (A), but on the cranial edge of the bone, changes in modulus have an important effect on predicted skeletal stresses; a difference of only 2.5 MPa in elastic modulus brings peak stresses from well below bone strength to stress magnitudes of fracture and beyond. (C) The model predicts that stresses in wing bones increase nearly in proportion to the mass of loads carried, suggesting that bats may need to alter wing kinematics or aerodynamics to maintain reasonable bone loads during weight carrying. (D) Bone stresses may be affected in unanticipated ways by redistributions of wing mass. Here, when the mass of the handwing, a small fraction of total wing mass, is doubled, peak stresses change significantly, but only during a fraction of the wingbeat cycle. (E) The model can be employed to partition the energetic cost of a wingbeat into components due to the individual mechanical and aerodynamic factors. Even for low values of skin stiffness, the greatest part of the energetic cost of flight as modeled is the energy required to produce and maintain the stretch of the wing membrane. (F) The total energetic cost of flight is highly sensitive to the stiffness of the wing skin. For all graphs, the horizontal axis gives time in seconds.

(figure 6.5D). In a more ecological context, we can model the effect of behaviors such as load carrying, relevant for understanding the mechanics and energetics of foraging in some frugivorous and carnivorous bats and of pregnancy in all bats. Model results suggest that loads of realistic magnitudes can have a very significant effect on the skeleton, bringing peak shear stresses in some cases from below to well above the shear strength of bone (35–40 MPa) (figure 6.5C). Similarly, model results suggest that the stiffness of wing membrane skin is important beyond the context of mechanics and aerodynamics. As wing membrane modulus increases, the total energetic cost of a wingbeat increases proportionately (figure 6.5E). When we employ the model to partition the energetic cost of a wingbeat into components related to each of the forces encountered by the wing, the reason for the strong relationship between skin modulus and flight energetics becomes clear: the energetic cost of physically extending the wing membrane and maintaining that extension during the wingbeat, particularly the downstroke, comprises the largest component of the energetic cost of flight, exceeding the energetic cost associated with wing inertia, generating lift, and overcoming drag severalfold (figure 6.5F). Thus, modeling reveals that energetic and mechanical considerations have likely played key roles in the evolution of the mechanical properties of the wing membrane and their variation among taxa.

THREE-DIMENSIONAL KINEMATIC ANALYSIS

High quality, three-dimensional kinematic data are necessary to model bat flight and its interspecific variation, and to examine how effectively aerodynamic theory developed for aircraft can be applied to bat flight. Until recently, such data were extremely difficult to obtain for several reasons. The rapid movements of bat wings require high imaging rates to accurately capture movements of the wingtips (typically greater than 250–500 images per second, and higher for smaller species with high wingbeat frequencies). With conventional film and video, these imaging rates require intense light that often prevents bats from flying normally or at all. Moreover, bats move several body lengths with each wingbeat cycle, hence it is impossible to simultaneously zoom in close enough to obtain good detail of wing motions and capture images from an entire wingbeat. We have attempted to address these issues by training subjects from a number of species of bats to fly in wind tunnels, the flying animal's equivalent of a treadmill, and imaging their flight with high-speed (500–1000 Hz) digital video under infrared lighting conditions. To date, we have obtained data from nine species of bats, ranging in body mass from 4 to 900 g, and including species from families of both Mega- and Microchiroptera. Although our analyses of these data are preliminary, they suggest that this approach may reveal a number of important features of bat flight.

Some potential limitations have been proposed for the use of wind tunnels in studies of animal flight. Rayner and Thomas (1991) considered the interaction of the wake behind a flying animal with the walls of a confined volume, such as a wind tunnel, and found that this interaction could lead to underestimates of the energetic cost of flight. According to their calculations, this effect was negligible if the size of the flight chamber was at least twice the wingspan of the animal. In addition, small differences have been found between the kinematics of cats walking on a treadmill versus walking on a stationary surface (Wetzel et al., 1975). There is no evidence that flying in a wind tunnel alters wing kinematics in any substantial way and wind tunnels have been widely used in the study of animal flight kinematics (e.g., Bruderer et al., 2001; Park et al., 2001; Tobalske, 1995; Tobalske and Dial, 1996).

Large-Scale Deformations of the Bat Wing Skeleton

Studies of vertebrate locomotion, including the flight of bats, typically assume that the bones remain quite stiff and rigid during movement, with few significant exceptions noted to date (see below). By tracking the movement of anatomical markers on the wing relative to fixed points on the body in three-dimensional space, we find that several elements of the bat wing skeleton undergo large deformations during flight. All species examined thus far display large deformations of the metacarpals and phalanges (figure 6.6). These deformations necessarily influence the three-dimensional conformation of the wing membrane; as the finger bones bend, the skin stretched between them will take on some characteristic three-dimensional curvature, determined not only by the aerodynamic forces exerted on the skin, but also by the deflection of the bones. This deflection, in turn, is determined by the structure of the bones, the aerodynamic forces on the wing, forces exerted by digital flexor and extensor muscles on the ventral and dorsal surfaces of the bone respectively, and by the force in the wing membrane. These data suggest that we must analyze wing shape during flight in new ways.

Angle of Attack

One aspect of airfoil design that significantly influences flow patterns, and hence the magnitude of lift and drag, is the angle of attack, the angle between a line connecting the leading and trailing edges of a wing and the direction of incoming airflow. Increasing angle of attack of a given airfoil in a given flow will increase the lift on the wing, up to a point at which lift drops precipitously and the stall ensues (Abbott and Doenhoff, 1999; Vogel, 1994). Aircraft typically fly at angles of attack less than 5°, although in extreme cases airplane wings may reach angles of attack as great as 10°. Moreover, angle of attack varies little during the flight of an engineered aircraft. These assumptions are not met in bat flight, so considerable caution may be necessary when applying standard aerodynamic theory to animal wings during flight (Aldridge, 1986; Hedrick et al., 2002).

Three-dimensional kinematic analysis is a particularly useful approach for making continuous measurements of angle of attack throughout a flight. We computed angle of attack from three-dimensional kinematics and corrected for the additional velocity of the wing's leading edge moving dynamically in

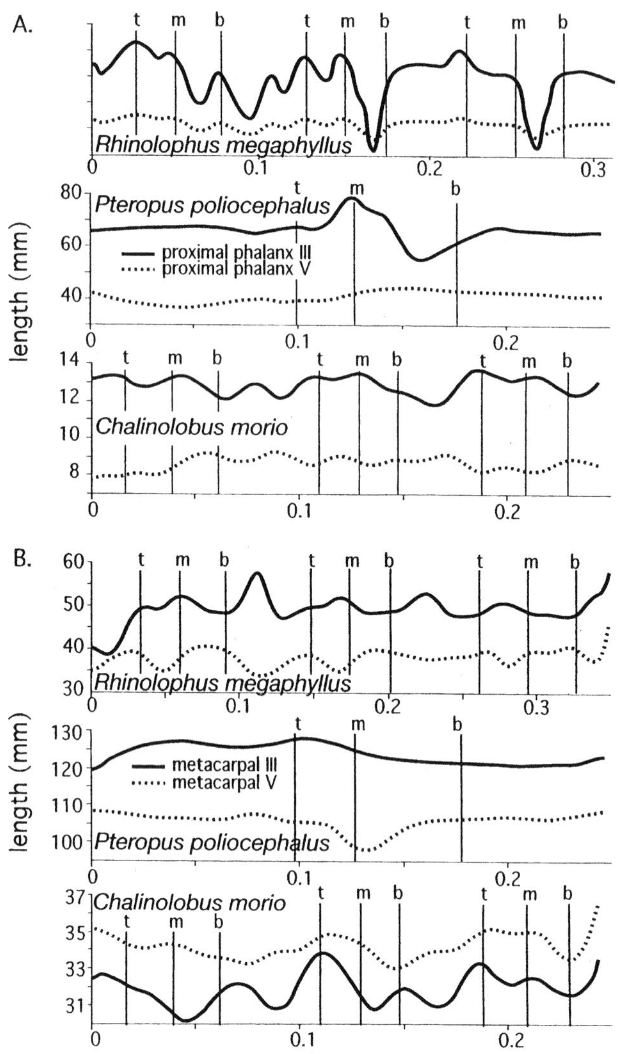

Figure 6.6 (A) Change in the distance between the proximal and distal interphalangeal joints during flight, as determined by three-dimensional kinematic analysis. The length of the third proximal phalanx is indicated by the continuous line, and that of the fifth proximal phalanx by the dotted line. (B) Change in the distance between the carpus and proximal interphalangeal joints during flight, as determined by three-dimensional kinematic analysis. The length of the third metacarpal is indicated by the continuous line, and that of the fifth metacarpal by the dotted line. For all graphs, the top, middle, and bottom of the downstroke are indicated by t, m, and b respectively; the horizontal axis gives time in seconds.

the airflow; this has been termed the "flapping velocity" (Aldridge, 1986). We found that the magnitudes of maximum and minimum angles of attack are far greater in bats than in any aircraft (−20° to +50°) (figure 6.7A), that the angle of attack is negative for a significant proportion of the lift-generating

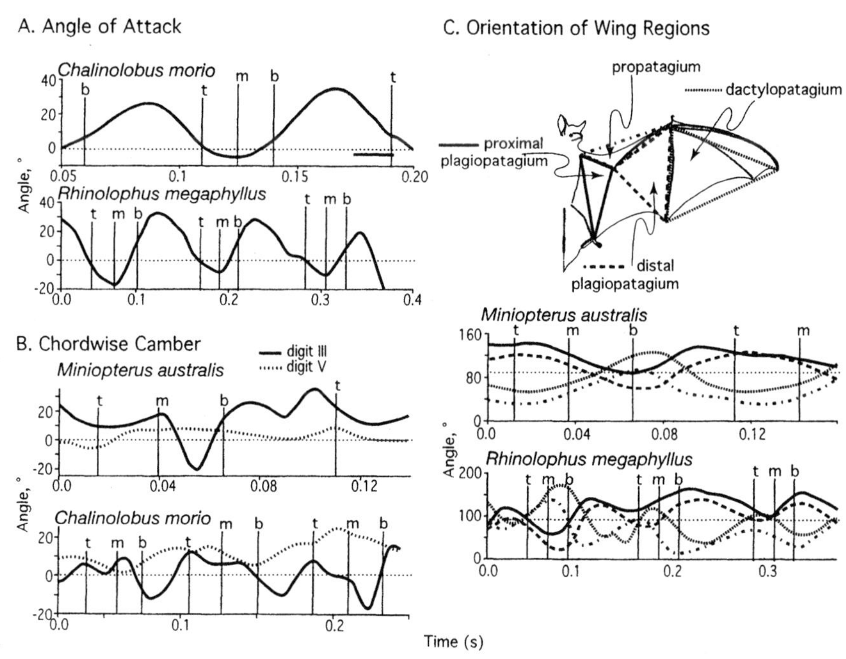

Figure 6.7 (A) Changes in the angle of attack during flight in *Chilinolobus morio* and *Rhinolophus megaphyllus* as determined by three-dimensional kinematic analysis. Zero angle of attack (coincidence of wing chord with incident velocity vector) is indicated by the horizontal dotted line. (B) Changes in the chordwise camber due to the third and fifth digits during flight in *Miniopterus australis* as determined by three-dimensional kinematic analysis. Chordwise camber induced by the positions of joints of the third digit is indicated by the continuous line, and by the fifth digit by the dotted line. (C) Orientation of wing regions (dactylopatagium, continuous line; proximal plagiopatagium, dashed line, distal plagiopatagium, dotted line, propatagium, dashed-dotted line) with respect to the horizontal plane of the wing in *Miniopterus australis* and *Rhinolophus megaphyllus*. When plane orientation equals 90° (horizontal dotted line), the plane of the wing is parallel to the ground. For all graphs, the top, middle, and bottom of the downstroke are indicated by T, M, and B respectively; the horizontal axis designates time in seconds.

downstroke, and that angle of attack changes continuously throughout the wingbeat, varying over a large range of angles (30–50° over a single wingbeat). This finding is significant because of the importance of angle of attack for the aerodynamic behavior of airfoils and other streamlined shapes. The distribution of air flow velocity, magnitude, and orientation around airfoils is sensitive to the orientation of the airfoil in the flow, and because of the direct inverse relationship between flow velocity and fluid pressure, the well-known Bernoulli effect, the distribution of positive and negative pressure around the airfoil reflects orientation. This leads to large changes in the magnitude and direction of the net aerodynamic force, the vector sum of the pressure acting around the entire surface of the airfoil. Lift is the upward component of the aerodynamic force, and will therefore change continuously as the angle of attack changes in an otherwise constant flow regime. Clearly, the mechanics of fixed-wing aircraft cannot be used to model these aspects of bat flight.

The very large angles of attack we have observed in bat wings suggest intriguing and as yet poorly understood features of bat flight aerodynamics. As noted, lift increases with angle of attack, but typically only until a certain relatively low stall value. Stall occurs because the airflow begins to separate from the dorsal surface of the wing, begins to form a turbulent wake, and ultimately no longer sustains a pattern of net air circulation around the wing. However, it is possible that bats employ novel mechanisms to maintain circulation despite an extremely high angle of attack. In at least some cases, insects appear to employ very large angles of attack to generate large amounts of lift without stall using specialized aerodynamic mechanisms (e.g., Sane and Dickinson, 2001). One such mechanism is the Kramer effect, in which a high angle of attack can transiently raise lift values above the maximum predicted by steady-state aerodynamics due to a separation of flow from the leading edge of the wing and an associated creation of leading edge vortices (Dudley, 2000; Ellington, 1984; Kramer, 1932; Wu et al., 1991). Once such leading edge vortices form, they may be stabilized by spanwise airflow along the wing, and such spanwise flows may be an important feature of flapping flight (Birch and Dickinson, 2001; Ellington et al., 1996; Maxworthy, 1979). Although these complex unsteady flow regimes have not been previously described for flying vertebrates, they are clearly of fundamental importance in the flapping flight of insects (Dickinson and Gotz, 1993). Insect wings also appear to generate additional lift by interacting with their own wake (Chow and Huang, 1982; Graham, 1983; Wagner, 1925; Walker, 1931). When the wing has not moved out of the range of the downstroke vortex of one wingbeat before another begins, one result can be "wake capture," in which the vorticity of the wingbeat results, in part, from a summation of effects of at least two wingbeats (Dickinson et al., 1999; Sane and Dickinson, 2001). This mechanism has proven to substantially increase lift production beyond predictions of simple aerodynamics in insects.

Most bats are probably too large and fast-flying to use the kind of wake capture seen in insects, but it increasingly appears likely that unsteady effects of some kind are important in vertebrate flight as well, particularly for bats and relatively small or slow-flying birds, and especially at flight speeds less than 5 m/s (Hedrick et al., 2002). The continuous change in angle of attack throughout the wingbeat cycle also suggests that important aspects of the aerodynamics of bat flight cannot be explained by fixed-wing aerodynamics. The continuously changing wing orientation may lead to large changes in the magnitude of lift forces during the wingbeat in direct relation to the angle of attack.

Wing Camber

In addition to angle of attack, flow around an airfoil and the resulting aerodynamic forces are strongly influenced by the degree to which an airfoil surface is cambered, or arched from front to back in dorsally convex manner. As in the case of angle of attack, we found that camber values in bats are large, ranging from −20° to +45°, and that camber varies continuously over a large range during all wingbeats (figure 6.7B). Values are often negative during the last third of the downstroke and the upstroke, indicating that the wing is bent in a dorsally concave fashion, at least locally along the span. These characteristics distinguish bats, once again, from fixed-wing aircraft and birds, but indicate possible similarities among bats and insects. Insect wings show a camber reversal between downstroke and upstroke (Dudley, 2000). Because insect wings lack intrinsic musculature, this change in conformation is a purely passive result of aerodynamic and inertial forces exerted on the wing interacting with the wing morphology and the mechanical properties of tissue. In bats, however, camber could be actively influenced by the degree of flexion or extension at carpal, finger joints, as well as by position of the thumb, and hence the propatagium, with respect to the plagiopatagium. In addition, the plagiopatagiales muscles, distinct chordwise bundles of intrinsic wing musculature with no direct skeletal attachments, have long been believed to play a role in regulating camber of the plagiopatagium (Strickler, 1978; Vaughan, 1959, 1970). Indeed, the net camber of a bat wing is likely a result of a combination of active control of wing conformation by intrinsic musculature and the interaction of wing tissue properties and overall geometry with complex aerodynamic forces exerted across the wing's surface.

Whatever the anatomical basis of the wing's camber, the curvature of the wing's surface, and the variation of this curvature along the wing's span and at any one location throughout the wingbeat cycle, may significantly influence airflow across the wing and the resulting aerodynamic forces. Camber clearly increases lift production in many insect species, even in groups in which the wings are predominantly planar, such as the Lepidoptera (Dudley, 2000; Ellington, 1984; Vogel, 1967). At the Reynolds numbers at which bats fly, the effect of camber on lift is even greater than for insects (Dickinson and Gotz, 1993). In fact, some of the most recent efforts to design more efficient airplanes have focused on approaches that allow the pilot to modulate both chordwise and spanwise camber in flight, given that wings of any fixed geometry represent optimal design only for a limited combination of altitudes and aircraft weights (Monner, 2001).

Are Bat Wings Fully Extended during the Downstroke?

Another central assumption of the application of aircraft aerodynamics to bat flight is that the wing is more or less fully extended through a substantial portion of the downstroke, and therefore maintains a constant wing area that is oriented in a single plane, perpendicular to local wing velocity; that is, for some important part of the downstroke, the wing moves like a paddle. In part, this assumption underlies the practice of assessing wing form by making tracings of fully extended wings; we typically accept that a fully extended wing posture, with the wing lying virtually flat, represents the form of the wing during the generation of aerodynamic forces. However, three-dimensional analysis again reveals surprising complexity. We partitioned the wing into four major triangular regions, each demarcated by three anatomical landmarks (joints or wingtip): the propatagium, the dactylopatagium, the proximal plagiopatagium, and the distal plagiopatagium. We then determined the three-dimensional spatial orientation of the planes connecting each of the groups of three landmarks, and computed the orientation of the vector perpendicular to each of these planes with respect to the horizontal; a measure of 90° thus indicates that a wing region is flat, extended, and parallel to the ground.

Plots of representative analyses demonstrate that the four regions only rarely coincide in orientation (all four lines intersect) (figure 6.7C). Furthermore, no part of the wing is regularly parallel to the ground at the middle of the downstroke. In fact, although three of four wing regions briefly approach horizontal in one species examined to date (figure 6.7C), most of our data show considerable three-dimensional complexity at all points in the wingbeat cycle, including the middle third of the downstroke. No animal in our sample closely approaches a fully extended wing in mid-downstroke, and different species differ from this expectation in a diversity of ways. As a consequence, it may be misleading to compare bat wing form among taxa based on tracings of fully extended wings, both because this wing posture may rarely be realized, and because the three-dimensional wing conformation, a primary determinant of airflow and thus aerodynamic forces, appears to differ significantly among species, leaving open the distinct possibility that two species with quite similar wing shapes might, by virtue of kinematic differences in their wing motions, employ quite different airfoils.

CONCLUSIONS AND FUTURE DIRECTIONS

Advances in the application of engineering approaches to understanding the tissues of the bat wing and the increased availability and decreased cost of sophisticated high-speed imaging technology together may profoundly affect our understanding of the mechanics and aerodynamics of bat flight, and of the relationship between wing shape and flight performance. In particular, until very recently it has been extremely difficult to obtain accurate three-dimensional information about the movements of bat wings in flight, and our knowledge of the three-dimensionality of wing conformation at a given moment in the wingbeat cycle, as well as the dynamically changing nature of wing shape has been limited. As bat researchers collect more and better three-dimensional data, it is possible that a new picture of wing function in flight will emerge.

Our studies of the mechanical and aerodynamic significance of wing tissue properties, computer modeling approaches, and recent preliminary analyses of three-dimensional kinematics in several bat species suggest that material complexity and dynamic aspects of bat flight are critically important. The materials of the wing, particularly wing membrane skin, but also wing bones, are unique among mammalian tissues, and possess mechanical properties that help drive the way wings deform when they experience aerodynamic loading. Different regions of the wing of any given bat, and the wings of distinct species with similar overall morphology, may deform in very different ways to identical loading conditions, thereby taking on distinctive three-dimensional form and aerodynamic performance. Aerodynamically relevant parameters such as angle of attack, wing camber, and orientation of regions of wing membrane relative to one another undergo large changes during the downstroke, and provide a picture of a wing that cannot be well understood as a simple airfoil.

Can we better specify the aerodynamic consequences of this picture of the morphology and behavior of the wing during flight? At present, we can say with confidence that the generation of lift by bat wings has yet to be fully understood, and that unsteady effects are likely extremely important. In this respect, we believe it will prove that most bats are quite different from most birds, and although the mechanisms involved are certainly different, the complex aerodynamics of insect wings may prove a better model for understanding bat flight aerodynamics.

In the context of this increasingly complex picture of the three-dimensionality of wing form during flight, it is also worthwhile to consider whether it is necessary for bats to actively control the wing shape, continually monitoring and adjusting angles of attack, camber, etc., to varying degrees along the wingspan. We suggest that this may not be the case, and that the dynamically changing nature of bat wing form suggests that bat wings may act as so-called reactive airfoils that are morphologically responsive to variable dynamic circumstances. In this view, the low stiffness of wing bones and the high anisotropy of wing skin interact with aerodynamic forces in a kind of feedback loop, such that increased loading increases the stiffness of the skin, changing, in turn, the way the wing responds to aerodynamic force, thereby changing the three-dimensional geometry in an adaptive manner.

Clearly, we have much work to do. The first important goal for studies of bat flight is to obtain much more detailed kinematic information from as many species as possible. The preliminary analyses presented here represent the "tip" of what we believe to be a very large "iceberg," and we predict that future work will uncover tremendous variation and complexity in the three-dimensionality of wing form in flight. A second important goal goes hand in hand with collecting three-dimensional data. Visualizing three-dimensional data is extremely difficult and requires that we go beyond traditional two-dimensional graphical presentation techniques. Computer graphics and animation techniques will be instrumental in interpreting spatially complex and dynamically changing forms. Finally, both physical modeling and computational fluid dynamic approaches may allow us to interpret the aerodynamic significance of wing form and its variation in ways that have not yet been possible. Difficult problems must be solved to go beyond the aerodynamics of fixed-wing aircraft to understand the functional performance of bat wings; the new insights we will gain will be well worth the effort.

ACKNOWLEDGMENTS

Many people have facilitated and contributed to the work described here. They include: Elizabeth Stockwell, Jennifer Skene, Les Hall, Nancy Irwin, Nicky Marcus, John Seyjaget and The Lubee Bat Conservancy, Ian Fischer, Mary Huber, The Civil Engineering Department of The University of Queensland, Andy Biewener, Tyson Hedrick, Emily Carrington, Kevin Middleton, David Baier, Leor Thomas, Eric and Emma Anderson, and Philip Watts. This work has been generously supported by National Science Foundation grants to S.M.S. and the Brown University Undergraduate Teaching and Research Assistantship Program.

LITERATURE CITED

Abbott, I.H., and A.E.V. Doenhoff. 1999. Theory of Wing Sections. Dover Publications, New York.

Aldridge, H.D.J.N. 1986. Kinematics and aerodynamics of the greater horseshoe bat, *Rhinolophus ferrumequinum*, in horizontal flight at various flight speeds. Journal of Experimental Biology, 126: 479–497.

Bartholomew, G.A., and R.E. Carpenter. 1973. Mechanics of flight in flying foxes. UCLA Media Center Productions, Los Angeles.

Birch, J.M., and M.H. Dickinson. 2001. Spanwise flow and the attachment of the leading-edge vortex on insect wings. Nature, 412: 729–733.

Bruderer, L., F. Liechti, and D. Bilo. 2001. Flexibility in flight behaviour of barn swallows (*Hirundo rustica*) and house martins (*Delichon urbica*) tested in a wind tunnel. Journal of Experimental Biology, 204: 1473–1484.

Chow, C.-Y., and M.-K. Huang. 1982. The initial lift and drag of an impulsively started airfoil of finite thickness. Journal of Fluid Mechanics, 118: 393–409.

Currey, J.D. 1979. Mechanical properties of bone tissues with greatly differing functions. Journal of Biomechanics, 12: 313–319.

Currey, J.D. 1984. The Mechanical Adaptations of Bones. Princeton University Press, Princeton, NJ.

Denny, M.W. 1993. Air and Water: The Biology and Physics of Life's Media. Princeton University Press, Princeton, NJ.

Dickinson, M.H., and K.G. Gotz. 1993. Unsteady aerodynamic performance of model wings at low Reynolds-numbers. Journal of Experimental Biology, 174: 45–64.

Dickinson, M.H., F.O. Lehmann, and S.P. Sane. 1999. Wing rotation and the aerodynamic basis of insect flight. Science, 284: 1954–1960.

Dudley, R. 2000. The Biomechanics of Insect Flight: Form, Function, Evolution. Princeton University Press, Princeton, NJ.

Ellington, C.P. 1984. The aerodynamics of hovering insect flight. IV. Aerodynamic mechanisms. Philosophical Transactions of the Royal Society of London, Series B, 305: 1–15.

Ellington, C.P., C. VandenBerg, A.P. Willmott, and A.L.R. Thomas. 1996. Leading-edge vortices in insect flight. Nature, 384: 626–630.

Gibson, T. 1977. The physical properties of skin. Pp. 69–77. In: Reconstructive Plastic Surgery: Principles and Procedures in Correction, Reconstruction and Transplantation (J.M. Converse, ed.). W.B. Saunders, Philadelphia.

Gibson, T., H. Stark and J.H. Evans. 1969. Directional variation in extensibility of human skin *in vivo*. Journal of Biomechanics, 2: 201–204.

Gordon, J.E. 1978. Structures: Or, Why Things Don't Fall Down. Plenum Press, New York.
Graham, J.M.R. 1983. The lift on an aerofoil in starting flow. Journal of Fluid Mechanics, 133: 413–425.
Hedrick, T.L., B.W. Tobalske, and A.A. Biewener. 2002. Estimates of circulation and gait change based on a three-dimensional kinematic analysis of flight in cockatiels (*Nymphicus hollandicus*) and ringed turtle doves (*Streptopelia risoria*). Journal of Experimental Biology, 205: 1389–1409.
Huber, M.H. 2000. Wings like springs: the structure and function of handwing bones in bats. Undergraduate Honors Thesis, Brown University, Providence, RI.
Kramer, M. 1932. Die Zunahme des Maximalauftriebes von Tragflügeln bei plötzlicher Anstellwinkelvergrösserung (Böeneffeckt). Zeitschrift für Flugtechnik und Motorluftschiffahrt 23: 185–189.
Liu, H., C.P. Ellington, K. Kawachi, C. Van den Berg, and A.P. Willmott. 1998. A computational fluid dynamic study of hawkmoth hovering. Journal of Experimental Biology, 201: 461–477.
Maxworthy, T. 1979. Experiments on the Weis-Fogh mechanism of lift generation by insects in hovering flight. Part 1. Dyanmics of the "fling." Journal of Experimental Biology, 93: 47–63.
Monner, H.P. 2001. Realization of an optimized wing camber by using form variable flap structures. Aerospace Science and Technology, 5: 445–455.
Nordin, M., T. Lorenz, and M. Campello. 2001. Biomechanics of tendons and ligaments. Pp. 102–125. In: Basic Biomechanics of the Musculoskeletal System (M. Nordin and V.H. Frankel, eds.). Lippincott Williams and Wilkins, Philadelphia.
Papadimitriou, H.M., S.M. Swartz, and T.H. Kunz. 1996. Ontogenetic and anatomic variation in mineralization of the wing skeleton of the Mexican free-tailed bat, *Tadarida brasiliensis*. Journal of Zoology (London), 240: 411–426.
Park, K.J., M. Rosen, and A. Hedenstrom. 2001. Flight kinematics of the barn swallow (*Hirundo rustica*) over a wide range of speeds in a wind tunnel. Journal of Experimental Biology, 204: 2741–2750.
Rayner, J.M.V., and A.L.R. Thomas. 1991. On the vortex wake of an animal flying in a confined volume. Philosophical Transactions of the Royal Society of London, Series B, 334: 107–117.
Sane, S.P., and M.H. Dickinson. 2001. The control of flight force by a flapping wing: lift and drag production. Journal of Experimental Biology, 204: 2607–2626.
Shadwick, R.B., A.P., Russell, and R.F. Lauff. 1992. The structure and mechanical design of rhinoceros dermal armour. Philosophical Transactions of the Royal Society of London, Series B, 337: 419–428.
Skene, J.A. 2000. The mechanical and energetic role of the wing membrane in bat flight. Undergraduate Honors Thesis, Brown University, Providence, RI.
Strickler, T.L. 1978. Functional Osteology and Myology of the Shoulder in Chiroptera. S. Karger, Basel, Switzerland.
Swartz, S.M. 1997. Allometric patterning in the limb skeleton of bats: implications for the mechanics and energetics of powered flight. Journal of Morphology, 234: 277–294.
Swartz, S.M., M.B. Bennett, and D.R. Carrier. 1992. Wing bone stresses in free flying bats and the evolution of skeletal design for flight. Nature, 359: 726–729.
Swartz, S.M., M.B. Bennett, J.A. Gray, and A. Parker. 1993. Bones built to bend: In vivo loading in the distal wing of fruit bats. American Zoologist, 33: 75A.

Swartz, S.M., M.D. Groves, H.D. Kim, and W.R. Walsh. 1996. Mechanical properties of bat wing membrane skin. Journal of Zoology (London), 239: 357–378.

Tobalske, B.W. 1995. Neuromuscular control and kinematics of intermittent flight in the European starling (*Sturnus vulgaris*). Journal of Experimental Biology, 198: 1259–1273.

Tobalske, B.W., and K.P. Dial. 1996. Flight kinematics of black-billed magpies and pigeons over a wide range of speeds. Journal of Experimental Biology, 199: 263–280.

Vaughan, T.A. 1959. Functional morphology of three bats: *Eumops, Myotis, Macrotus*. Publications, Museum of Natural History, University of Kansas, 12: 1–153.

Vaughan, T.A. 1970. The muscular system. Pp. 140–194. In: Biology of Bats, Vol. 1 (W.A. Wimsatt, ed.). Academic Press, New York.

Vogel, S. 1967. Flight in *Drosophila*. III. Aerodynamic characteristics of fly wings and wing models. Journal of Experimental Biology, 46: 431–443.

Vogel, S. 1994. Life in Moving Fluids: The Physical Biology of Flow. Princeton University Press, Princeton, NJ.

Wagner, H. 1925. Über die Entstehung des dynamischen Aufstriebes von Tragflügeln. Zeitschrift für angewandte Mathematik und Mechanik, 5: 17–35.

Wainwright, S.A., W.D. Biggs, J.D. Currey, and J.M. Gosline. 1976. Mechanical Design in Organisms. Edward Arnold, London.

Walker, P.B. 1931. A new instrument for the measurement of fluid motion; with an application to the development of the flow around a wing started impulsively from rest. Report of the Memoirs of the Aeronautical Research Council, No. 1402.

Watts, P., E.J. Mitchell, and S.M. Swartz. 2001. A computational model for estimating the mechanics of horizontal flapping flight in bats: model description and validation. Journal of Experimental Biology, 204: 2873–2898.

Wetzel, M.C., A.E. Atwater, J.V. Wait, and D.G. Stewart. 1975. Neural implications of different profiles between treadmill and overground locomotion timings in cats. Journal of Neurophysiology, 38: 492–501.

Wu, J.Z., A.D. Vakili, and J.M. Wu. 1991. Review of the physics of enhancing vortex lift by unsteady excitation. Progress in Aerospace Science, 28: 73–131.

7

Performance Analysis as a Tool for Understanding the Ecological Morphology of Flower-Visiting Bats

Christopher W. Nicolay & York Winter

Performance analysis provides a tool for understanding the functional and ecological correlates of morphological variation. We have examined two measures of nectar-feeding performance within phyllostomid bats: tongue extension and rate of nectar extraction. Tongue extension provides a measure of maximal capacity and potentially limits the lengths of flowers that a bat can visit. Rate of nectar extraction at different feeders provides a measure of performance at a task within the capacity of the bats. We found strong relationships between apparent morphological specialization for nectarivory and tongue extension. However, morphological specialization is not strongly associated with nectar extraction rates. Tongue extension is more closely tied to apparent morphological specialization for nectarivory in phyllostomids than is rate of nectar extraction, which is influenced by feeding styles and behavioral plasticity. Other factors, such as the ability to extract small amounts of nectar from flowers, probably also play an important role in the evolution of nectarivory.

INTRODUCTION

Performance analysis has emerged as a powerful tool in functional morphology for examining the connection between morphology and behavioral capacity. In functional morphology, "performance" is defined as the quantifiable score at some behavioral or physiological task (Arnold, 1983; Wainwright, 1991). Performance analysis investigates the functional consequences of anatomical variation. Although the discussion of performance in this chapter is directed toward the feeding apparatus of flower-visiting bats, with slight modification this approach can be tailored to fit any morphological or physiological system.

The Conceptual Framework for Performance Analysis

Morphology rarely influences fitness directly, but instead is subject to selection only when a particular morphology enables an individual to do something that others with different morphologies cannot. For example, selection for long legs cannot be achieved unless long-legged animals can do something (such as elude predators, acquire resources, attract mates) "better" than can their shorter-legged counterparts. Cryptic coloration is subject to selection only to the degree that variation in color and pattern helps an animal avoid detection by predators. As formulated by Arnold (1983), performance analysis is used to address the relationship between morphology and fitness by breaking the problem into two steps: (1) establishing the relationship between morphology and performance, and (2) evaluating the effect of variation in performance on fitness. A complete analysis of evolutionary functional morphology ideally addresses both gradients.

The link between morphology and performance is typically evaluated in the laboratory using controlled behavioral experiments coupled with detailed morphological observations. Establishing the association between performance and fitness requires appropriately designed ecological investigations. In practice, demonstrating the relationships between morphology and performance is easier than establishing the relationships between performance and fitness. This seems especially evident in bats, which are long-lived, slow-reproducing, far-ranging, and mobile animals. Conducting appropriately designed ecological experiments with samples great enough to powerfully test for significance presents a substantial challenge. Even when experimental designs are excellent, episodes of directional selection may be limited to periods of annual, supra-annual, or catastrophic events, and therefore may not be detectable during "normal" periods.

Several other factors potentially complicate this seemingly simple research paradigm. First, the degree to which morphology is heritable limits the potential for selection to operate on morphology. Ultimately, a measure of performance must have some degree of heritability (through its linkage with heritable morphology) and contribute to variation in reproductive success (at least during some point in the evolutionary history of a lineage) to evolve through natural selection (Shaffer and Formanowicz, 2000). Although factors such as diet and biomechanical feedback affect morphology and performance, the variation produced by these processes is not heritable, but the result of phenotypic plasticity (Carter et al., 1991; Frost et al., 1998).

Second, the association between morphology, performance, and fitness is complicated because most anatomical systems serve multiple functions. Performance usually results from the complex interactions of multiple organs (muscles, bones) and anatomical systems (e.g., muscular, skeletal, nervous), which can independently vary in morphology, histology, and physiology (Bauwens et al., 1995; Garland et al., 1995). As a result, conflicting functional demands can be placed on any single system. However, similarity in performance is possible through different combinations of morphological features or different behaviors that accomplish the same goal (Aerts et al., 2000).

Any morphological system of interest may be affected by selection operating on apparently unrelated systems. For example, a change in neurocranial volume (brain size) may alter the size and leverage of the masticatory muscles originating from the skull. The bite forces generated may increase or decrease as a result, but not due to selection operating directly on the masticatory system. In bats, different modes of echolocation may influence many aspects of cranial morphology, including the masticatory system (Pedersen, 1998).

Additionally, behavioral plasticity mediates and helps overcome the constraints imposed by morphology. For example, some bats alter the placement of food items along the toothrow to increase bite power in response to hardness of the food (Dumont, 1999). These issues present challenges for the investigator and potentially confound results. However, rather than detracting from the applicability of performance analysis, these issues underscore the need for rigorous examination of fundamental assumptions about the relationships between form and function.

Pragmatic Application of Performance Analysis

Beyond fitting into a theoretical framework for understanding the evolution of morphology, performance analysis is used in a more pragmatic and exploratory manner to dissect form–function problems into manageable parts. Performance analysis can be employed to address three broad levels of inquiry: (1) the functional correlates of morphological variation, (2) the ecological correlates of variation in morphology and performance, and (3) the evaluation of scenarios concerning how/why a particular morphology evolved. These three levels build upon each other to form a hierarchy of increasing complexity, where our understanding of each higher level builds upon the accuracy of our knowledge of the levels below.

A fundamental goal of performance analysis is to establish the link between morphology and performance. Performance data can then be employed to examine how the variation in performance is manifest in ecology and evolution. Performance plays a major role in sculpting patterns of resource use and ecology (Irschick and Losos, 1998; Wainwright, 1994, 1996). In this sense, resource use includes habitats occupied (Irschick and Losos, 1998; Vanhooydonck et al., 2000), prey and food items consumed (Herrel et al., 1999; Wainwright, 1996), and combinations of the two (Norberg, 1994; Saunders and Barclay, 1992; Schluter, 1996).

Types of Performance

Behavioral variables that reflect performance usually attempt to either capture a measure of maximal capacity, or measure the ability to use resources within this range (Reilly and Wainwright, 1994; Wainwright, 1996). By placing absolute limits on the range of activities an organism can accomplish, maximal performance helps circumscribe the *fundamental* niche of a species—the range of resources and habitats potentially utilized. The ability to use resources within the fundamental niche—along with other factors such as resource availability, choice, and competition—helps define the *realized* niche of a species (the actual patterns of resource use) (Reilly and Wainwright, 1994).

METHODS

Our work with flower-visiting phyllostomids has focused on variables reflecting both of these measures of performance. Tongue extension was chosen as a measure of maximal capacity related to nectarivory. The maximum distance the tongue can be extended ultimately places upper limits on the lengths of flowers a bat can visit (Winter and Helversen, 2003), and consequently defines part of the fundamental niche of flower-visiting bats (the flowers from which a bat is physically able to extract nectar).

We measured tongue extension by allowing bats to feed ad libitum from a clear, graduated 9-mm-diameter test tube without replacing the nectar solution it contained. The bats fed from the tubes until they could no longer reach the nectar. Videotaping visits using an infrared camera enabled observation of the distance the tongue was extended past the tip of the snout.

Rate of nectar extraction provides a measure of feeding performance at a task within the capacity of the bats. In the wild, glossophagine bats visit flowers extremely quickly (< 1 second) and take amounts of nectar that typically range from 0.01 to 0.30 g per visit). Increasing feeding rate can be argued to both increase the net return of energy per unit of time spent foraging and reduce the risk of predation while visiting a flower (Heinrich, 1975; Helversen, 1993).

We measured nectar-feeding performance using artificial flowers of three different lengths (20, 30, 40 mm) at each of two diameters (19 mm, 26 mm). Feeder diameters produced no significant effect on feeding performance, so data from the two diameters were pooled for each length. The following discussion is limited to feeder length (20, 30, 40 mm). The apparatus used for measuring feeding performance is shown in figure 7.1. Bats were filmed while feeding from a clear plastic "flower" connected to a large nectar reservoir, to ensure minimal change in the nectar level as the bats fed. The nectar supply was placed on a balance connected to a computer, which recorded the amount of nectar taken per visit. A light sensor connected to the computer recorded the occurrence of visits to the feeder and facilitated the calibration of computer-recorded data with video replay. The computer also controlled a curtain that swung shut immediately after a visit, which excluded the bat from the feeder and allowed the balance to stabilize between visits. The duration of each visit (time) was analyzed by counting frames on video playback (30 frames/second). Rate of nectar extraction (amount/time) was calculated for each visit. Because the artificial flowers were constructed from clear plastic, the number of licks per visit and the distance the head was inserted into the feeder could also be recorded.

NECTAR-FEEDING PERFORMANCE IN PHYLLOSTOMID BATS

Flower visitation is common in bats of the families Phyllostomidae and Pteropodidae, and within each of these families several species are specialized

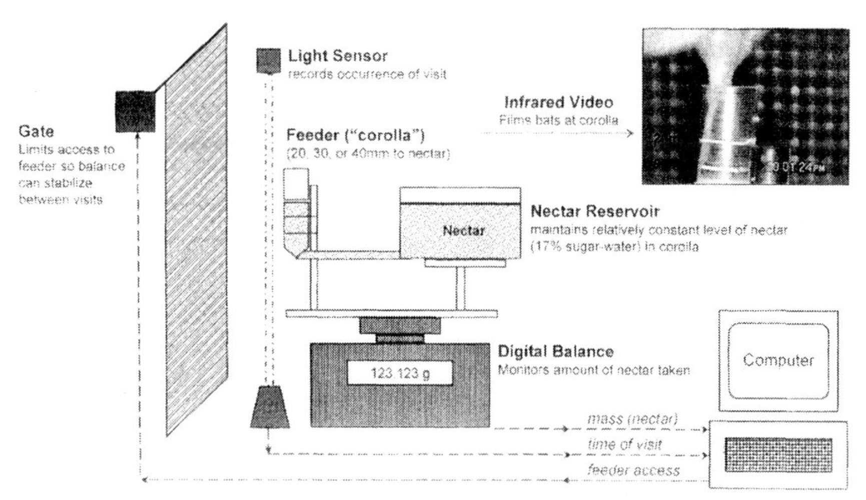

Figure 7.1 Schematic representation of feeder design. Bats were only able to access nectar through the artificial flower. Three different lengths of flowers were used (20, 30, 40 mm). The gate was normally open, but was automatically closed for 12 seconds after each visit (detected by interruption of the light sensor) to exclude the bat, allowing the balance time to stabilize between visits. The computer was housed in a separate room away from the bats.

for feeding from flowers (Dobat, 1985; Heithaus, 1982). Nectar also comprises part of the diet of the New Zealand endemic species *Mystacina tuberculata*, Mystacinidae (Arkins et al., 1999), and incidental consumption of flower products occurs in *Antrozous pallidus*, Vespertilionidae (and probably other species) as they glean arthropods from flowers (Herrera et al., 1993). Of all these groups, the family Phyllostomidae has the greatest diversity of flower-visiting species (in both morphology and number of species) and contains the species that possess the most extreme degree of morphological specialization toward nectarivory (Koopman, 1981).

In any functional analysis, the potential for confounding factors increases as phylogenetic distance increases (see chapter 9 for further discussion of these issues and approaches for dealing with them). To reduce the effect of phyletic distance, our studies have been restricted to a single, diverse family (Phyllostomidae). We measured nectar-feeding performance in six phyllostomid species: three members of the subfamily Glossophaginae (*Glossophaga soricina*, *Leptonycteris curasoae*, *Choeronycteris mexicana*), and three phyllostomids from other subfamilies (*Artibeus jamaicensis*, Stenodermatinae; *Carollia perspicillata*, Carolliinae; *Phyllostomus discolor*, Phyllostominae) (figure 7.2).

Within phyllostomids, flower-visitation characterizes two subfamilies (Glossophaginae and Lonchophyllinae), and has been documented in at least some species of all the other subfamilies (with the exception of the vampire bats, Desmodontinae) (Dobat, 1985; Gardner, 1977). Flower-visiting

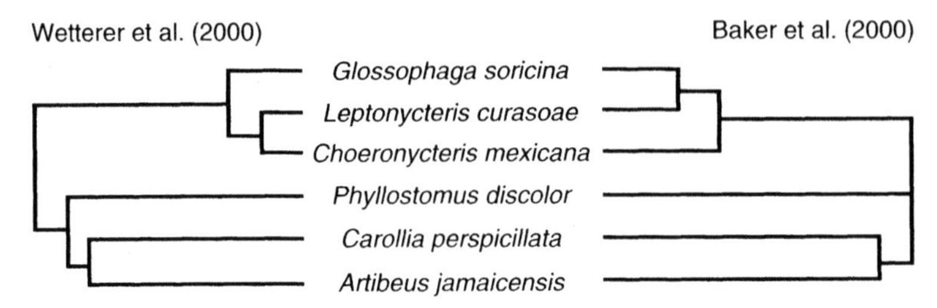

Figure 7.2 Phylogenetic relationships among species included in this study. Two separate phylogenies are shown. The phylogeny of Wetterer et al. (2000) is a "total-evidence" phylogeny, while that of Baker et al. (2000) is based on RAG-2 DNA sequences.

styles of the nectarivorous "glossophagines"* differ greatly from other phyllostomids. Most non-glossophagine phyllostomids land on flowers to feed, spend at least 1–3 seconds on the flower, and appear to drain a flower entirely of its nectar before leaving (Heithaus et al., 1974, 1975; Helversen, 1993). Glossophagines typically feed while hovering, and their visits to flowers last less than 1 second (e.g., Helversen, 1993; Sazima et al., 1989). Glossophagines apparently do not entirely drain flowers of nectar at each visit (Helverson, 1993). Within glossophagines, there is a trend for increasing specialization of the skull, teeth, and tongue related to increasing reliance on nectar in the diet (Freeman, 1995, 1998; Koopman, 1981; Phillips, 1971; Solmsen, 1999). Our sample includes members of Glossophaginae that have been classified as primitive regarding the degree of morphological specialization for nectarivory (*G. soricina*), moderately derived (*L. curasoae*), and highly derived (*C. mexicana*) (Koopman, 1981).

Tongue Extension

Sympatric species of glossophagines tend to feed on similar sets of flowers, and the relationships between flower shape and bat morphology are diffuse (Heithaus, 1982). Different species of plants depend on bats as pollinators to varying degrees (Sazima et al., 1994; Fleming et al., 2001), and the importance of bats as pollinators of a particular species can vary with season and geographic location (Fleming et al., 2001). However, flowers pollinated by glossophagines are often easily distinguished from flowers pollinated by other phyllostomids by their smaller nectar rewards, and small flower size, and presentation of the flower that requires hovering flight (Helversen, 1993; Vogel, 1968, 1969a,b). Within Brazilian bat-pollinated assemblages,

* Recent phylogenetic analyses (Baker et al., 2000; Wetterer et al., 2000) agree that Glossophaginae and Lonchophyllinae are distinct clades which are both predominantly nectarivorous but differ in aspects of tongue morphology (Griffiths, 1982). At least two other genera (*Erophylla* and *Phyllonycteris*) appear to be somewhat specialized for feeding on nectar, although there is currently no consensus on their taxonomic placement. Whether all nectar-feeding phyllostomids form a unified clade (Wetterer et al., 2000) or not (Baker et al., 2000) is currently unresolved. Because of the current disagreement on the relationships among species, the term "glossophagine" in this chpater refers to the traditional ecological and functional grouping of nectarivores, irrespective of phylogeny, although the three nectarivores we studied belong to the subfamily Glossophaginae.

one species often combines a moderately long flower (at least) with the requirement of hovering flight, which probably discourages visitation by unspecialized bats (M. Sazima et al., 1999). However, this pattern does not necessarily characterize all neotropical bat-pollinated flower assemblages.

Tongue extension ultimately places upper limits on the lengths of flowers a bat can visit, and determines how close a bat must approach and how deeply the head must be inserted to reach the nectar. We found significant correlations between maximum tongue extension and apparent specialization for nectarivory indicated by the skull (Nicolay, 2001; Winter and Helversen, 2003). Specialized nectarivores have smaller teeth, long and relatively narrow skulls, snouts, and jaws (Freeman, 1995, 1998), and also possess modifications of the extrinsic tongue musculature and hyoid apparatus that facilitate tongue extension (Griffiths, 1982). These factors probably reflect a trade-off of masticatory power for increased ability to feed with the tongue. Because the tongue fills the oral cavity (often protruding slightly), the length of the snout is indicative of resting tongue length. Cranial specialization in part reflects the relative size of the resting tongue. Additionally, the extrinsic tongue muscles migrate toward the back of the tongue, increasing their ability to extend and retract the tongue (Griffiths, 1982). Longer resting tongues, placement of the extrinsic tongue muscles, and possibly hydrostatic mechanisms for tongue elongation are all associated with the ability to extend the tongue, and modified in morphologically specialized bats (Griffiths, 1982; Winkelmann, 1971; Winter and Helversen, 2003).

Glossophagines are distinguished from other phyllostomids by their highly protrusible tongues (figure 7.3). Glossophagines can extend the tongue past the tip of the snout to a distance greater than the length of the skull. In contrast, *Carollia perspicillata* and *Artibeus jamaicensis* can extend the tongue only a minimal distance (less than 7 mm) past the tip of the snout. *Phyllostomus discolor* has long and mobile tongue compared with other non-glossophagines, although it cannot extend its tongue proportionately as far as can glossophagines. In many parts of its range, *P. discolor* is highly nectarivorous during the dry season (Heithaus et al., 1974, 1975; Bonaccorso, 1979), which may be associated with its tongue mobility. Unlike glossophagines, the tongue of *P. discolor* is not covered with enlarged lingual papillae to extract nectar (Griffiths, 1982), and *P. discolor* is probably less dependent on nectar in the diet than are glossophagines.

Rate of Nectar Extraction

The extremely rapid visits of glossophagines to flowers suggest that rate of nectar extraction is a potentially useful measure of performance. Our experiments revealed several differences in feeding behavior that distinguish glossophagines from other phyllostomids (table 7.1). During feeding trials, glossophagines always hovered while feeding, visited feeders more quickly, and extracted less nectar per visit. In contrast, other phyllostomids either landed on the feeders by holding on with the thumbs

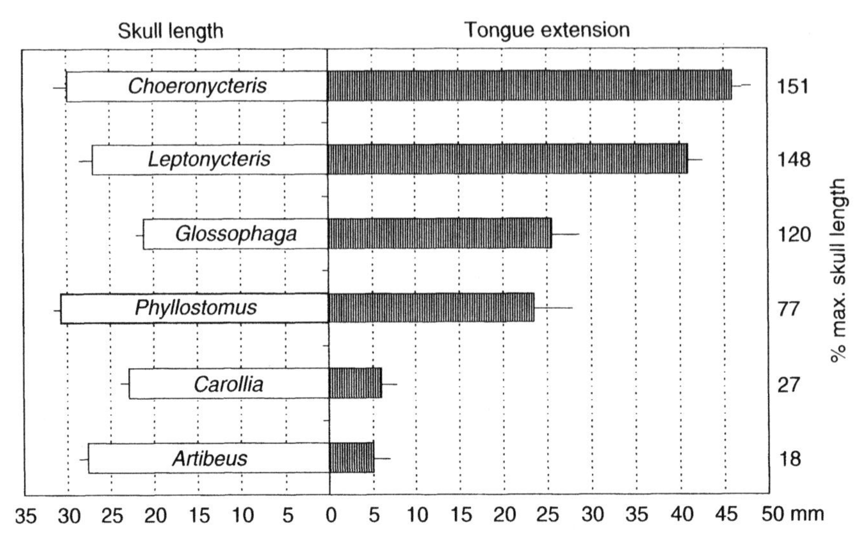

Figure 7.3 Maximum skull length (left) and maximum tongue extension (right; distance the tongue is extended past the tip of the snout). Bars represent means ± standard deviations. Percentages express maximum tongue extension relative to maximum skull length (data from Nicolay, 2001). Labels indicate genera and species shown in figure 7.2.

(*Carollia perspicillata* and *Artibeus jamaicensis*) or used an anchored-hovering strategy, grabbing the feeder with the hind feet while flapping the wings (*Phyllostomus discolor*). Non-glossophagines remained at the feeders for longer periods of time, and took greater amounts of nectar per visit. These results correspond to the flower-visiting habits of these bats in the wild (Helversen, 1993).

These experiments also demonstrated that increasing the length of the flower significantly reduces the rate of nectar extraction for all species (table 7.1; statistical analysis in Nicolay, 2001). Because of their limited tongue extension, *Carollia perspicillata* and *Artibeus jamaicensis* were unable to reach the nectar and feed from the 40-mm-long feeders. Ultimately flower length does impose limits on the sizes of flowers a bat can potentially visit. Within the range of flowers from which the bats were able to reach the nectar, nectar extraction rates were not associated with morphological specialization. Of the three glossophagines, *Leptonycteris curasoae* (the moderately derived species) always had the highest size-adjusted rate of nectar extraction, *Choeronycteris mexicana*, the highly derived nectarivore was second, and *Glossophaga soricina* had the lowest rate of nectar extraction. Thus, morphological specialization is not necessarily associated with an increased rate of nectar extraction. Nor do glossophagines necessarily extract nectar more quickly than do other phyllostomids. At the shortest feeders (20 mm), *Phyllostomus discolor* removed nectar more quickly than did *G. soricina*.

Table 7.1 Measures of Nectar-Feeding Performance in Phyllostomid Bats

	Diet and Description	Body Mass (g)	Max. Tongue Extension (mm)	Feeder Length (mm)	Time per Visit (s)	Amount of Nectar per Visit (mg)	Rate (mg/s)	Size-adjusted Rate (mg/s/g)
Choeronycteris mexicana (*n* = 8)	Highly derived nectarivore	17–20	43–46	20	0.45 ± 0.11	113 ± 23	261 ± 44	14.4 ± 1.8
				30	0.51 ± 0.10	106 ± 22	214 ± 49	11.1 ± 2.0
				40	0.58 ± 0.11	93 ± 23	162 ± 41	8.9 ± 1.5
Leptonycteris curasoae (*n* = 7)	Moderately derived nectarivore	23–29	39–41	20	0.35 ± 0.08	197 ± 42	559 ± 63	21.0 ± 3.3
				30	0.43 ± 0.05	196 ± 27	453 ± 57	17.6 ± 3.2
				40	0.51 ± 0.08	192 ± 34	379 ± 77	13.8 ± 3.6
Glossophaga soricina (*n* = 10)	Primitive nectarivore–omnivore	10–12	24–26	20	0.73 ± 0.18	108 ± 21	154 ± 36	13.2 ± 3.3
				30	0.86 ± 0.16	83 ± 26	97 ± 25	8.0 ± 1.8
				40	0.96 ± 0.11	55 ± 18	57 ± 16	5.0 ± 1.2
Phyllostomus discolor (*n* = 5)	Omnivore with seasonal nectarivory	40–46	21–24	20	1.56 ± 0.54	1103 ± 585	710 ± 233	16.8 ± 5.4
				30	1.37 ± 0.30	517 ± 128	403 ± 152	9.5 ± 3.6
				40	1.80 ± 0.88	322 ± 128	225 ± 140	5.4 ± 3.4
Carollia perspicillata (*n* = 6)	Understory frugivore	18–22	5–7	20	1.94 ± 0.71	272 ± 98	142 ± 18	7.0 ± 1.0
				30	4.17 ± 2.6	364 ± 165	103 ± 50	5.0 ± 2.3
				40	–	–	–	–
Artibeus jamaicensis (*n* = 6)	Canopy frugivore	31–41	4–6	20	21.4 ± 19.1	892 ± 489	72 ± 44	2.0 ± 1.2
				30	10.5 ± 9.4	419 ± 276	56 ± 29	1.5 ± 0.8
				40	–	–	–	–

Species are listed in order from most morphologically specialized (*Choeronycteris mexicana*) to least specialized (*Artibeus jamaicensis*) for feeding on nectar. Descriptions derived primarily from Koopman (1981) and Bonaccorso (1979). All data are from Nicolay (2001). Dash (–) indicates range, plus-minus (±) indicates standard deviation. Three different lengths of artificial flowers were used; *A. jamaicensis* and *C. perspicillata* were unable to feed from the 40-mm-long feeder. Size-adjusted rate is the rate divided by body mass for each individual.

Much of the variation in rate of nectar extraction can be attributed to differences in feeding styles. Species differed not only in duration and amount of nectar extracted per visit; they also varied in the average number of licks taken per visit and how deeply they inserted the snout into the feeder. Of all the bats, the most specialized species, *C. mexicana*, consistently inserted its snout *least* deeply into the flower, instead relying on its long tongue to reach the nectar. In experimental settings (especially at short flowers), *C. mexicana* did not use its long snout as a probe, although it did insert its snout completely into the feeders during the maximum tongue extension experiments. *C. mexicana* appears to stay away from short flowers so that nectar can easily be reached by the long, mobile tongue. Since *C. mexicana* tended to feed at a greater distance from the flower, the rate of nectar extraction was reduced because the tongue must travel a greater distance between the mouth and the nectar, reducing licking frequency.

In contrast, *L. curasoae* inserted its snout into the feeders the most deeply of the glossophagines, regardless of feeder length. This reduces the distance the tongue has to travel, increasing the rate of licking because the tongue covers a smaller distance from the mouth to the nectar, and potentially increases the percent of the tongue that is immersed in the nectar. These factors all contribute to a higher rate of nectar extraction.

The species that landed on the feeders almost always inserted their heads so deeply into the feeder that the snout contacted the nectar. These bats licked very quickly, and spent more time on the feeders overall than did glossophagines. Because they landed on the feeders, handling times (time spent landing on the feeder, reaching the nectar, and taking off) were naturally greater than in the hovering glossophagines. Landing on the flowers may incur a greater risk of predation than hovering, due to longer visits and the difficulty of pushing away from the flower to escape. However, landing on flowers may increase the success of locating nectar and extracting a larger percentage of it contained within the flower.

Other Aspects of Performance

One of the greatest challenges of performance analysis is to design behavioral experiments that accurately mimic behavior in the wild yet enable quantification of the intended behavior under constant conditions to increase accuracy, reliability, and comparability of results. While the experiments described here capture certain fundamental aspects of nectar-feeding performance, they do not encompass all that is involved with feeding on nectar (Winter and Helversen, 2001). Nectar in a flower is quite limited, whereas our feeders contained an effectively unlimited nectar supply. In the wild, flowers produce only a finite amount of nectar per night, which is continually depleted by visitors to the flower. Thus, species relying on nectar as a dietary staple are theoretically under selective pressure for the ability to remove small quantities of nectar (Winter and Helversen, 2003). This pressure probably drove the evolution of the lengthened papillae on the tongues of highly nectarivorous species (but not other bats), which function as mops for extracting small amounts of nectar (Griffiths, 1982; Koopman, 1981).

Natural selection may act strongly on nectarivorous species during periods when nectar is in short supply, such as seasonal fluctuations or supra-annual catastrophic events. These periods may favor the ability to visit a flower and extract some nectar, rather than removing nothing. In such situations, the rate or efficiency at which nectar is extracted may be relatively unimportant compared with the ability to utilize a resource not available to other bats. The ability to extract a minimal amount of nectar may also influence when a bat switches its diet to alternative food resources (insects, fruits, or possibly novel flowers).

Quantifying this ability requires measuring the smallest amount of nectar that can be extracted, while still verifying that some nectar was indeed removed. This type of performance has proven difficult to measure; however, we suggest that the ability to extract miniscule amounts of nectar is another important aspect of nectar-feeding performance, and has probably been an important factor during the evolution of glossophagines.

SUMMARY

Although measuring performance is not always an easy task, performance analysis holds the potential to reveal much about the functional, ecological, and evolutionary importance of morphology. We have found that morphological specialization is related to tongue extension (maximal capacity) but is not strongly associated with nectar extraction rates. This is not particularly surprising since tongue extension has a more obvious, direct relationship to anatomy (resting tongue length) than rate of nectar extraction. Nectar extraction is influenced not only by the morphology of the feeding apparatus, but is also strongly affected by the locomotor ability, feeding styles, and behavioral plasticity. Understanding the relationship between morphology and nectar-feeding performance requires the realization that nectarivory may involve elements of performance not fully captured by traditional or easily measured variables.

Performance analyses inevitably underscore the need for collection of ecological data relevant to understanding the greater impact of morphology on ecology and evolution. Such data include year-round data on resource availability (e.g., phenology, resource density, fruit and insect availability), specific information concerning plant biology (e.g., flower shape and presentation, nectar amount and composition, anthesis, nectar defenses, pollen amount, pollen placement), and diets of the bats (flowers visited, seasonal variation, differences between sexes, alternate resources used). Together, performance experiments and ecological investigations hold the potential to reveal much about the evolution of nectarivory, niche partitioning within flower-visiting bat assemblages, and the mutualism between bats and the plants they pollinate.

ACKNOWLEDGMENTS

We would like to thank A. Herrel and B. Jayne for critical review and insightful commentary that greatly improved this chapter. E.R. Dumont and O. von Helversen have provided valuable input and assistance to this research throughout its development. The University of Erlangen (O. von Helversen), University of Munich—LMU (G. Müller), and Northeastern Ohio Universities College of Medicine (J. Wenstrup) all generously provided access to captive animals and facilities to conduct this research. This work was supported by an NSF Dissertation Improvement Grant (IBN 0073146 to E.R. Dumont and C.W. Nicolay), the German Academic Exchange Service (DAAD), and the American Museum of Natural History (Theodore Roosevelt Fund).

LITERATURE CITED

Aerts, P., R. van Damme, B. Vanhooydonck, A. Zaaf, and A. Herrel. 2000. Lizard locomotion: how morphology meets ecology. Netherlands Journal of Zoology, 50: 261–277.

Arkins, A.M., A.P. Winnington, S. Anderson, and M.N. Clout. 1999. Diet and nectarivorous foraging behaviour of the short-tailed bat (*Mystacina tuberculata*). Journal of Zoology (London), 247: 183–187.

Arnold, S.J. 1983. Morphology, performance, and fitness. American Zoologist, 23: 347–361.

Baker, R.J., C.A. Porter, J.C. Patton, and R.A. van den Bussche. 2000. Systematics of bats of the family Phyllostomidae based on *RAG2* DNA sequences. Occasional Papers of the Museum of Texas Tech University, 202: 1–16.

Bauwens, D., T. Garland, Jr., A.M. Castilla, and R. van Damme. 1995. Evolution of sprint speed in lacertid lizards: morphological, physiological, and behavioral covariation. Evolution, 49: 848–863.

Bonaccorso, F.J. 1979. Foraging and reproductive ecology in a Panamanian bat community. Bulletin of the Florida State Museum, Biological Science, 24: 359–408.

Carter, D.R., M. Wong, and T.E. Orr. 1991. Musculoskeletal ontogeny, phylogeny, and functional adaptation. Journal of Biomechanics, 24: 3–16.

Dobat, K. 1985. Blüten und Fledermäuse: Bestäubung durch Fledermäuse und Flughunde (Chiropterophilie). Waldemar Kramer, Frankfurt.

Dumont, E.R. 1999. The effect of food hardness on feeding behaviour in frugivorous bats (Phyllostomidae): an experimental study. Journal of Zoology (London) 248: 219–229.

Fleming, T.H., C.T. Sahley, J.N. Holland, J.D. Nason, and J.L. Hamrick. 2001. Sonoran Desert columnar cacti and the evolution of generalized pollination systems. Ecological Monographs, 71: 511–530.

Freeman, P.W. 1995. Nectarivorous feeding mechanisms in bats. Biological Journal of the Linnean Society, 56: 439–463.

Freeman, P.W. 1998. Form, function, and evolution in skulls and teeth of bats. Pp. 140–156. In: Bat Biology and Conservation (T.H. Kunz and P.A. Racey, eds.). Smithsonian Institution Press, Washington, DC.

Frost, H.M., J.L. Ferretti, and W.S.S. Jee. 1998. Perspectives: some roles of mechanical usage, muscle strength, and the mechanostat in skeletal physiology, disease and research. Calcified Tissue International, 62: 1–7.

Gardner, A.L. 1977. Feeding habits. Pp. 293–350. In: Biology of Bats of the New World Family Phyllostomatidae, Part II (R.J. Baker, J.K. Jones, Jr., and D.C. Carter, eds.). Special Publications of the Museum of Texas Tech University, 13: 1–364.

Garland, T. Jr., T.T. Gleeson, B.A. Aronovitz, C.S. Richardson, and M.R. Dohm. 1995. Maximal sprint speeds and muscle fiber composition of wild and laboratory house mice. Physiology and Behavior, 58: 869–876.

Griffiths, T.A. 1982. Systematics of the New World nectar-feeding bats (Mammalia, Phyllostomidae) based on the morphology of the hyoid and lingual regions. American Museum Novitates, 2742: 1–45.

Heinrich, B. 1975. Energetics of pollination. Annual Review of Ecology and Systematics, 6: 139–170.

Heithaus, E.R. 1982. Coevolution between bats and plants. Pp. 327–367. In: Ecology of Bats (T.H. Kunz, ed.). Plenum Press, New York.

Heithaus, E.R., P.A. Opler, and H.G. Baker. 1974. Bat activity and pollination of *Bauhinia pauletia:* plant–pollinator coevolution. Ecology, 55: 412–419.

Heithaus, E.R., T.H. Fleming, and P. Opler. 1975. Foraging patterns of resource utilization in seven species of bats in a seasonal tropical forest. Ecology, 56: 841–853.

Helversen, O. von. 1993. Adaptations of flowers to the pollination by glossophagine bats. Pp. 41–59. In: Plant–Animal Interactions in Tropical Environments (W. Barthlott, ed.). Museum Alexander Koenig, Bonn.

Herrel, A., P. Aerts, J. Fret, and F. de Vree. 1999. Morphology of the feeding system in agamid lizards: ecological correlates. Anatomical Record, 254: 496–507.

Herrera, L.G., T.H. Fleming, and J.S. Findley. 1993. Geographic variation in carbon composition of the pallid bat, *Antrozous pallidus,* and its dietary implications. Journal of Mammalogy, 74: 601–606.

Irschick, D.J., and J.B. Losos. 1998. Do lizards avoid habitats where performance is submaximal? The relationship between sprinting capabilities and structural habitat use in Caribbean Anoles. The American Naturalist, 154: 293–305.

Koopman, K.F. 1981. The distributional patterns of New World nectar-feeding bats. Annals of the Missouri Botanical Garden, 68: 352–369.

Nicolay, C.W. 2001. Ecological Morphology and Nectar-Feeding Performance in Flower-Visiting Bats. Ph.D. Dissertation, Kent State University, Kent, OH.

Norberg, U.M. 1994. Wing design, flight performance, and habitat use in bats. Pp. 205–239. In: Ecological Morphology (P.C. Wainwright and S.M. Reilly, eds.). University of Chicago Press, Chicago.

Pedersen, S.C. 1998. Morphometric analysis of the chiropteran skull with regard to mode of echolocation. Journal of Mammalogy, 79: 91–103.

Phillips, C.J. 1971. The dentition of glossophagine bats: development, morphological characteristics, variation, pathology and evolution. Miscellaneous Publications, Museum of Natural History, University of Kansas, 54: 1–138.

Reilly, S.M., and P.C. Wainwright. 1994. Conclusion: ecological morphology and the power of integration. Pp. 339–354. In: Ecological Morphology (P.C. Wainwright and S.M. Reilly, eds.). University of Chicago Press, Chicago.

Saunders, M.B., and R.M.R. Barclay. 1992. Ecomorphology of insectivorous bats: a test of predictions using two morphologically similar species. Ecology, 73: 1335–1345.

Sazima, I., S. Vogel, and M. Sazima. 1989. Bat pollination of *Encholirium glaziovii,* a terrestrial bromeliad. Plant Systematics and Evolution, 168: 167–179.

Sazima, M., I. Sazima, and S. Buzato. 1994. Nectar by day and night: *Siphocampylus sulfurous* (Lobeliaceae) pollinated by hummingbirds and bats. Plant Systematics and Evolution, 191: 237–246.
Sazima, M., S. Buzato, and I. Sazima. 1999. Bat-pollinated flower assemblages and bat visitors in two Atlantic forest sites in Brazil. Annals of Botany, 83: 705–712.
Schluter, D. 1996. Ecological causes of adaptive radiation. The American Naturalist, 148: S40–64.
Shaffer, L.R., and D.R. Formanowicz, Jr. 2000. Sprint speeds of juvenile scorpions: among-family differences and parent–offspring correlations. Journal of Insect Behavior, 13: 45–54.
Solmsen, E.-H. 1999. New World nectar-feeding bats: biology, morphology and craniometric approach to systematics. Bonner Zoologische Monographien, 44: 1–118.
Vanhooydonck, B., R. van Damme, and P. Aerts. 2000. Ecomorphological correlates of habitat partitioning in Corsican lacertid lizards. Functional Morphology, 14: 358–368.
Vogel, S. 1968. Chiropterophilie in der neotropischen Flora. Neue Mitteilungen I. Flora, 157: 562–602.
Vogel, S. 1969a. Chiropterophilie in der neotropischen Flora. Neue Mitteilungen II. Flora, 158: 185–222.
Vogel, S. 1969b. Chiropterophilie in der neotropischen Flora. Neue Mitteilungen III. Flora, 158: 289–323.
Wainwright, P.C. 1991. Ecomorphology: experimental functional anatomy for ecological problems. American Zoologist, 31: 680–693.
Wainright, P.C. 1994. Functional morphology as a tool in ecological research. Pp. 42–59. In: Ecological Morphology (P.C. Wainwright and S.M. Reilly, eds.). University of Chicago Press, Chicago.
Wainwright, P.C. 1996. Ecological explanation through functional morphology: the feeding biology of sunfishes. Ecology, 77: 1336–1343.
Wetterer, A.L., M.V. Rockman, and N.B. Simmons. 2000. Phylogeny of phyllostomid bats (Mammalia: Chiroptera): data from diverse morphological systems, sex chromosomes, and restriction sites. Bulletin of the American Museum of Natural History, 248: 1–200.
Winkelmann, J.R. 1971. Adaptations for Nectar-Feeding in Glossophagine Bats. Ph.D. Dissertation, University of Michigan, Ann Arbor, MI.
Winter, Y., and O. von Helversen. 2001. Bats as pollinators: foraging energetics and floral adaptations. Pp 148–170, In: Cognitive Ecology of Pollination (L. Chittka and J.D. Thomson, eds.). Cambridge University Press, Cambridge.
Winter, Y., and O. von Helversen. 2003. Operational tongue length in phyllostomid nectar-feeding bats. Journal of Mammalogy, 84: 886–896.

8

Quadrupedal Bats: Form, Function, and Evolution

William A. Schutt, Jr., & Nancy B. Simmons

Although bats (Chiroptera) are unique among mammals because of their ability to fly, they also exhibit quadrupedal locomotion during a variety of nonflight behaviors. This chapter concerns the evolution, morphology, and behavior associated with quadrupedal locomotion in bats. Although many bat families contain members that exhibit some degree of quadrupedal locomotion, we concentrate on three groups of bats that exhibit extensive quadrupedal behavior: vampire bats (Phyllostomidae: Desmodontinae); naked bulldog bats, *Cheiromeles* (Molossidae); and the New Zealand short-tailed bat, *Mystacina tuberculata* (Mystacinidae). Numerous morphological traits that apparently facilitate quadrupedal locomotion have evolved independently in these three lineages.

INTRODUCTION

When bats are not flying or resting in their roosts, they often exhibit some type of quadrupedal behavior while moving around on various substrates. The awkward appearance of quadrupedal locomotion in most bats is primarily a reflection of the fact that chiropteran limbs are markedly different from those of a typical quadrupedal mammal. Obviously, the forelimbs of bats have evolved into wings. Less obvious, but just as important, are the changes in the morphology and orientation that have evolved in their hindlimbs.

Vaughan (1970: 135) divided bats into three groups based on "variation in the posture and proportions of the hind limbs." Such variation was thought to be "related principally to the great differences that occur in roosting habits and in modes of terrestrial locomotion." One group of bats, typified by phyllostomids such as *Glossophaga* and *Macrotis*, have hindlimbs that are rotated approximately 180° from the condition in a typical quadrupedal mammal. In these forms, which we refer to as "Type 1" bats, the thin femora project caudally and slightly dorsad, and the head of the femur is in line with the long axis of the bone. The distal section of the leg (the shank) is directed caudally and ventrad. The feet are directed caudally (figure 8.1A). This limb

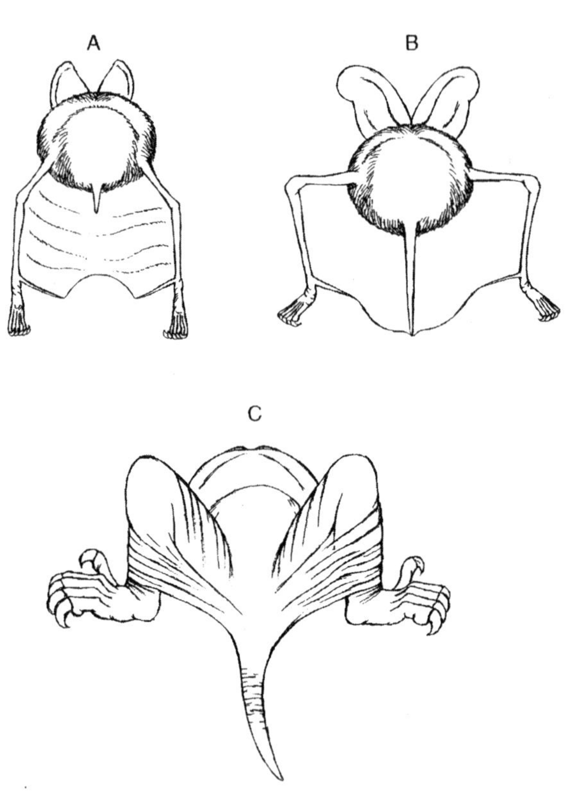

Figure 8.1 Examples of variation in the orientation of bat hindlimbs. (A) "Type 1" bats are nonquadrupedal; (B) "Type 2" bats are capable of crawling; (C) "Type 3" bats are capable of agile and/or specialized quadrupedal locomotion.

orientation gives these bats a spider-like stance (Vaughan, 1959, 1970). The tibia is also lightly built and the fibula is incomplete (in that it spans only the distal portion of the tibia). Because of the lightweight structure of the hindlimb bones and the unique arrangement of the hindlimbs in Type 1 bats, they are unable to crawl forward and "never alight on horizontal surfaces" (Vaughan, 1970: 136).

In many bats, hindlimb bones appear delicate because they have greater length-to-diameter ratios or have undergone reduction. These changes may have evolved in bats as a way to reduce overall body mass as an adaptation for flight (Vaughan, 1959). In many noncursorial mammals, for example, the fibula acts as the origin for muscles that manipulate the digits or rotate the distal segment of the hindlimb (pes). In cursorial mammals, the mass of the limbs is often reduced through bone fusion or reduction. In some cursors (e.g., jack rabbits, *Lepus*), the fibula is fused to the tibia, whereas in others (e.g., the pronghorn antelope, *Antilocapra*) the fibula is reduced to a splint-like nub at the ankle (Hildebrand, 1995). Similarly, the incomplete fibula in some bats may represent a trade-off between foot twisting and limb rotation and requirements for a lightweight musculoskeletal system. Further study is needed to determine whether the absence of a complete fibula plays a role in limiting quadrupedal locomotor ability (e.g., by limiting the rotation of the pes).

Another important consequence of delicate hindlimb bones (especially the femora) is that the typical bat hindlimb skeleton does not appear well suited to undergo compressive loads associated with quadrupedal locomotion (Howell and Pylka, 1977). Furthermore, these lightweight bones may be unable to resist the bending forces to which they would normally be subjected during quadrupedal locomotion. As in the role of a complete versus incomplete fibula, we suggest that biomechanical analysis of structures such as the femur and tibia (as well as the hip, knee, and ankle joints) will shed light on the comparative quadrupedal locomotion in bats.

In most bats (including some phyllostomids and most or all members of the families Emballonuridae, Nycteridae, Natalidae, and Vespertilionidae), the hindlimbs are rotated laterally approximately 90°, with the shank directed ventrad (figure 8.1B). This arrangement of limb elements, which we term "Type 2," gives these bats a somewhat reptilian stance. Although the hindlimb bones are also generally reduced in diameter relative to length (and the fibula is often incomplete), these bats can crawl quite well (Vaughan, 1970) and some of them exhibit extensive quadrupedal locomotion during roosting and feeding behavior. For example, Greenhall (1968) maintained specimens of *Vampyrum spectrum* (Phyllostomidae) in captivity and reported them to be careful, slow stalkers.

In the final group of bats ("Type 3"; e.g., vampire bats, molossids such as *Cheiromeles*, and *Mystacina*), the femur is directed anterolaterally, with the shank directed almost vertically (figure 8.1C). The feet are not generally in line with the shank (as in Types 1 and 2) but, instead, are directed laterally. Although there is considerable variation in the morphology of the hindlimb bones, they are generally robust. With few exceptions (e.g., *Diphylla ecaudata*, Phyllostomidae, Desmodontinae), the fibula is complete. Although bats in this group are capable of extremely agile quadrupedal locomotion (some of it quite spectacular) only a few researchers have studied this phenomenon (e.g., Vaughan, 1959; Altenbach, 1979; Schutt, 1998; Schutt et al., 1997) and there remains much to be learned about the subject. The primary goals of this chapter are to review what is currently known about quadrupedal locomotion in bats and to stimulate future study on this subject.

VAMPIRE BATS

The order Chiroptera contains over 1100 extant species (Simmons, 2005) but only three of these are vampire bats. These bats are widely distributed within the tropical and subtropical regions of the New World from Northern Mexico south to northern Argentina, Uruguay, and Chile (Koopman, 1988). Their unique feeding habits have been reported with varying degrees of accuracy for over four hundred years but the group remained relatively unstudied by biologists until an outbreak of rabies in the 1930s in Trinidad (Ditmars and Greenhall, 1935; for a review see Greenhall and Schmidt, 1988). Since then, behavioral studies on vampire bats have revealed an elaborate social structure (reviewed by Wilkinson, 1988) as well as the demonstration of altruistic behavior such as reciprocal blood sharing between related and unrelated individuals (Wilkinson, 1984). Researchers have also learned that, as a consequence of their unique feeding habits, vampire bats are the most agile quadrupedal bats in the world (Beebe, 1927; Ditmars and Greenhall, 1935; Dalquist, 1955; Wimsatt, 1959; Altenbach, 1979; Schutt et al., 1997). "Alighting near its intended prey, *Desmodus rotundus* uses a quadrupedal gait that varies between walking, spider-like scrambling, jumps into short flights and hopping" (Schutt, 1998: 159). Not surprisingly, vampire bats have a number of unique morphological adaptations that are closely involved with their blood-feeding lifestyle (Altenbach, 1979; for reviews, see Bhatnagar, 1988; Schutt, 1998).

Locomotor Performance and Morphology

While many bat species have limited quadrupedal locomotor capabilities, vampire bats exhibit a degree of complexity and agility during quadrupedal locomotion that is unparalleled in the order Chiroptera (Ditmars and Greenhall, 1935; Altenbach, 1979, 1988; Schutt et al., 1997). Alighting some distance from their intended prey, the common vampire bat, *Desmodus rotundus*, makes its approach using a quadrupedal gait that varies between walking, spider-like scrambling, jumps into short flights, and hopping. *Desmodus* feeds primarily from the ground but often feeds while situated atop, or hanging from its prey.

Altenbach (1979) used high-speed still photography and cinetography and electromyography to examine the mechanisms of walking, hopping, jumping, and climbing in *Desmodus*. These maneuvers were determined to be extremely variable, not only in terms of direction and velocity, but also with regard to the relative contributions of the forelimbs and hindlimbs. For example, Altenbach determined that most of the force behind the jump is generated by contraction of the powerful pectoralis muscles which adduct the wings. Extension of the elbows and ventral flexion of the elongated thumbs also supply components of upward thrust. During the jump, the hindlimbs do not act in force production. Rather, they aid in balance and function primarily in jump preparation. Just before the jump is initiated, extension of the knees tips the animal forward—shifting the bat's mass anteriorly—above the thrust-producing wings and into the line of the jump (Altenbach, 1979, 1988; Schutt et al., 1997). Jumping in *Desmodus* serves several roles. Since fresh blood contains no fat and almost no carbohydrates, *Desmodus* must consume between 50% and 100% of its body mass in blood each night to survive (Wimsatt and Guerriere, 1962; Wimsatt, 1969; McNab, 1973). Jumping functions to get the heavily loaded vampire bat airborne from a horizontal surface, a critical first step in initiating flight back to the roost following a blood meal. Jumping may also be useful to vampire bats in avoiding terrestrial predators and probably prevents them from being trampled by large prey such as cattle.

Variation exists in quadrupedal locomotor performance between *Desmodus rotundus* and the white-winged vampire bat, *Diaemus youngi*. This variation may be related to terrestrial versus arboreal feeding modes in these taxa (Schutt et al., 1997). The diet of *Diaemus* consists primarily of avian blood (Gardner, 1977; Goodwin and Greenhall, 1961; Sazima and Uieda, 1980; Uieda et al., 1992), which is obtained from perching birds, stalked as they sleep, although these bats also have been reported to feed on mammalian blood (Goodwin and Greenhall, 1961; Uieda, 1986; Greenhall, 1988; Schutt et al., 1999). In captivity, *Diaemus* is a stealthy and agile arboreal hunter that exhibits terrestrial feeding behavior similar to that of *Desmodus*, although *Diaemus* has never been observed to run or perform the flight-initiating jumps observed in *Desmodus*. The third vampire species, *Diphylla ecaudata*, is an arboreal hunter that feeds entirely on avian blood (Uieda, 1986; Uieda et al., 1992), although little is known about its quadrupedal behavior.

Differences in the feeding behavior of vampire bats are also reflected in their limb morphology (Schutt, 1998). Koopman (1988) reported that the thumbs of *Diaemus* and *Diphylla* are only one eighth the length of the third digit and possess a single metacarpal pad. The thrust-generating thumb of *Desmodus* is longer (one fifth the length of the third digit) and has two metacarpal pads. Schutt (1998) demonstrated that hindlimb skeletal elements of *Desmodus* (figure 8.2A) and *Diaemus* (figure 8.2B) are more robustly built than the same elements in *Diphylla* (figure 8.2C). It has been hypothesized that the sturdier bones of *Desmodus* may allow the hindlimb of this species to safely withstand the compressive loading that typifies terrestrial locomotion (Schutt, 1998), although this has not been proven experimentally and thus requires further study. There are additional morphological differences between vampire bat species. For example, in *Desmodus* the tibia is roughly T-shaped in cross-section and the fibula is complete (Vaughan, 1970; Schutt, 1998) (figure 8.2A). In *Diaemus* and *Diphylla*, the tibia and fibula are more rounded in cross-section and in *Diphylla* the fibula is incomplete (figure 8.2B,C). Future studies should be designed to examine the variation in vampire bat hindlimb morphology from a functional perspective.

Interestingly, *Diphylla* has a unique digitiform calcar (absent in *Desmodus* and *Diaemus*) that is employed like a sixth digit to facilitate grasping (Schutt and Altenbach, 1997). Similar to roosting behavior exhibited by molossids such as *Eumops perotis* (Vaughan, 1959), *Diphylla* uses its hindlimbs to grope about in search of a solid foothold before pulling itself backward (W.A. Schutt, pers. obs.). *Diphylla* has never been reported to exhibit terrestrial locomotion during feeding behavior. Rather, it is purely an arboreal hunter, feeding exclusively upon avian blood (see Greenhall, 1988 for a review). As in *Desmodus* and *Diaemus*, the limb morphology of *Diphylla* reflects further specialization within a unique feeding mode (Schutt, 1998). An interesting biomechanical question is whether the unique calcar of *Diphylla* enables it to

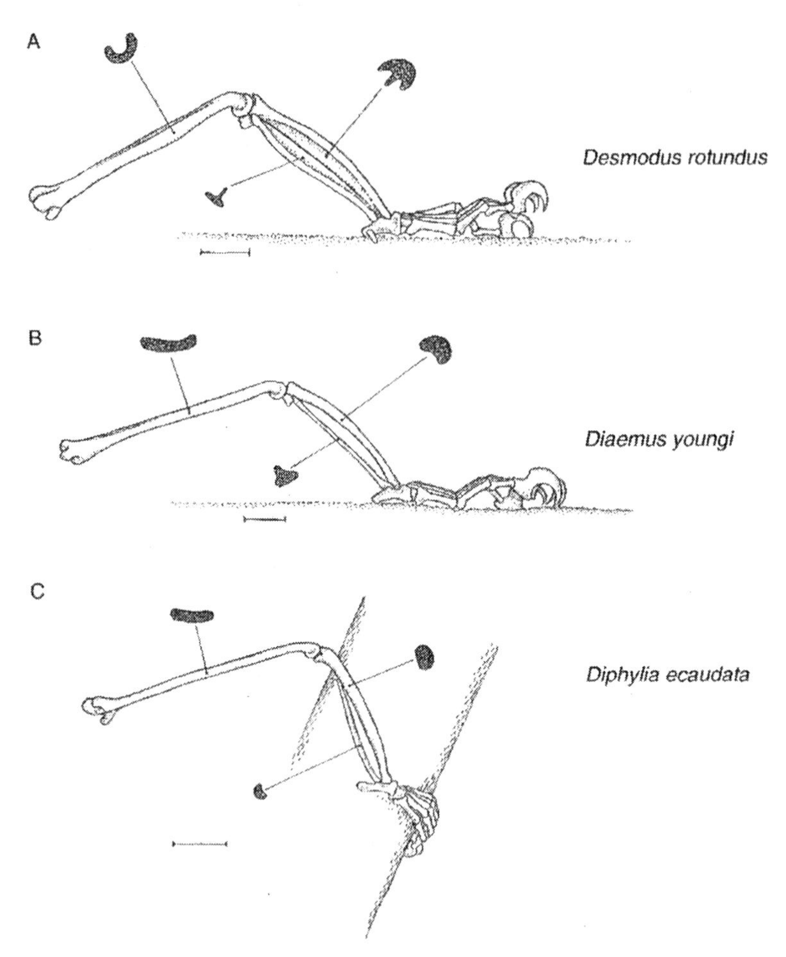

Figure 8.2 Hindlimb bones of (A) *Desmodus rotundus*, (B) *Diaemus youngi*, and (C) *Diphylla ecaudata*. Dark shaded inserts represent cross-sectional views of the bone at the point indicated.

compensate functionally for its reduced fibula and relatively delicate hindlimb bones.

Evolution of Quadrupedal Locomotion

Vampire bats (Desmodontinae) belong to a large Neotropical clade of leaf-nosed bats (Phyllostomidae) that are generally characterized by poor or moderately developed quadrupedal locomotion (i.e., Type 1 or Type 2). Closely related families (e.g., Mormoopidae, Noctilionidae) exhibit similar limb architecture. Accordingly, it seems likely that the specializations seen in vampires evolved from one of these ground plans, probably a Type 2 arrangement. Altenbach (1979) suggested that vampire bats initially fed on small terrestrial mammals or other vertebrates and that terrestrial locomotion developed in response to this mode of feeding. Schutt (1998) reviewed several hypotheses on vampire bat origins before proposing his own—the arboreal omnivore hypothesis. According to this, Miocene protovampires exploited arboreal food sources such as insects, small reptiles, birds, and mammals, much as do some extant phyllostomids (e.g., *Vampyrum spectrum*). During the mid- to late Miocene, protovampires would have encountered larger and increasingly diverse arboreal vertebrates. Under these conditions, protovampires would have undergone dietary and behavioral changes which allowed them to exploit larger animals as a food source—by feeding only on their blood. Derived conditions for terrestrial blood-feeding (e.g., robust hindlimb bones) may have evolved as the *Desmodus/Diaemus* lineage came down from the trees to exploit terrestrial vertebrates.

CHEIROMELES

Two species of *Cheiromeles* (Molossidae) are recognized by most authors—*C. torquatus* and *C. parvidans* (Lekagul and McNeely, 1977; Honacki et al., 1982; Corbet and Hill, 1991, 1992; Ingle and Heaney, 1992; Flannery, 1995)—although these taxa are sometimes treated as subspecies of *C. torquatus* (e.g., Koopman, 1989, 1993, 1994). *Cheiromeles torquatus* is found in peninsular Malaysia, southern Thailand, Sumatra, Java, Borneo, and Palawan Island in the Philippines (Sanborn, 1952; Corbett and Hill, 1992; Ingle and Heaney, 1992). *Cheiromeles parvidens* is found in Sulawesi and throughout the Philippines except Palawan Island (Corbet and Hill, 1992; Ingle and Heaney, 1992).

Locomotor Performance and Morphology

Although *Cheiromeles* appears to be devoid of body hair, these bats actually have an abundance of short, fine hairs, particularly on the underside of the head and body (figure 8.3A). A tuft of stiff bristles is found on the lateral surface of the first hindlimb digit. Yet another unusual morphological characteristic of *Cheiromeles* is a pair of subaxillary "pouches" (25–50 mm deep), which open posteriorly. These cavities are formed by the plagiopatagial attachment to the sides of the body, upper arm, and thigh. *Cheiromeles* uses its hindfeet (which have highly modified first digits) to insert the distal ends of the folded wings into these cavities (figure 8.3B,C) (Medway, 1969; Freeman, 1981; Nowak, 1994). With its wings tucked away, *Cheiromeles* is said to employ all four limbs during climbing and arboreal locomotion (Allen, 1886; Burton, 1955). Schutt and Simmons (2001) described a number of morphological features found in both species of *Cheiromeles* that are unique within Chiroptera. Their morphology and possible functions are briefly reviewed below.

The hallux (digit I) in *Cheiromeles* is positioned at a right angle to digits II–V (figures 8.3, 8.4). The muscles in the region (e.g., m. lumbricales) appear characteristic except for m. abductor hallucis brevis (figure 8.4). This muscle is quite large and its origin is a well-developed ridge on the medial surface of metatarsal I. Metatarsal I is a sturdy bone, with a shaft diameter approximately

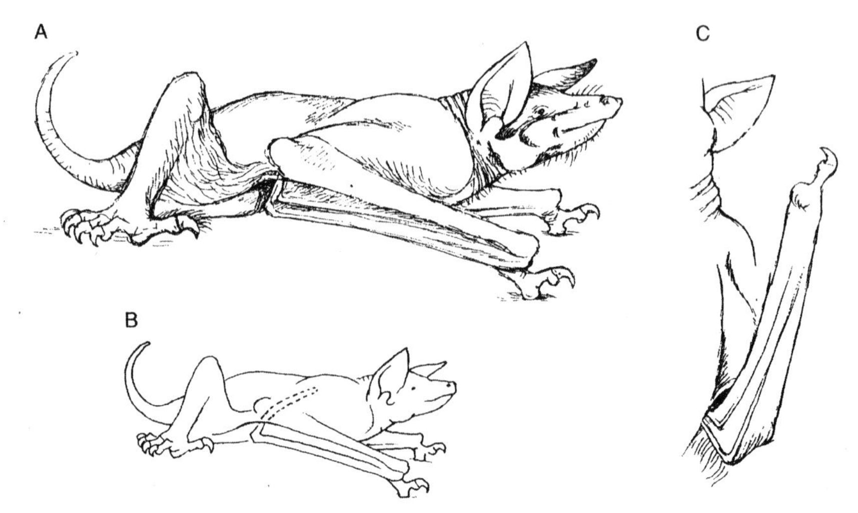

Figure 8.3 (A) Quadrupedal stance of *Cheiromeles torquatus*. (B) Lateral view showing the position of the phalanges III and IV inserted into the subaxillary "pouch." (C) Ventral view.

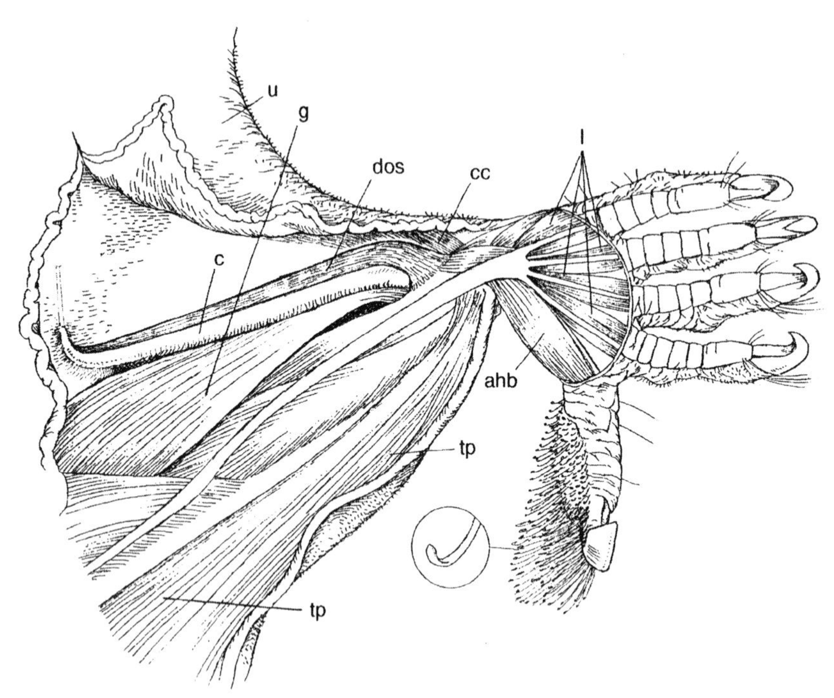

Figure 8.4 Plantar view of the dissected hindlimb of *Cheiromeles parvidens* (AMNH 241941). Note the position of the calcar, the opposable hallux (digit I), and the spatulate bristles on digit I. The fascia binding the calcar to m. gastrocnemius has been removed. ahb, m. abductor hallucis brevis; c, calcar; cc, m. calcaneo-cutaneus; dos, m. depressor ossis styliformis; g, m. gastrocnemius; l, mm. lumbricales; tp, m. tensor plagiopatagii; u, uropatagium.

twice that of any of the other metatarsals. The distal articulating surface (head) of metatarsal I is broad with a pronounced lip that extends around the entire end of the bone. The claw on digit I is flattened on its lateral surface but this does not appear to be a consequence of wear (figure 8.4). A patch of tissue on the medial aspect of digit I (which extends from the base of the proximal phalanx onto the claw) bears stiff bristles whose ends are spatulate (figure 8.4). The bristles and the unique digit that bears them are apparently used for grooming. Additionally, this bat employs its hindfeet, prior to undertaking quadrupedal locomotion, to help insert the folded wings into the subaxillary "pouches" (Medway, 1969; Freeman, 1981; Nowak, 1994).

The calcar of *Cheiromeles* does not project into the uropatagium (interfemoral membrane) as this element does in other bats (figure 8.4). Instead, the calcar is directed anteriorly (approximately parallel to the shank). Most of the calcar shaft lies in a slight depression along the lateral surface of the gastrocnemius muscle. The calcar is bound to m. gastrocnemius by fine fibers of connective tissue and both structures are overlain by a thick sheath of fascia. The result is that the calcar of *Cheiromeles* appears to be incapable of abduction toward the rudimentary uropatagium. The morphology of this structure suggests that it may serve to strengthen and support the shank, presumably during quadrupedal locomotion. Alternately, m. depressor ossis styliformis, which is typically employed to pull the calcar laterally, may have been co-opted for a new function. Given the mass of the femur in *Cheiromeles*, m. depressor ossis styliformis may serve a role in mutual stabilization of both the limb and patagia during the swing phase of quadrupedal locomotion.

While structures such as the subaxillary pouches, uniquely positioned hallux, and nonmobile calcar appear to be related to quadrupedal locomotion (Schutt and Simmons, 2001), future studies on *Cheiromeles* should be designed to better understand their function. Another characteristic of this bat that requires a more definitive explanation is its apparently hairless body. Freeman (1999) suggested that the sparse body hair of *Cheiromeles* is related to its feeding behavior, which she believes involves bat spending a considerable amount of time rooting through plant litter in search of large insects such as beetles. There are, however, no published accounts of *Cheiromeles* foraging on the ground. Medway (1978) reported that the diet of *Cheiromeles torquatus* consists of termites and other insects captured on the wing, most commonly over clearings or rice paddies. Recently, feeding buzzes have been recorded (T. Kingston, pers. comm.), lending support to Medway's claim that *Cheiromeles* is an aerial insectivore. In a recent video by Brown and Berry (pers. comm.), a specimen of C. *torquatus* captured in peninsular Malaysia exhibited the ability to walk, run, and climb vertically (which it does in a backward direction), but it did not attempt to initiate flight from the ground (nor has this behavior been reported). *Cheiromeles* are large, sturdy bats. Reported body mass for *C. torquatus* is 150–196 g (Medway, 1969; Payne et al., 1985) and 75–100 g for C. *parvidens* (Flannery, 1995). Like other molossids, these bats appear unable to initiate flight from a level surface, but instead must launch themselves from a position above the ground. *Cheiromeles* roosts primarily in caves, rock crevices, and large hollow trees (Nowack, 1994). We hypothesize that quadrupedal locomotion in *Cheiromeles* functions primarily during roosting behavior and is not related to terrestrial or arboreal feeding. Likewise, we propose that the hairless appearance of *Cheiromeles* is related to roosting habits, not feeding.

Evolution of Quadrupedal Locomotion

Cheiromeles are the largest members of the molossid family, a group characterized by moderately well developed quadrupedal locomotion (Type 3). The unique specializations exhibited by *Cheiromeles* seem to be associated with the large size of these bats, coupled with their roosting habits. While smaller molossids are well adapted for crevice-dwelling (Vaughan, 1959, 1970), larger species would obviously face greater difficulty moving within tight spaces. In *Cheiromeles*, the subaxillary pouches likely function by helping to get the distal portion of the wings out the way while increasing the stability of the forelimb as a weight-bearing appendage (figure 8.3). Similarly, the unique morphology of the calcar may increase the efficiency of the hindlimb during quadrupedal locomotion. However, quadrupedal locomotor efficiency has not been proven experimentally in these extremely large insectivorous species.

MYSTACINA

The New Zealand short-tailed bat, *Mystacina tuberculata* (Mystacinidae) is one of New Zealand's two extant indigenous mammals. A second species, *M. robusta*, was once widespread throughout New Zealand but in recent times it was found only on Solomon and Big South Cape islands. The arrival of black rats in the early 1960s quickly led to the extinction of *M. robusta*, which was last captured in 1965 (Flannery and Schouten, 2001). The phylogenetic affinities of the family Mystacinidae have been somewhat controversial, but most workers now agree that Mystacinidae belongs in the superfamily Noctilionoidea (Pierson et al., 1986; Kirsch et al., 1998; Kennedy et al., 1999; Van den Bussche and Hoofer, 2000; Simmons and Conway, 2001).

Mystacina tuberculata is a small (10–19.7 g, pre-feeding body mass), omnivorous bat (Lloyd, 2001). It feeds on forest fruits (Daniel, 1976, 1979), pollen, nectar (Arkins et al., 1999), as well as aerial and terrestrial arthropods (Dwyer, 1962; Daniel, 1976). One of the characteristics that makes this bat so remarkable is its rodent-like agility during terrestrial locomotion (Daniel, 1979).

Although a number of bats (e.g., *Antrozous*, *Nycteris*, *Megaderma*, *Cardioderma*, *Macroderma*) obtain insects by gleaning from various substrates, most bats capture insects in flight. *Mystacina* is unique among bats because, in addition to hunting aerial prey, it also forages for both plant material and insects while on the ground and on tree trunks and branches. It is reported to be agile and fast-moving while hunting on the ground or climbing (Dwyer, 1962; Daniel, 1979; Parsons, 1998; Lloyd, 2001). Jones et al. (2003) reported that while *Mystacina* uses echolocation to locate prey in the air, it listens for prey-generated noise and uses olfaction when hunting on the ground. Ecroyd (1993, 1994) demonstrated that while obtaining pollen and nectar from the endangered, ground-dwelling plant *Dactylanthus taylorii* (Balanophoaceae), *Mystacina* pollinated the plant. *Dactylanthus* (wood rose), is New Zealand's only flowering parasitic plant, and the only bat-pollinated plant in the world to produce flowers at ground level (Ecroyd, 1993). Unfortunately, *Dactylanthus*, like *Mystacina*, is under severe threat from introduced mammals and habitat destruction. Presently, the continued survival of these unique and rare species is in considerable doubt. Preservation efforts should concentrate on maintaining habitats that are free from introduced mammals such as rats and opossums (Ecroyd, 1993).

Locomotor Performance and Morphology

Mystacina is reported to roost in hollow trees and in rock crevices, but there are also claims that these bats roost in burrows that they excavate with their teeth (Daniel, 1979). Even though the occurrence of this seemingly unique behavior has recently been called into question (S. Parsons and G. Jones, pers. comm.), *Mystacina* exhibits remarkable quadrupedal locomotion that has yet to be fully studied.

Mystacina exhibits a number of morphological specializations apparently related to terrestrial locomotion. The claws on the thumb and hindlimb digits are unique in that they have basal talons (or denticles) (Daniel, 1979). The patagia are tough and leathery and the wings can be folded tightly against the body during quadrupedal locomotion (Daniel, 1979). Examination of preserved specimens revealed a stiffened and elongated strip of cartilage-like material embedded within the plagiopatagium and running roughly parallel to the radius and ulna. We speculate that this structure is involved in securing the folded phalanges of digits III–V during quadrupedal locomotion. Further study should be undertaken to investigate the development, composition, and function of this structure. The hindlimb bones of *Mystacina* are robust and the fibula is complete. As in agile quadrupedal bats such as *Desmodus* and *Cheiromeles*, the head of the femur is offset from the femoral shaft and there is a pronounced femoral neck. Detailed analysis of the functional morphology and biomechanics of quadrupedal locomotion in *Mystacina* await future study.

Evolution of Quadrupedal Locomotion

The relationships of Mystacinidae within the superfamily Noctilionoidea suggest that *Mystacina* evolved from ancestors characterized by a Type 1 or Type 2 stance. The absence of mammalian predators is thought to have had an important influence on the evolution of quadrupedal locomotion in *Mystacina* (Daniel, 1979). Additionally, lack of competition from rodents and insectivores appears to have enabled *Mystacina* to exploit the terrestrial niches normally filled by these small mammals. As in other bats that exhibit extremely agile quadrupedal locomotion (e.g., *Desmodus*), *Mystacina* has retained its ability to fly, but it is relatively slow, cannot sustain prolonged flight, and it generally flies within a few meters of the ground (Daniel, 1979).

CONCLUSIONS

Many bats exhibit some degree of quadrupediality, but only vampire bats, molossids, and *Mystacina* are though to have evolved specializations for quadrupedal locomotion. Although several generalizations can be made about the anatomy of quadrupedal bats (e.g., they have sturdy hindlimb bones), quadrupedal locomotion in bats is relatively unstudied (especially in nonvampire bats). In this chapter we have reviewed a number of interesting morphological adaptations that appear to be related to quadrupedal locomotion, and many of these require further investigation. For example, it would be informative to compare the functional morphology of locomotion in *Mystacina* with that of *Desmodus rotundus*. Although both these bats are able to balance the requirements for complex quadrupedal locomotion with those for flight, only *Desmodus* has been sufficiently studied. Future studies should employ interdisciplinary approaches and new technology (such as computer modeling) to investigate quadrupedal locomotion from new perspectives (e.g., ontogeny). While traditional techniques such as electromyography may help to better elucidate the function of muscles involved in quadrupedal locomotion, new levels of synthesis and integration are needed to carry functional morphology beyond the study of structure–function relationships.

ACKNOWLEDGMENTS

The authors wish to thank Robert Adamo, Bob Berry, Pat Brown, Pat Brunauer, Elizabeth Dumont, Neil Duncan, Brock Fenton, Trish Freeman, Jonathan Geisler, Kim Grant, John Hermanson, Tigga Kingston, Tom Kunz, Darrin Lunde, Geraldine Moore, Guy Musser, Stuart Parsons, Steve Rossiter, Janet Schutt, William Robert Schutt, Richard Sinclair, and Eric Stiner for their help and support in this and related projects. Special thanks go to Patricia Wynne for her fine artwork. W.A.S. thanks Southampton College of Long Island University for its continued support. This study was supported by Southampton College of Long Island University (Research Release Time granted to W.A.S.) and NSF Research Grant DEB-9873663 to N.B.S.

LITERATURE CITED

Allen, H. 1886. Muscles in the hind-limb of *Cheiromeles torquatus*. Science, 7: 506.

Altenbach, J.S. 1979. Locomotor morphology of the vampire bat *Desmodus rotundus*. Special Publication. The American Society of Mammalogists, 6: 1–137.

Altenbach, J.S. 1988. Locomotion. Pp. 71–83. In: Natural History of Vampire Bats (A.M. Greenhall and U. Schmidt, eds.). CRC Press, Boca Raton, FL.

Arkins, A.M., A.P. Winnington, S. Anderson and M.N. Clout. 1999. Diet and insectivorous foraging behaviour of the short-tailed bat (*Mystacina tuberculata*). Journal of Zoology (London), 247: 183–187.

Beebe, W. 1927. The vampire's bite. Bulletin of the New York Zoological Society, 30: 113–115.

Bhatnagar, K. 1988. Anatomy. Pp. 31–40. In: Natural History of Vampire Bats (A.M. Greenhall and U. Schmidt, eds.). CRC Press, Boca Raton, FL.

Burton, M. 1955. Bulldog Bats. III. London News, 226: 28.

Corbet, G.B., and J.E. Hill. 1991. A World List of Mammalian Species. Oxford University Press, Oxford.

Corbet G.B., and J.E. Hill. 1992. The Mammals of the Indomalayan Region: A Systematic Review. Oxford University Press, Oxford.

Dalquist, W.W. 1955. Natural history of the vampire bats of eastern Mexico. American Midland Naturalist, 53: 79–87.

Daniel, M.J. 1976. Feeding by the short-tailed bat (*Mystacina tuberculata*) on fruit and possibly nectar. New Zealand Journal of Zoology, 3: 391–398.

Daniel, M.J. 1979. The New Zealand short-tailed bat, *Mystacina tuberculata*: a review of present knowledge. New Zealand Journal of Zoology, 6: 357–370.

Ditmars, R.L., and A.M. Greenhall. 1935. The vampire bat—a presentation of undescribed habits and review of its history. Zoologica, 19: 53–76.

Dwyer, P.D. 1962. Studies on the two New Zealand bats. Zoology Publications from Victoria University of Wellington, 28: 1–28.

Ecroyd, C.E. 1993. In search of the wood rose. Forest and Bird, February: 24–28.

Ecroyd, C.E. 1994. Location of short-tailed bats using *Dactylanthus*. Conservation Advisory Science Notes. Department of Conservation, Wellington, New Zealand. Moluccan Islands. Cornell University Press, Ithaca, NY.

Flannery, T. 1995. Mammals of the South-west Pacific and Moloccan Islands. Cornell University Press, Ithaca, NY.

Flannery, T., and P. Schouten. 2001. A Gap in Nature. Atlantic Monthly Press, New York.

Freeman, P. 1981. A multivariate study of the family Molossidae (Mammalia, Chiroptera): morphology, ecology, evolution. Fieldiana (Zoology), 7: 1–173.

Freeman, P. 1999. Nudity in the naked bulldog bat *Cheiromeles*. Bat Research News, 40: 170–171.

Gardner, A.L. 1977. Feeding habits, Pp. 293–350. In: Biology of Bats of The New World family Phyllostomatidae, Part II (R.J. Baker, J.K. Jones, Jr., and D.C. Carter, eds.). Special Publications of the Museum of Texas Tech University, 13: 1–364.

Goodwin, G.W., and A.M. Greenhall. 1961. A review of the bats of Trinidad and Tobago. Bulletin of the American Museum of Natural History, 122: 187–302.

Greenhall, A.M. 1968. Notes on the behavior of the false vampire bat. Journal of Mammalogy, 49: 337–340.

Greenhall, A.M. 1988. Feeding behavior. Pp. 111–131. In: Natural History of Vampire Bats (A.M. Greenhall and U. Schmidt, eds.). CRC Press, Boca Raton, FL.

Greenhall, A.M., and U. Schmidt (eds.). 1988. Natural History of Vampire Bats. CRC Press, Boca Raton, FL.

Hildebrand, M. 1995. Analysis of Vertebrate Structure. Wiley, New York.

Honacki, J.H., K.E. Kinman, and J.W. Koeppl. 1982. Mammal Species of the World: A Taxonomic and Geographic Reference. Allen Press, Lawrence, KS.

Howell, D., and J. Pylka. 1977. Why bats hang upside-down: a biomechanical hypothesis. Journal of Theoretical Biology, 69: 625–631.

Ingle, N.R., and L.R. Heaney. 1992. A key to the bats of the Philippine Islands. Fieldiana (Zoology), 69: 1–44.

Jones, G., P.I. Webb, J.A. Sedgeley, and C.F.J. O'Donnell. 2003. Mysterious *Mystacina*: how the New Zealand short-tailed bat (*Mystacina tuberculata*) locates insect prey. Journal of Experimental Biology, 206, 4209–4216.

Kennedy, M., A.M. Paterson, J.C. Morales, S. Parsons, A.M. Winnington, and H.G. Spencer. 1999. The long and the short of it: branch lengths and the problem of placing the New Zealand short-tailed bat, *Mystacina*. Molecular Phylogenetics and Evolution, 3: 405–416.

Kirsch, J.A.W., J.M. Hutcheon, D.G.P. Byrnes, and B.D. Lloyd. 1998. Affinities and historical zoogeography of the New Zealand short-tailed bat, *Mystacina tuberculata* Gray 1843, inferred from DNA-hybridization comparison. Journal of Mammalian Evolution, 5: 33–64.

Koopman, K.F. 1988. Systematics and distribution. Pp. 7–17. In: Natural History of Vampire Bats (A.M. Greenhall and U. Schmidt, eds.). CRC Press, Boca Raton, FL.

Koopman, K.F. 1989. Distributional patterns of Indo-Malaysian bats (Mammalia: Chiroptera). American Museum of Natural History, Novitates, 2942: 1–19.

Koopman, K.F. 1993. The Order Chiroptera. Pp. 137–241. In: Mammal Species of the World: A Taxonomic and Geographic Reference (D. Wilson and D.E. Reeder, eds.). Smithsonian Institution Press, Washington, DC.

Koopman, K.F. 1994. Chiroptera: Systematics. Walter de Gruyter, New York.

Lekagul, B., and J.A. McNeely. 1977. Mammals of Thailand. Association for Conservation of Wildlife, Bangkok.

Lloyd, B.D. 2001. Advances in New Zealand Mammalogy 1990–2000: short-tailed bats. Journal of the Royal Society of New Zealand, 31: 59–81.

McNab, B.K. 1973. Energetics and the distribution of vampires. Journal of Mammalogy, 54: 131–144.

Medway, Lord. 1969. The Wild Mammals of Malaya and Offshore Islands Including Singapore. Oxford University Press, Kuala Lumpur, Malaysia.

Medway, Lord. 1978. The Wild Mammals of Malaysia (Peninsular Malaysia) and Singapore. Oxford University Press, Kuala Lumpur, Malaysia.

Nowak, R.M. 1994. Walker's Bats of the World. The Johns Hopkins University Press, Baltimore.

Parsons, S. 1998. The effect of recording situation on the echolocation calls of the New Zealand lesser short-tailed bat (*Mystacina tuberculata* Gray). New Zealand Journal of Zoology, 25: 147–156.

Payne, J., C.M. Francis, and K. Phillips. 1985. A Field Guide to the Mammals of Borneo. WWF-Malaysia, Kuala Lumpur, Malaysia.

Pierson, E.D., V.M. Sarich, J.M. Lowenstein, M.J. Daniel, and W.E. Rainey. 1986. A molecular link between the bats of New Zealand and South America. Nature, 6083: 60–63.

Sanborn, C.C. 1952. Philippine zoological expedition 1946–1947, mammals. Fieldiana (Zoology), 33:89–158.

Sazima, I., and W. Uieda. 1980. Feeding behavior of the white-winged vampire bat, *Diaemus youngi*, on poultry. Journal of Mammalogy, 61: 102–104.

Schutt, W.A., Jr. 1998. The chiropteran hindlimb morphology and the origin of blood feeding in bats. Pp. 157–168. In: Bat Biology and Conservation (T.H. Kunz and P.A. Racey, eds.). Smithsonian Institution Press, Washington, DC.

Schutt, W.A., Jr., and J.S. Altenbach. 1997. A sixth digit in *Diphylla ecaudata*, the hairy legged vampire bat. Mammalia, 6: 280–285.

Schutt, W.A., Jr., and N.B. Simmons. 2001. Morphological specializations of *Cheiromeles* (naked bulldog bats; Family Molossidae) and their role in quadrupedal locomotion. Acta Chiropterologica, 3: 225–236.

Schutt, W.A., Jr., J.W. Hermanson, Y.H. Chang, D. Cullinane, J.S. Altenbach, F. Muradali, and J.E.A. Bertram. 1997. Functional morphology of the common vampire bat, *Desmodus rotundus*. Journal of Experimental Biology, 200: 3003–3012.

Schutt, W.A., Jr., F. Muradali, K. Mondol, K. Joseph, and K. Brockmann. 1999. The behavior and maintenance of captive white-winged vampire bats, *Diaemus youngi* (Phyllostomidae: Desmodontinae). Journal of Mammalogy, 80: 71–81.

Simmons, N.B. 2001. Reassessing bat diversity: how many species are there in the world? Bat Research News, 42: 179–180.

Simmons, N.B. 2005. Order Chiroptera. Pp. 312–529. In: Mammal Species of the World: A Taxonomic and Geographic Reference. 2nd Ed. (D.E. Wilson and D.M. Reeder, eds.) Johns Hopkins University Press, Baltimore.

Simmons, N.B., and T. Conway. 2001. Phylogenetic relationships of mormoopid bats (Chiroptera: Mormoopidae) based on morphological data. Bulletin of the American Museum of Natural History, 258: 1–98.

Uieda, W. 1986. Aspectos da morfologia lingual das tres especies de morcegos hematofagos (Chiroptera, Phyllostomidae). Revista Brasileira de Biologia, 46: 581–587.

Uieda, W., S. Buck, and I. Sazima. 1992. Feeding behavior of the vampire bats, *Diaemus youngi* and *Diphylla* ecaudata, on smaller birds in captivity. Ciencia e Cultura, 44: 410–412.

Van Den Bussche, R.A., and S.R. Hoofer. 2000. Further evidence for inclusion of the New Zealand short-tailed bat (*Mystacina tuberculata*) within Noctilionoidea. Journal of Mammalogy, 81: 865–874.

Vaughan, T.A. 1959. Functional morphology of three bats: *Eumops*, *Myotis*, *Macrotis*. Publications of the Museum of Natural History, University of Kansas, 12: 1–153.

Vaughan, T.A. 1970. The skeletal system. Pp. 97–138. In: Biology of Bats, Vol. 1 (W.A. Wimsatt, ed.). Academic Press, New York.

Wilkinson, G.S. 1984. Reciprocal food sharing in the vampire bat. Nature, 308: 181–184.

Wilkinson, G.S. 1988. Social organization and behavior. Pp. 85–97. In: Natural History of Vampire Bats (A.M. Greenhall and U. Schmidt, eds.). CRC Press, Boca Raton, FL.

Wimsatt, W.A. 1959. Portrait of a vampire. Ward's Natural Science Bulletin, 32: 35–63.

Wimsatt, W.A. 1969. Transient behavior, nocturnal activity patterns, and feeding efficiency of vampire bats (*Desmodus rotundus*) under natural conditions. Journal of Mammalogy, 50: 233–244.

Wimsatt, W.A., and A. Guerriere. 1962. Observations on the feeding capacities and excretory functions of captive vampire bats. Journal of Mammalogy, 43: 17–27.

9

The Correlated Evolution of Cranial Morphology and Feeding Behavior in New World Fruit Bats

Elizabeth R. Dumont

Plant-visiting phyllostomids exhibit remarkable diversity in cranial design. This chapter seeks to explain the functional basis of this diversity by evaluating the correlated evolution of cranial morphology and feeding behavior within the lineage. Several significant associations between morphology and behavior are documented using two modern comparative techniques: independent contrasts and squared-change parsimony. Based on models of mammalian mastication and results of previous experimental studies, variation in the magnitude of masticatory stress is a common theme in the evolution of the morphology/behavior associations. Inspection of reconstructed ancestral character states suggests that the evolution of fig-feeding is associated with shifts to morphologies and behaviors that reflect increased masticatory stress. This conclusion is provisional and requires testing with additional fig-feeding clades. Ultimately, explicitly evolutionary analyses such as this one are powerful tools for elucidating patterns of morphological and behavioral integration in chiropteran evolution.

INTRODUCTION AND BACKGROUND

Functional studies are fundamentally comparative, focusing on associations expressed by species that are closely related and/or fill similar ecological niches. In accordance with modern thoughts concerning the fundamental importance of phylogeny (e.g., Brooks and McLennan, 1991; Felsenstein, 1985; Harvey and Pagel, 1991), comparative biologists are increasingly accounting for phylogenetic structure in analyzing these associations (e.g., Eggleton and Vane-Wright, 1994; Martins, 1996; Wainwright and Reilly, 1994). Prior to the development of modern comparative methods, functional morphologists lacked adequate tools for evaluating the evolution of continuous morphological and behavioral data. The development of techniques to accomplish this task (for reviews, see Harvey and Pagel, 1991;

Rohlf, 2001) has enabled functional morphology to truly progress in its search for evolutionary patterns in form–function relationships. This study employs two comparative techniques to investigate the correlated evolution of continuous morphological and behavioral characters: independent contrasts (Felsenstein, 1985; Garland et al., 1992; Harvey and Pagel, 1991) and squared-change parsimony analysis (Garland et al., 1991; Huey and Bennett, 1987; Maddison, 1991; Martins and Garland, 1991). Independent contrasts can also be used with discrete data and with combinations of discrete and continuous characters. Independent contrasts and squared-change parsimony have very different philosophical and statistical approaches, but each can be used to evaluate evolutionary correlations between continuous variables within a clearly defined phylogenetic context.

The goal of this chapter is to evaluate the correlated evolution of cranial morphology and feeding among plant-visiting phyllostomids. The structural diversity within this group is remarkable, yet few studies have investigated its functional implications. Although the question asked here is relatively simple, the answer has significant implications for our understanding of form–function relationships in the evolution of the phyllostomid feeding apparatus. Moreover, this study will set the stage for elucidating broader patterns in the evolution of cranial morphology and feeding behavior in mammals.

METHODS

Morphological Data

The first step in this analysis was to identify morphological features of the skull and dentary that vary among phyllostomid frugivores with divergent trophic adaptations. Building on previous work (Dumont, 1997; Freeman, 1988), a dataset containing 23 continuous variables describing the shape of the skull and dentary was assembled for 19 plant-visiting species of phyllostomids (table 9.1, figure 9.1). Variables were chosen to reflect variation in skull form that suggest differing biomechanical properties. These include estimates of muscle lever arms, load arms, and gape, as well as elements of cranial shape often associated with resistance to twisting and bending stresses encountered during mastication.

For each specimen, measurements for each variable were divided by the geometric mean of all variables measured for that individual and then log transformed (e.g., Darroch and Mosiman, 1985; Dumont, 1997; Falsetti et al., 1993). This procedure generates individual-based shape values that avoid the problem of sample dependence inherent in regression-based size adjustments. A principal components analysis (with Varimax rotation) of species means for the size-adjusted variables demonstrates significant morphological variation among these species (figure 9.2, table 9.2). Variables with factor loadings greater than 0.70 are considered to contribute meaningfully to variation within the sample (Tabachnick and Fidell, 1996).

Table 9.1 Species Included in Morphometric Analyses and Their Sample Sizes

Species	Females (*N*)	Males (*N*)
Ametrida centurio	1	1
Anoura geoffroyi	5	5
Ardops nichollsi	–	2
Ariteus flavescens	2	2
Artibeus jamaicensis	5	5
Artibeus phaeotis	5	5
Brachyphylla cavernarum	3	3
Carollia perspicillata	5	5
Centurio senex	4	4
Chiroderma villusom	3	3
Ectophylla alba	3	3
Erophylla sezekorni	1	1
Glossophaga soricina	5	5
Phylloderma stenops	–	2
Phyllonycteris poeyi	2	2
Platyrrhinus helleri	3	3
Pygoderma bilabiatum	4	4
Rhinophylla pumilio	4	4
Sturnira lilium	7	7

The first principal component (PC1) summarizes 46% of the variation among species and is negatively correlated with dentary length (TDL) and positively correlated with condyle width (CW), palate width (PM1), the depth of the dentary (MDD), and the occlusal area of the lower second molar (AREA). Species with high scores on PC1 have relatively short, deep dentaries with wide condyles, wide palates, and large molars. The second principal component (PC2) summarizes an additional 16% of the variation among species and is positively correlated with skull height (SKH) and width (PSW). The skulls of species with high scores on PC2 are relatively tall and wide both posteriorly and across the palate at the canines. Two variables (CH, condyle height; CPH, coronoid process height) strongly contribute to a third principal component (PC3), but this component explains only a small proportion of variation within the sample.

This simple morphometric analysis identifies a subset of variables that drive the structural diversity in the skulls and dentaries of plant-visiting phyllostomids. In this case, the variables entered into the analysis reflect aspects of the masticatory apparatus that are hypothesized to be correlated

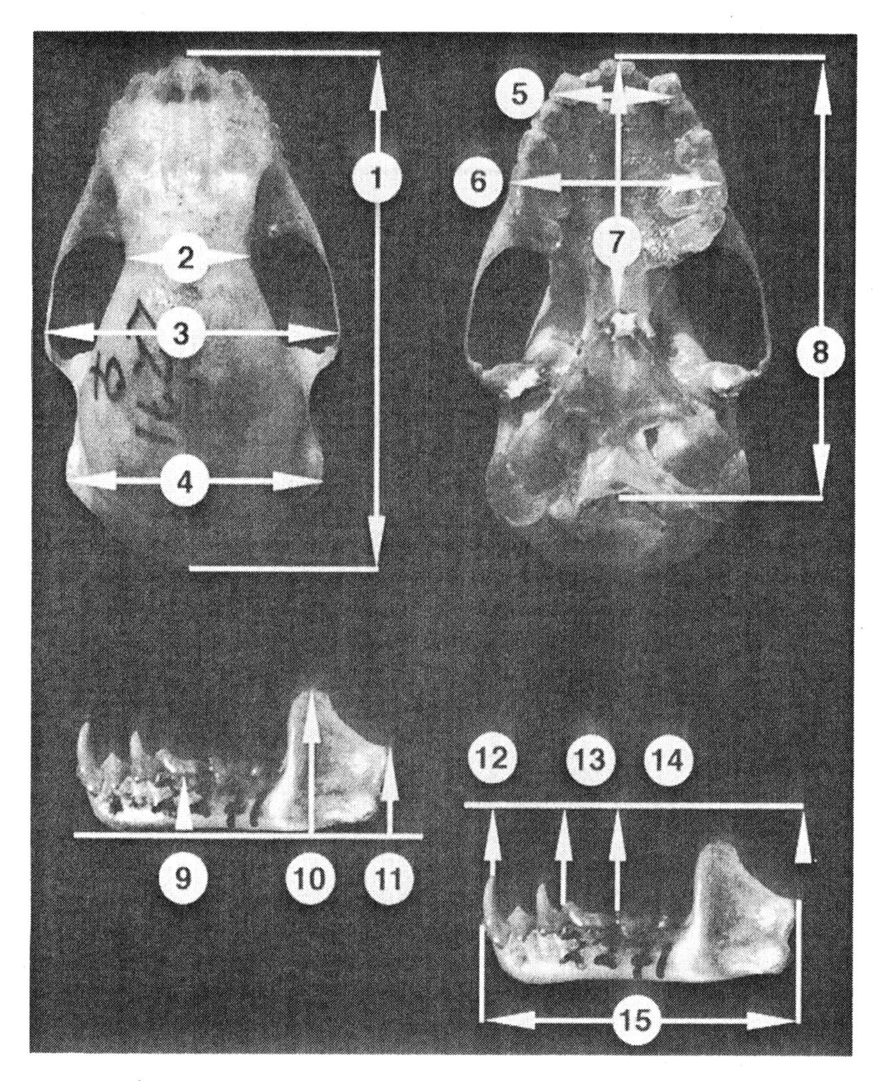

Figure 9.1 Morphological variables used in the principal components analysis. Dorsal and ventral views of a skull and lateral views of a dentary of *Artibeus jamaicensis* illustrate: 1, total skull length (TSL); 2, middle skull width (MSW); 3, maximum zygomatic breadth (MZB); 4, posterior skull width (PSW); 5, palate width at canines (PC); 6, palate width at M1 (PM1); 7, total palate length (TPL); 8, anterior skull length (ASL); 9, dentary depth under M1 (MDD); 10, coronoid process height (CPH); 11, condyle height (CH); 12, condyle to canine bite point (CC); 13, condyle to M1 bite point (CM1); 14, condyle to M3 bite point (CM3); 15, total dentary length (TDL). Linear measurements not illustrated here include: dentary condyle length (CL), condyle width (CW), and M2 area (Area, M2 length × M2 width). Calculations for estimating gape angle, masseter origin/insertion ratio, masseter lever arm, and pterygoid lever arm are described in the literature (Herring and Herring, 1974; Spencer and Demes, 1993).

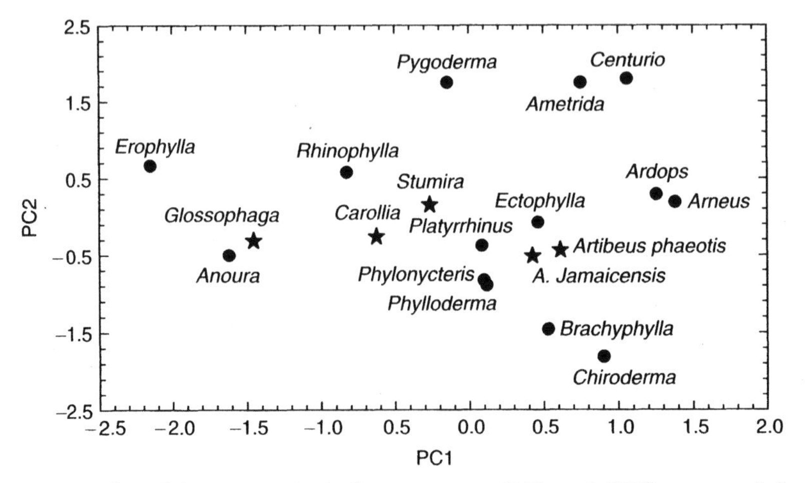

Figure 9.2 Plot of first two principal components (PC1 and PC2) generated from size-adjusted cranial shape data for 19 species of plant-visiting phyllostomids. Species with high scores on PC1 have relatively short, deep dentaries with wide condyles, wide palates, and large molars. Species with high scores on PC2 have relatively tall skulls that are wide both posteriorly and across the palate at the canines. Focal species used in the analysis of evolutionary correlations between morphology and behavior are identified by a star.

with variation in feeding behavior. Variables from the first two principal components with factor loadings greater than or equal to 0.70 are used as morphological characters in subsequent evolutionary analyses.

Behavioral Data

Data summarizing biting and chewing behavior for five phyllostomid species (*Glossophaga soricina, Carollia perspicillata, Sturnira lilium, Artibeus phaeotis,* and *Artibeus jamaicensis*) are drawn from Dumont (1999) (table 9.3). This small sample represents the first installment of a larger database summarizing feeding behavior in frugivorous bats. In this study, behavioral data were collected under two experimental conditions designed to elicit a broad range of feeding behaviors: soft-fruit feeding and hard-fruit feeding. Variation in food texture has been cited as a basis for morphological and behavioral divergence in many vertebrate communities (e.g., Freeman, 1979; Herring, 1985; Wainwright, 1987). Here, fruit hardness was varied to evaluate the level of behavioral plasticity displayed by different species during fruit processing.

Several different behavioral characters are used in this analysis. The most basic are the average number of bites used to detach a mouthful of fruit (SBITES, bites during soft-fruit feeding; and HBITES, bites during hard-fruit feeding) and the average number of chews used to process mouthfuls of fruit (SCHEWS, chews during soft-fruit feeding; and HCHEWS, chews during hard-fruit feeding). These characters provide estimates of the effort expended in breaking down food. (Here "effort" implies relatively more or

Table 9.2 Factor Loadings for Variables on the First Three Principal Components (PC1, PC2, and PC3)

Variable	**PC1**	**PC2**	**PC3**
TSL (Total skull length)	0.44	0.54	−0.43
MSW (Middle skull width)	−0.25	0.67	0.22
MZB (Maximum zygomatic breadth)	0.44	0.54	−0.43
PSW (Posterior skull width)	0.08	**0.93**	0.04
PC (Palate width at canines)	0.07	**0.80**	−0.25
PM1 (Palate width at M1)	**0.72**	0.50	−0.17
TPL (Total palate length)	−0.68	−0.59	−0.32
ASL (Anterior skull length)	**−0.80**	−0.40	−0.07
MDD (Dentary depth under M1)	**0.72**	0.46	0.40
CPH (Coronoid process height)	0.47	0.07	**0.76**
CH (Condyle height)	−0.10	−0.14	**0.83**
CC (Condyle to canine bite point)	**−0.90**	−0.37	−0.01
CM1 (Condyle to M1 bite point)	**−0.82**	−0.36	0.05
CM3 (Condyle to M3 bite point)	**−0.79**	−0.26	0.01
TDL (Total dentary length)	**−0.91**	−0.35	−0.01
CL (Condyle length)	−0.64	0.12	−0.01
CW (Condyle width)	**0.83**	−0.06	0.14
AREA (m2 length × width)	**0.73**	−0.10	−0.20
GAPE (Gape angle)	−0.26	−0.59	0.29
SKH (Skull height)	0.14	**0.86**	−0.07
MOI (Masseter origin:insertion ratio)	−0.67	0.39	0.16
MTPL (Masseter lever arm:palate length ratio)	0.55	0.54	0.29
PTPL (Pterygoid lever arm:palate length ratio)	0.14	0.62	0.01
Eigenvalue	10.54	3.70	2.24
Proportion of explained variance (%)	45.81	16.08	9.75

Factor loadings > 0.70 (in bold) are considered high. Eigenvalues and the proportion of explained variance for each component are also reported.

less time spent on a task and does not reflect a physiological measurement of work.)

Two other, more complex characters describe differences in the mechanical efficiency of bites used to detach pieces of fruit. These categories of efficiency are based on a simple model of the dentary as a class III lever (Herring, 1993), where the temporomandibular joint is the fulcrum, the distance from the joint to the anterior masseter muscle attachment is the muscle lever arm, and the distance from the joint to the food item is the bite load arm. Using this model, bites that are centered over the anterior

Table 9.3 Behavioral Data

	Soft-fruit Feeding					Hard-fruit Feeding				
	SCHEWS	**SBITES**	**%I**	**%E**	**%UNI**	**HCHEWS**	**HBITES**	**%IH**	**↑E**	**↑UNI**
Glossophaga soricina	10	11	8	0	0	0	0	0	0	0
Carollia perspicillata	7	4	9.2	0	1	9	11	8.9	3	6
Sturnira lilium	28	10	43.4	0	17	30	9	14	17	6
Artibeus phaeotis	45	6	47	2	12	14	5	7.8	60.7	74
Artibeus jamaicensis	54	5	29.1	19.9	60	42	6	0	60.1	29

Data include the number of chews per mouthful (SCHEWS and HCHEWS), the numbers of bites used to detach a mouthful of fruit (SBITES and HBITES), and the proportion of mechanically efficient bites (%E, ↑E), inefficient bites (%I, %IH), and unilateral bites (%UNI and ↑ UNI) used by each species while feeding on soft and hard fruits (Dumont, 1999).

teeth (incisors and canines) are termed "inefficient" (%I and %IH) because the length of the bite point load arm far exceeds the length of the muscle lever arm. In contrast, bites centered over the posterior teeth (premolars and molars) on one side of the jaw (i.e., unilateral) are termed "efficient" (%E) as the bite point load arms more closely approximate masticatory muscle lever arms and concentrate bite forces over a smaller area. The total proportion of all unilateral bites (%UNI, which includes %E) was also used as a character as studies have documented that unilateral loading results in predictable strain patterns in the dentary (Hylander, 1979a,b) and may be associated with the shape of the palate and rostrum (Covey and Greaves, 1994). The proportional increases in efficient and unilateral bites during hard-fruit feeding (↑E and ↑UNI) are analogous to %E and %UNI, and document the degree to which species switch to more efficient biting strategies when they encounter hard fruits.

Correlation Analysis

The ascendancy of parsimony and maximum likelihood techniques for reconstructing phylogenetic history during the 1980s provided the foundation for the development of modern comparative statistical techniques. Phylogenies are the backbone against which modern comparative analyses are built. Because phylogenetic structure significantly impacts analyses based on it, selecting the most well supported phylogeny is critical.

Fortunately, the relationships among the focal species used here (figure 9.3) have been studied extensively. This cladogram is extracted from a supertree

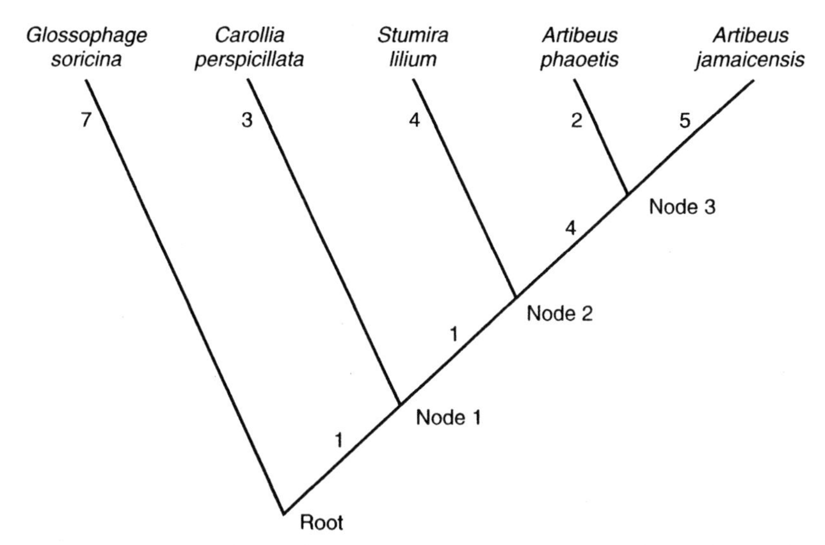

Figure 9.3 Cladogram of relationships among species used in this analysis. Branch lengths and topology derived from Jones et al. (2002). Node 2 unites members of the subfamily Sternodermatinae. Species derived from node 1 regularly consume figs.

that presents the consensus topology from more than 100 phylogenies published since 1970 (Jones et al., 2002). Within this tree, relationships among phyllostomids are based on 36 molecular and morphological studies. Despite a history of controversies in phyllostomid systematics, the relationships among the focal taxa used in this study are well supported. The branch lengths reported here are derived from the structure of the supertree and reflect numbers of branching events rather than time. (Branch lengths were multiplied by two prior to analysis to meet the requirements of the independent contrasts software; Purvis and Rambaut, 1995.)

The techniques used here to assess the evolutionary correlation between morphological and behavioral characters, independent contrast and squared-change parsimony, make similar assumptions about the process of evolution. Each assumes that evolution proceeds through a series of small, random changes that accumulate at a constant rate (i.e., a Brownian motion model). Experiments demonstrate that under most models of evolution, independent contrasts and squared-change parsimony yield similar results and exhibit low rates of type I error (Martins and Garland, 1991). Therefore, several studies have employed both techniques as a means of validating results (Garland et al., 1991; Walton, 1993; Westneat, 1995). Despite their similarities, the two techniques differ markedly in their statistical and philosophical approaches to the study of evolution (Garland et al., 1997; Harvey and Pagel, 1991; Pagel, 1993).

The goal of independent contrasts (Felsenstein, 1985) is to produce a set of independent, phylogeny-adjusted variables ("contrasts") that can be analyzed using traditional statistical techniques. Among the programs available to calculate independent contrasts, I selected the program CAIC (Comparative Analysis by Independent Contrasts; Purvis and Rambaut, 1995) for its ability to test for violations of statistical assumptions. Following Felsenstein's (1985) original method, the program begins generating contrasts by calculating the difference between values of variables from adjacent crown tips. This step is repeated for each node. All contrasts are calculated using two variables at a time with one (independent) variable being used to determine the direction of subtraction between pairs of values at each node. Raw contrasts are then transformed into standardized linear contrasts through division by the square root of the sum of their branch lengths, a measure of expected variance (for detailed descriptions see Garland et al., 1992; and Purvis and Rambaut, 1995). $N-1$ contrasts, one representing each node, are generated from any fully bifurcating tree ($N=$number of tip species/taxa). The correlated evolution of contrasts can be analyzed accurately using parametric correlation or regression statistics with the intercept forced through the origin (for discussions of this requirement see Garland et al., 1992; and Harvey and Pagel, 1991).

In this analysis, independent contrasts were used to assess the correlation between each of 80 pairs of morphological and behavioral characters. In each case, the morphological characters were treated as independent variables. In 23 of these comparisons, one or more assumptions of the method were violated (homogeneity of variance in residuals, and independence of contrasts from their expected variance and age of the node; Purvis and Rambaut, 1995; www.bio.ic.ac.uk/evolve/software/caic/assumptions). Violations were corrected for 16 contrasts by transforming the raw data by ($\mathrm{Ln}(X+10)$ or $-1/x$). The remaining nine contrasts could not be adequately transformed and were assessed using an algorithm (brunch) designed for the analysis of discrete characters (Purvis and Rambaut, 1995) which makes no assumptions about mode of evolution and yields more conservative results.

The correlated evolution of each pair of morphological and behavioral characters was investigated by calculating a Pearson product–moment correlation coefficient with an intercept forced through the origin (SPSS Base 10 for Macintosh, SPSS Inc., Chicago, USA; Garland et al., 1992; Purvis and Rambaut, 1995). Although the variables in this dataset are almost certainly not independent, *P* values were not adjusted for experiment-wise, as each pairing of morphological and behavioral data was conceived a priori as independent hypotheses of bivariate relationship.

Unlike independent contrasts, squared-change parsimony (also known as minimum evolution) seeks to reconstruct absolute values for all ancestral nodes and, as a result, the pattern of character evolution along all branches of a cladogram (Garland et al., 1991; Huey and Bennett, 1987; Maddison, 1991; Martins and Garland, 1991; Pagel, 1992). The squared-change parsimony algorithm calculates nodal values by minimizing their sum of squared changes over the cladogram. For this analysis I selected the collection of comparative programs entitled PDAP (Phenotypic Diversity Analysis Program; Garland et al., 1993) for its ease of use and thorough documentation. In this software package, the squared-change parsimony algorithm finds nodal values using an iterative approach. First, each node is assigned the mean of adjacent tip/node values weighted by branch length. These values are then re-adjusted over a user-defined number of iterations (100 in this study) until the values stabilize at their minimum (for a detailed description of the method see Garland et al., 1997). Interestingly, the value of the root node generated by squared-change parsimony and independent contrasts is exactly the same (Garland and Ives, 2000; Garland et al., 1997).

It is important to point out that nodal values generated using squared-change parsimony are not statistically independent. Rather they are calculated to reconstruct the evolution of individual continuous characters. Indeed, the most common application of squared-change parsimony is in reconstructing values at ancestral nodes (Garland et al., 1997; Losos, 1990; Miles and Dunham, 1996). However, the evolutionary correlation between two characters can be assessed using the estimates of character change along each branch as paired variables in a correlation analysis (Garland et al., 1991; Martins and Garland, 1991).

For this study, squared-change parsimony analysis was used to reconstruct nodal values and the amount of change along each branch of the cladogram for all 18 morphological and behavioral characters.

Because branch lengths in this analysis are equal and do not reflect time, this analysis assumes a punctuated model of evolution (Martins and Garland, 1991). Parametric correlations of changes along each branch were calculated for all 80 pairs of morphological with behavioral characters (SPSS Inc., Chicago, USA; Garland et al., 1991). A criticism of this method is that fully bifurcating trees result in values for $2N-2$ branches, far exceeding the initial number of data points (Nunn and Barton, 2001; Pagel, 1993). As suggested by Martins and Garland (1991), a more conservative $N-2$ degrees of freedom was used to assess the significance of the correlation coefficients.

RESULTS

Independent contrasts and squared-change parsimony identified a virtually identical pattern of significant correlations and trends in the data (tables 9.4, 9.5), although significance values tend to be lower under independent contrasts. Associations between morphology and behavior are widespread, with 14 of 18 variables involved in either significant correlations or trends. However, there is undoubtedly some degree of interdependence among characters within both the morphological and behavioral datasets, and the total number of associations may ultimately be reduced.

Both analyses yielded the following significant correlations between cranial morphology and feeding behavior: increased dentary depth (MDD) is associated with increased efficient and unilateral biting during soft-fruit eating (%E and %UNI), increased condyle width (CW) is associated with a shift to unilateral biting during hard-fruit feeding (↑UNI), and increased palate width at the canines (PC) is correlated with both increased numbers of chews per mouthful during soft-fruit feeding (SCHEWS) and a shift toward a higher proportion of efficient bites during hard-fruit feeding (↑E).

Because this dataset is relatively small, it is reasonable to mention several trends in associations between morphology and behavior that emerged from both analyses. One is the tendency for the number of inefficient bites during soft-fruit feeding (%I) to increase with the width of the palate at M1 (PM1) and to decrease with the length of the dentary (TDL). Another is the tendency for increased dentary depth (MDD) to be associated with increased chewing during hard-fruit feeding (HCHEWS). Finally, while both analyses identify a negative relationship between skull height (SKH) and number of bites during hard-fruit feeding (HBITES), the association is not significant under independent contrasts. Though not statistically significant, the fact that these trends emerge from both analyses suggests that they are not random and should be explored further with larger datasets.

DISCUSSION

This is the first study of mammals to assess the correlated evolution of cranial shape and detailed accounts of feeding behavior gathered in the field. The significant, correlated evolution of these elements among plant-visiting

Table 9.4 Direction and Significance Values for Correlations between Morphological (column) and Behavioral (row) Variables Determined Using Independent Contrasts

	TDL	PM1	MDD	CW	Area	PSW	SKH	PC
Soft-fruit feeding								
SCHEWS	ns	ns	ns	ns	ns	ns	ns	+ **≤0.05**
SBITES	ns	ns	ns	ns	ns	ns	ns	ns
%I	– **≤0.1**	+ **≤0.1**	ns	ns	ns	ns	ns	ns
%E	ns	ns	+ **≤0.01**	ns	ns	ns	ns	ns
%UNI	ns	ns	+ **≤0.01**	ns	ns	ns	ns	ns
Hard-fruit feeding								
HCHEWS	ns	ns	+ **≤0.1**	ns	ns	ns	ns	ns
HBITES	ns	ns	ns	ns	ns	ns	– **≤0.1**	ns
%IH	ns	ns	ns	ns	ns	ns	ns	ns
↑E	ns	ns	ns	ns	ns	ns	ns	+ **≤0.01**
↑UNI	ns	ns	ns	+ **≤0.01**	ns	ns	ns	ns

The directions of significant correlations (in bold) and trends ($P \leq 0.1$) are provided (df = 3). See figure 9.1 and table 9.2 for descriptions of variables.

phyllostomids suggests strong links between form and function. While rigorous biomechanical testing and modeling of bat crania are necessary to demonstrate the underlying mechanisms by which morphology translates into behavior, the associations documented here are supported by functional analyses of the skull and dentary in other mammals.

Classic studies of primates demonstrate that both biting and chewing impose bending and torsional stresses on the dentary (Hylander, 1979a,b). Bilateral biting (%I and %IH) vertically bends both dentaries, while unilateral biting (%E, ↑E, %UNI, and ↑UNI) and chewing twist the working side and vertically bend the balancing side. Building on this, simple models of dentary cross-sectional shape (e.g., Bouvier, 1986a,b; Daegling, 1992; Ravosa, 1991) predict that tall, narrow dentaries are more resistant to bending while dentaries with rounder cross-sections are more resistant to torsion.

Table 9.5 Direction and Significance Values for Correlations between Morphological (column) and Behavioral (row) Variables Determined Using Squared-Change Parsimony

	TDL	PM1	MDD	CW	Area	PSW	SKH	PC
Soft-fruit feeding								
SCHEWS	ns	ns	ns	ns	ns	ns	ns	+ **≤0.05**
SBITES	ns	ns	ns	ns	ns	ns	ns	ns
%I	– ≤ 0.1	+ ≤ 0.1	ns	ns	ns	ns	ns	ns
%E	ns	ns	+ **≤0.01**	ns	ns	ns	ns	ns
%UNI	ns	ns	+ **≤0.01**	ns	ns	ns	ns	ns
Hard-fruit feeding								
HCHEWS	ns	ns	+ ≤0.1	ns	ns	ns	ns	ns
HBITES	ns	ns	ns	ns	ns	ns	– **≤0.05**	ns
%IH	ns	ns	ns	ns	ns	ns	ns	ns
↑E	ns	ns	ns	ns	ns	ns	ns	+ **≤0.05**
↑UNI	ns	ns	ns	+ **≤0.01**	ns	ns	ns	ns

The directions of significant correlations (in bold) and trends ($P \leq 0$ .1) are provided (df = 3). See figure 9.1 and table 9.2 for descriptions of variables.

Additional documentation of cortical bone thickness can provide even more precise information about strain patterns (e.g., Biknivicius and Ruff, 1992; Daegling and Hylander, 1998, 2000). In this study, dentary depth is significantly associated with high numbers of unilateral bites (%E and %UNI) as well as a trend toward increased chewing. Documenting the width and the distribution of cortical bone in these dentaries awaits analysis of cross-sectional data. Nevertheless, our understanding of dentary function during feeding supports a functional association between dentary depth and biting behavior.

The positive correlation between condyle width (CW) and increased unilateral biting (↑UNI) may be associated with increased stress at the temporomandibular joint during hard-fruit feeding. Again in primates, Hylander (1979c) demonstrated that strain in the subcondylar region of the (contralateral) mandibular condyle tends to be higher during isometric, unilateral molar biting than during either incision or mastication. Subsequent allometric studies concluded that increased condyle width is associated with diets that emphasize posterior tooth use (Bouvier, 1986a,b; Smith et al., 1983), and presumably serves to distribute increased joint reaction forces. These studies support a functional interpretation of the relationship between condyle width and unilateral biting discovered here.

Finally, a model of palate shape suggests that relative increase in palate width is associated with resistance to torsional stress (Covey and Greaves, 1994). Although this model has not been tested, the significant correlation between palate width at the canines (PC) and both high numbers of chews (SCHEWS) and a shift to efficient biting (↑E) fits with its predictions. In sum, all the significant associations between morphology and behavior described in this study revolve around the theme of resisting and generating stresses during food processing.

From a broader perspective, the data presented here can also be used to investigate the ecological context of evolution in morphological and behavioral characters. For example, it would be of interest to know whether shifts in morphologies and behaviors associated with increased masticatory stress are associated with the evolution of a particular dietary strategy. Fig-feeding is proposed to be a form of hard-object feeding in bats (Dumont, 1999, 2003). Thus, one could predict that evolution of fig-feeding in the *Artibeus* clade (node 3, figure 9.3) is linked with marked changes in morphology and behavior.

Hypotheses such as this can be addressed using contrasts with a discrete character (e.g., the presence or absence of figs in the diet) as a predictor variable (Purvis and Rambaut, 1995). Unfortunately, the technique cannot be used with this dataset because fig-feeding is limited to the two crown taxa, thus allowing the calculation of a single contrast. As a purely heuristic alternative, nodal values reconstructed using squared-change parsimony analysis can offer some, albeit limited, insights into the morphological and behavioral shifts that accompany fig-feeding.

The reconstructed nodal values for morphological and behavioral characters highlighted by correlation analyses are presented in table 9.6. The three patterns of change in nodal values all show marked shifts at the fig-feeding node (node 3, figure 9.3 and table 9.6). First, an increase in nodal value at node 3 is characteristic of the morphological data. The second pattern of nodal value reconstructions additionally includes an increase at the node uniting stenodermatines (node 2, figure 9.3 and table 9.6). Finally, two behavioral characters (%E and %UNI) exhibit a marked increase in value at the sternodermatine node followed by an immediate decrease at node 3. It is tempting to suggest that these patterns of change point to morphological and behavioral shifts that accompany the evolution of fig-feeding in phyllostomids. Unfortunately, there are no techniques for generating confidence intervals for nodes reconstructed using squared-change parsimony and, moreover, the accuracy of ancestral node reconstruction using

Table 9.6 Nodal Values of Significantly Correlated Morphological and Behavioral Characters Reconstructed Using Squared-Change Parsimony (Garland et al., 1997)

	Morphology			**Behavior**				
	MDD	**CW**	**PC**	**SCHEWS**	**%E**	**%UNI**	**↑E**	**↑U**
Root	−0.84	−2.07	−0.28	22.12	4.20	14.39	20.31	14.64
Node 1	−0.75	−2.04	−0.26	23.85	4.80	16.45	23.22	16.73
Node 2	−0.45	−1.97	−0.24	31.20	7.0	23.65	32.86	23.73
Node 3	−0.99	−1.62	−0.20	37.79	2.89	16.12	44.17	46.45

See figure 9.3 for a graphical summary of node numbers. Definitions of morphological characters are given in figure 9.2 and behavioral characters are described in the text.

any method is questionable (e.g., Losos, 1999; Webster and Purvis, 2002). Patterns of nodal change must be interpreted with discretion.

In sum, this analysis demonstrates significant correlations between cranial morphology and feeding behavior in the evolution of plant-visiting phyllostomids. The presence of these correlations begins to identify the functional implications of long-recognized patterns of morphological diversity seen within the group. Specifically, variation in the patterns of stress encountered during food processing appears to be a significant factor underlying structural and behavioral evolution. Although implied by these data, a robust test of the relationship between morphology and behavior in the evolution of fig-feeding awaits data from additional fig-feeding clades. Importantly, the approach to the study of mammalian feeding used here is unique and is a significant step toward understanding the functional interdependence of cranial morphology and feeding behavior in mammalian evolution.

ACKNOWLEDGMENTS

I thank Kate Jones for access to the dataset on which the phylogeny used in this study is based and for many, many discussions about independent contrasts. Both comparative programs used in this study (CAIC and PDAP) are available from web sites maintained by their authors (A. Purvis and T. Garland, respectively). This work was supported by a grant from the National Science Foundation (IBN-9507488).

LITERATURE CITED

Biknivicius, A.R., and C.B. Ruff. 1992. The structure of the mandibular corpus and its relationship to feeding behaviours in extant carnivorans. Journal of Zoology (London), 228: 479–507.

Bouvier, M. 1986a. A biomechanical analysis of mandibular scaling in Old World monkeys. American Journal of Physical Anthropology, 69: 473–482.

Bouvier, M. 1986b. Biomechanical scaling of mandibular dimensions in New World monkeys. International Journal of Primatology, 7: 551–567.

Brooks, D.R., and D.A. McLennan, 1991. Phylogeny, Ecology, and Behavior: A Research Program in Comparative Biology. University of Chicago Press, Chicago.

Covey, D.S.G., and W.S. Greaves. 1994. Jaw dimensions and torsion resistance during canine biting in the Carnivora. Canadian Journal of Zoology, 72: 1055–1060.

Daegling, D.J. 1992. Mandibular morphology and diet in the genus *Cebus*. International Journal of Primatology, 13: 545–570.

Daegling, D.J., and W.L. Hylander. 1998. Biomechanics of torsion in the human mandible. American Journal of Physical Anthropology, 105: 73–88.

Daegling, D.J., and W.L. Hylander. 2000. Experimental observation, theoretical models, and biomechanical inference in the study of mandibular form. American Journal of Physical Anthropology, 112: 541–552.

Darroch, J.N., and J.N. Mosiman. 1985. Canonical and principle components of shape. Biometrika, 72: 241–252.

Dumont, E.R. 1997. Cranial shape in fruit, nectar and exudate feeding mammals: implications for interpreting the fossil record. American Journal of Physical Anthropology, 102: 187–202.

Dumont, E.R. 1999. The effect of food hardness on feeding behavior in frugivorous bats (Family Phyllostomidae): an experimental study. Journal of Zoology (London), 248: 219–229.

Dumont, E.R. 2003. Bats and fruit: an ecomorphological approach. Pp. 398–429. In: Bat Ecology (T.H. Kunz and M.B. Fenton, eds.). University of Chicago Press, Chicago.

Eggleton, P., and R.I. Vane-Wright. 1994. Phylogenetics and Ecology. Academic Press, San Diego.

Falsetti, A.B., W.L. Jungers, and T.M. Cole. 1993. Morphometrics of the callitrichid forelimb: a case study in size and shape. International Journal of Primatology, 14: 551–572.

Felsenstein, J. 1985. Phylogenies and the comparative method. The American Naturalist, 125: 1–15.

Freeman, P.W. 1979. Specialized insectivory: beetle-eating and moth-eating molossid bats. Journal of Mammalogy, 60: 467–479.

Freeman, P.W. 1988. Frugivorous and animalivorous bats (Microchiroptera): dental and cranial adaptations. Biological Journal of the Linnean Society, 33: 249–272.

Garland, T., Jr., and A.R. Ives. 2000. Using the past to predict the present: confidence intervals for regression equations in phylogenetic comparative methods. The American Naturalist, 155: 346–364.

Garland, T., Jr., R.B. Huey, and A.F. Bennett. 1991. Phylogeny and thermal physiology in lizards: a reanalysis. Evolution, 45: 1969–1975.

Garland, T., Jr., P.H. Harvey, and A.R. Ives. 1992. Procedures for the analysis of comparative data using phylogenetically independent contrasts. Systematic Biology, 41: 18–32.

Garland, T., Jr., P.E. Midford, A.W. Dickerman, J.A. Jones, and R. Diaz-Uriarte. 1993. PDAP: Phenotypic Diversity Analysis Programs, version 5.0.

Garland, T., Jr., K.L.M. Martin, and R. Diaz-Uriarte. 1997. Reconstructing ancestral trait values using squared-change parsimony: plasma osmolarity at the origin of amniotes. Pp. 425–501. In: Amniote Origins: Completing the Transition to Land (S.S. Sumida and K.L.M. Martin, eds.). Academic Press, San Diego.

Harvey, P., and M.D. Pagel. 1991. The Comparative Method in Evolutionary Biology. Oxford University Press, New York.
Herring, S.W. 1985. Morphological correlates of masticatory patterns in peccaries and pigs. Journal of Mammalogy, 66: 603–617.
Herring, S.W. 1993. Functional morphology of mammalian mastication. American Zoologist, 33: 289–299.
Herring, S.W., and S.E. Herring. 1974. The superficial masseter and gape in mammals. The American Naturalist, 108: 561–576.
Huey, R.B., and A.F. Bennett. 1987. Phylogenetic studies of coadaptation: preferred temperatures versus optimal performance temperatures of lizards. Evolution, 41: 1098–1115.
Hylander, W.L. 1979a. The functional significance of primate mandibular form. American Journal of Physical Anthropology, 106: 223–240.
Hylander, W.L. 1979b. Mandibular function in *Galago crassicaudatus* and *Macaca fasicularis*: an in vivo approach to stress analysis of the mandible. Journal of Morphology, 159: 253–296.
Hylander, W.L. 1979c. An experimental analysis of temporomandibular joint reaction forces in macaques. American Journal of Physical Anthropology, 51: 433–456.
Jones, K.E., A. Purvis, A. MacLarnon, O.R.P. Bininda-Emonds, and N.B. Simmons. 2002. A phylogenetic supertree of the bats (Mammalia: Chiroptera). Biological Reviews, 77: 233–259.
Losos, J.B. 1990. Ecomorphology, performance capability, and scaling of West Indian *Anolis* lizards: an evolutionary analysis. Ecological Monographs, 60: 369–388.
Losos, J.B. 1999. Uncertainty in the reconstruction of ancestral character states and limitations on the use of phylogenetic comparative methods. Animal Behaviour, 58: 1319–1324.
Maddison, W.P. 1991. Squared-change parsimony reconstructions of ancestral states for continuous-valued characters on a phylogenetic tree. Systematic Zoology, 40: 304–314.
Martins, E.P. 1996. Phylogenies and the Comparative Method in Animal Behavior. Oxford University Press, New York.
Martins, E.P., and T. Garland, Jr. 1991. Phylogenetic analyses of the correlated evolution of continuous characters: a simulation study. Evolution, 45: 534–557.
Miles, D.B., and A.E. Dunham. 1996. The paradox of phylogeny: character displacement of analyses of body size in island *Anolis*. Evolution, 50: 594–603.
Nunn, C.L., and R.A. Barton. 2001. Comparative methods for studying primate adaptation and allometry. Evolutionary Anthropology, 10: 81–98.
Pagel, M.D. 1992. A method for the analysis of comparative data. Journal of Theoretical Biology, 156: 431–442.
Pagel, M. 1993. Seeking the evolutionary regression coefficient: an analysis of what comparative methods measure. Journal of Theoretical Biology, 164: 191–205.
Purvis, A., and A. Rambaut. 1995. Comparative analysis by independent contrasts (CAIC): an Apple Macintosh application for analyzing comparative data. Computer Applications in Biosciences, 11: 247–251.
Ravosa, M.J. 1991. Structural allometry of the prosimian mandibular corpus and symphysis. Journal of Human Evolution, 20: 3–20.
Rohlf, J.F. 2001. Comparative methods for the analysis of continuous variables: geometric interpretations. Evolution, 55: 2143–2160.

Smith, R.J., C.E. Petersen, and D.P. Gipe. 1983. Size and shape of the mandibular condyle in primates. Journal of Morphology, 177: 59–68.
Spencer, M.A., and B. Demes. 1993. Biomechanical analysis of masticatory system configuration in Neanderthal and Inuits. American Journal of Physical Anthropology, 91: 1–20.
Tabachnik, B.G., and L.S. Fidell. 1996. Using Multivariate Statistics. 2nd Ed. Harper and Row, New York.
Wainwright, P.C. 1987. Biomechanical limits to ecological performance: mollusc-crushing by the Caribbean hogfish, *Lachnolaimus maximus* (Labridae). Journal of Zoology (London), 213: 283–297.
Wainwright, P.C., and S.M. Reilly. 1994. Ecological Morphology: Integrative Organismal Biology. University of Chicago Press, Chicago.
Walton, B.M. 1993. Physiology and phylogeny: the evolution of locomotor energetics in hylid frogs. The American Naturalist, 141: 26–50.
Webster, A.J. and A. Purvis. 2001. Testing the accuracy of methods for reconstructing ancestral states of continuous characters. Proceedings of the Royal Society of London, Series B, 269: 143–149.
Westneat, M.W. 1995. Feeding, function, and phylogeny: Analysis of historical biomechanics in labrid fishes using comparative methods. Systematic Biology, 44: 361–383.

III

Roosting Ecology and Population Biology

Gary F. McCracken, Linda F. Lumsden, &
Thomas H. Kunz

Like all mammals, bats must find a place to live, mate, and raise their young. As recently as 30 years ago, it was thought that these basic needs of bats were met in a relatively simple fashion. A common perception of the time was that bats aggregated to satisfy their physiological needs, and that they formed large groups because there were only a limited number of suitable roosts available to satisfy these requirements. Because aggregations were forced by the need to stay warm, roosting associations and social interactions among individuals within groups, including mating and even the care of offspring, were believed to be largely passive and random. Our knowledge of bat population biology and social organization has expanded enormously since Peter Dwyer (1971) questioned this common perception in a prophetic essay entitled "Are bats socially conservative?" in which he predicted with regard to sociality in bats that "surely much more awaits discovery."

We now know that rather than being "socially conservative" the over 1100 species of extant bats present a diverse and exciting panoply of animals for studies of social evolution, roosting ecology, and population biology. Many exciting discoveries in this area of bat biology have been reviewed (Bradbury, 1977; Kunz, 1982; Kunz and Lumsden, 2003; Lewis, 1995; McCracken and Wilkinson, 2000; Burland and Worthington-Wilmer, 2001; Wilkinson and McCracken, 2003). Our goals in this introduction are to link the eight studies presented here with major themes in the earlier literature, and to focus on the advances and new discoveries that this new research provides.

Early studies on social organization in bats came from documenting nonrandom structure in roosting associations of bats of different age, sex, and reproductive condition. These "structural" associations suggested monogamy in a few cases, but more typically various forms of polygyny with mating systems based on harems, mating territories, and dominance hierarchies (e.g., Bradbury, 1977). Notwithstanding, the lifestyles of many species belied obvious structure and have continued to defy our abilities to document their behaviors. This is true for many bats, especially those temperate-zone species that are known primarily from maternity colonies and for which mating associations are essentially unknown, and for species that live and mate in mixed-sex colonies but also mate at alternative sites. Temperate and tropical species of bats that roost in foliage or tree cavities (Kunz and Lumsden, 2003) also confound description by frequently moving among roost sites in an apparently haphazard fashion. Six of the eight chapters in part III (chapters 10–13, 16 and 17) concern such challenging species.

In contrast to earlier studies, which have mostly concerned tropical bats (80 of 118 and 55 of 66 species reviewed by Bradbury (1977) and McCracken and Wilkinson (2000), respectively, are tropical), these same six studies involve bats that are exclusively temperate or temperate/tropical in their distributions. Because almost 75% of all bat species are found in the tropics, these earlier studies do not reflect a "bias" toward tropical bats. Rather the studies presented here illustrate that behavior and population structure of temperate species have increasingly been subjects of detailed study.

Notable, also, is that six of the eight chapters in part III (chapters 10–15) use molecular gene markers to describe social systems or population structure. As recently as 1998, when McCracken and Wilkinson (2000) completed a thorough review of mating systems in bats, there were fewer than 10 published studies in which molecular markers were employed. The studies presented here illustrate the increasing use of molecular assays in research on social evolution and population structure in bats, and demonstrate some of the finest such applications to date.

The chapters on *Plecotus auritus* (chapter 10), *Myotis bechsteinii* (chapter 11), and *Rhinolophus ferrumequinum* (chapter 12) couple field studies with analysis of genetic markers to describe not only the structure of social units, but also their function with regard to fitness and microgeographic population structure. All three species live in highly structured maternity colonies that are characterized by strong natal philopatry that is strict for females in *M. bechsteinii* (no marked females switched colonies), extremely strong for female *R. ferrumequinum*, and strong in both sexes for *P. auritus*, with only a few documented movements among hundreds of bats banded in the latter two species. As a consequence of strong philopatry, colonies of all three species consist of matrilineal groups, and striking structure among adjacent colonies of *M. bechsteinii* is demonstrated by marked variation in mitochondrial DNA haplotypes inherited through female lineages. However, despite natal philopatry, analysis of nuclear microsatellite variation shows that average relatedness among colony members is low, approaching zero, in all three species. This surprising result is attributed to gene flow due to male dispersal in *M. bechsteinii* and *R. ferrumequinum*, and because 80% of documented paternities in *P. auritus* result from matings with males outside the natal colony. The population consequences of these findings are that genetic structure among adjacent colonies is low, and there is no evidence that colonies are inbred as measured by nuclear gene markers. In addition to dissecting the fitness and population consequences of social systems, these studies also consider selective factors that act to shape social systems, including roost site use, roost site location, the location of foraging sites, and foraging associations among bats that roost together. Possibilities of information exchange at roost sites and babysitting of offspring are considered in chapter 11.

The chapters on *Cynopterus sphinx* (chapter 14), *Saccopteryx bilineata* (chapter 15), and *Tadarida brasiliensis* (chapter 15) also build on field data and the dissection of mating systems or analysis of population structure using molecular markers. Storz et al. in chapter 14 present a broader evolutionary and biogeographic analysis of the effects of social structure and demography on genetically effective population size (N_e), and consider implications for the importance of gene flow and genetic drift in structuring populations and affecting rates of morphological evolution. These authors demonstrate that social structure reduces N_e in *C. sphinx*, raising the possibility for an increased importance of genetic drift. However, analysis of population structure shows little effect of drift as values of standardized genetic variances (F_{ST}) among populations are low. While evidence exists for isolation by distance across a large area (peninsular India), effects of overlapping generations counter the reductions in N_e due to social structure, and the effects of drift appear to be mitigated by extensive gene flow. These authors suggest that such effects may be expected as the general pattern in bats. From this analysis, they conclude that substantial morphological variation in body size in *C. sphinx* across India is best explained by selection rather than the stochastic effects of genetic drift coupled with reduced gene flow.

Gene flow and effective population size also are the focus of Russell and McCracken's analysis of population structure in *T. brasiliensis* in chapter 13, and Storz et al.'s suggestion that genetic drift is likely to have little effect, and gene flow large effect, in bat populations is supported in the extreme by this study. *Tadarida brasiliensis* has one of the widest distributions of any bat species in the Western Hemisphere and is undoubtedly one of the most abundant mammals on earth, frequently living in colonies that exceed millions of individuals. Many earlier studies have described a host of differences among populations that suggest possible population structuring, with restriction in gene flow among populations, and perhaps the differentiation of populations into separate taxa. These differences among populations include hibernation versus migration, the possible use of different migratory routes, the use of different types of roosts, and differences in morphology. From sequence variation in the mitochondrial DNA control region, Russell and McCracken present a large-scale analysis of population structure that documents significant differences between North American and South American populations, but a striking absence of geographic structure among populations throughout the large range of this species in North America. The results of this study indicate that the several hundred million bats that occupy North America are members of one large panmictic population. Correspondingly, the genetically effective size of this North American population is estimated from molecular sequence diversity to exceed 200 million female bats, a number that is orders of magnitude larger than estimates similarly obtained for other mammalian species. The authors conclude that these bats exhibit substantial plasticity with regard to important behavioral traits, and that any recognition of taxonomic or population units based on these traits is unwarranted.

Working at a very different scale, Voigt et al. in chapter 15 provide new data and a new interpretation of the already well-studied social system of *Saccopteryx bilineata*. The authors integrate molecular data with detailed studies on the variability in display behaviors and the behavioral interactions of males and females to dissect the effects that these behaviors have on mating success and fitness. Harem size and the complexity of display calls by males positively influence their fitness, but males display less and attempt to copulate less with higher ranking females than with females of lower rank. Females of higher rank also roost in warmer locations, which enhances their fitness. Voigt et al. conclude that females are dominant in their encounters with males and that this is not a "classical" harem system because female choice plays the major role in determining male mating success. They also conclude that female choice is the driving selective force in the elaborate courtship displays in this species. Causal descriptions of the selective forces that shape social systems in bats and other mammals have long been based (perhaps "pigeonholed") on considerations of what males can economically defend (females, resources), with the role of female choice recieving little attention. This study is at the vanguard of appreciating the role of female choice in bats.

Several contributors (Burland et al., Kerth, Voigt et al.) consider physical and thermal attributes of the roost sites used by bats and how these attributes may affect fitness by influencing roost site selection, roost site location, and the location of foraging sites. These are topics of major focus in the chapters by O'Donnell and Sedgeley (chapter 17) and Lumsden and Bennett (chapter 16). We began this introduction by discounting the view that physiological needs act as contraints to the evolution of social complexity. However, this should not be equated with dismissing the possible importance of physiology in influencing the evolution of social complexity in bats. The chapters by O'Donnell and Sedgeley and Lumsden and Bennett present a sophisticated analysis of these issues. Both studies concern bats that typically roost in tree cavities and change roost sites frequently, often on a daily basis. O'Donnell and Sedgeley summarize their 9 years of detailed work on *Chalinolobus tuberculatus*, whereas Lumsden and Bennett summarize their own multi-year studies and those of several other researchers on *Nyctophilus geoffroyi*. Both studies demonstrate that thermal characteristics play important roles in the choice and use of roost sites, and that roost sites are selected with regard to fine features of thermal stability, optimal orientation, cavity size and cavity entrance. Lumsden and Bennett also emphasize the importance of roost site locations and opening dimensions in providing protection from predators. These studies document that choice of roost site, movements between roosts, and the associations among the individual bats using multiple roosts are anything but haphazard. Incredible fine-tuning of roost site choice in *C. tuberculatus* is documented. Colonies of this species move almost daily among hundreds of different roost sites, but were found to use the same sites within the same few days in subsequent years, apparently to capitalize on the seasonally optimal thermal characteristics of the available roosts.

These studies have important conservation implications by documenting that a network of diverse roost sites is essential to meet the needs of these and other forest-dwelling bats. Intact forest habitat is needed to provide this network, and fragmentation of habitat forces bats to roost in less optimal sites and to commute further from roost sites to foraging areas. O'Donnell and Sedgeley also document that bats roosting in larger native tree species with specific cavity dimensions have higher rates of reproduction and survival.

An emerging and important discovery in the last few years is that many bats live in "fission–fusion" societies that are nonetheless highly social and highly structured. In these systems, colonies are fragmented into numerous, smaller subgroups that intermingle and cycle among dozens or even hundreds of roosts, but colonies are nonetheless closed systems in which colony-mates interact with one another both in roosts and on foraging grounds, but not with other bats from other colonies. Using capture/recapture and radiotelemetry, O'Donnell and Sedgeley document separate fission–fusion colonies consisting of 29, 53, 72, 99, and 131 adult females and young of *C. tuberculatus*. Using the same techniques, Kerth in chapter 11 documents four fission–fusion colonies ranging from 15 to 45 adult females/colony in *Myotis bechsteinii*, and coupling this with mitochondrial DNA analysis documents that these colonies are, indeed closed, matrilineal groups. Kerth's study in Europe provides the greatest detail, to date, on the complexity of social interactions among fission–fusion colony members. O'Donnell and Sedgeley in New Zealand provide the greatest detail on roost site characteristics and the physiological factors that drive these complex social systems. Other examples of fission–fusion societies have recently been documented in bats in North America and in the Neotropics, and it appears that such unexpected systems are common and another example of the discoveries that await us.

A major commonality shared by these studies is that all have involved intensive fieldwork, either by the authors themselves or by previous researchers whose work allows the detailed interpretations presented. As examples, Rossiter et al.,'s work on *R. ferrumequinum* is built on continuous studies of the same colony that date back to the mid-1950s, Burland et al.'s research on *P. auritus* is based on studies begun 18 years previously, Kerth's work on *M. bechsteinii* is based on 12 years of continuous research, and O'Donnell and Sedgeley's studies of *C. tuberculatus* began in 1992. Russell and McCracken's study interprets extensive banding data from *T. brasiliensis* that date back to the 1950s, and detailed behavioral studies of *S. bilineata* (Voigt et al.) began in the late 1960s.

While research on population biology and social evolution in bats has lagged behind such work on other mammals (see chapter 11), this has been due to logistic and technical difficulties of studying long-lived, highly vagile, nocturnal animals with cryptic lifestyles. Through the diligence of researchers, the maturation of several long-term studies, and the development of new technologies (smaller and smaller radiotransmitters, passive integrated transponders (PIT-tags), night vision devices, infrared video, thermal imaging, and molecular techniques, to mention a few), bats are increasingly taking their place among the most exciting animals for studies of social evolution, roosting ecology, and population biology. The eight chapters in part III adequately illustrate that there is no inherent limitation to the richness of bats as systems for such studies.

LITERATURE CITED

Bradbury, J.W. 1977. Social organization and communication. Pp. 1–73. In: Biology of Bats, Vol. 3 (W.A. Wimsatt, ed.). Academic Press, New York.

Burland, T.M., and J. Worthington-Wilmer. 2001. Seeing in the dark: molecular approaches to the study of bat populations. Biological Reviews, 76: 389–409.

Dwyer, P.D. 1971. Are bats socially conservative? Fauna 1: 31–35.

Kunz, T.H. 1982. Roosting ecology of bats. Pp. 1–55. In: Ecology of Bats (T.H. Kunz, ed.). Plenum Press, New York.

Kunz, T.H., and L.F. Lumsden. 2003. Ecology of cavity and foliage roosting bats. Pp. 3–89. In: Bat Ecology (T.H. Kunz and M.B. Fenton, eds.). University of Chicago Press, Chicago.

Lewis, S.E. 1995. Roost fidelity in bats: a review. Journal of Mammalogy, 76: 481–496.

McCracken, G.F., and G.S. Wilkinson. 2000. Bat mating systems. Pp. 321–362. In: Reproductive Biology of Bats (E.G. Crichton and P.H. Krutzsch, eds.). Academic Press, New York.

Wilkinson, G.S., and G.F. McCracken. 2003. Bats and balls: sexual selection, sperm competition, and female choice in bats. Pp. 128–155. In: Bat Ecology (T.H. Kunz and M.B. Fenton, eds.). University of Chicago Press, Chicago.

10

Social and Population Structure in the Brown Long-Eared Bat, *Plecotus auritus*

Tamsin M. Burland, Abigail C. Entwistle, & Paul A. Racey

The brown long-eared bat (*Plecotus auritus*) is a Palaearctic species with low wing loading and low aspect ratio, which feeds primarily by gleaning. Like most temperate-zone bat species, *P. auritus* forms colonies during summer. However, this species is unusual in that adult males are present within these colonies. Using both ecological and genetic techniques, the pattern of social organization and population structure was determined for this species close to the northern border of its distribution. A ringing study conducted at 30 summer roost sites determined that *P. auritus* formed small, mixed-sex colonies, with natal recruitment and long-term association to the colony in both sexes. Little exchange among roosts was identified; just six movements were recorded from 1138 recaptures and no individual was caught for a second time at a new site. *P. auritus* was found to be highly selective of its summer roost site at this latitude, which may explain the high natal philopatry and colony stability identified. Mean colony relatedness was low. This is probably a result of low skew in male reproductive success and a high incidence of inter-colony mating; over 80% of colony offspring were fathered by males from other summer colonies. Such a pattern of mating suggests that neither inbreeding avoidance nor competition for mates is likely to force males to disperse from their natal colony. The strong spatial distribution of individuals in the summer population was not reflected in patterns of genetic structure. Low levels of genetic differentiation among colonies were identified, probably as a result of inter-colony mating. However, genetic isolation by distance was evident within the population, suggesting that mating among colonies is restricted to those located close by. It is proposed that, at high latitudes, social organization and population structure in *P. auritus* are strongly influenced both by the climate and by the species' wing morphology.

INTRODUCTION

An accurate understanding of a species' behavior and ecology requires detailed knowledge of its social organization and population structure.

For the purposes of this chapter, social organization refers to the extent to which individuals are partitioned into social groups, group size, sexual composition and levels of kinship within the group, patterns of dispersal among groups and the species' mating system (after Wilson, 1975). In turn, population structure describes the level of variation in the spatial distribution and genetic composition of individuals (Hewitt and Butlin, 1997), and thus can be highly influenced by social organization.

The study of social organization and population structure is best achieved by combining direct (ecological) and indirect (genetic) techniques. Ecological studies provide basic compositional data and an estimation of the behavior and distribution of individuals (Begon et al., 1996; Krebs and Davies, 1987). However, important dispersal events can be missed (Slatkin, 1987) and estimates of reproductive success or genetic relatedness based on direct observation alone may be inaccurate and/or unfeasible (Hughes, 1998). By contrast, genetic studies are particularly effective in identifying parentage and relatedness. However, patterns of genetic variation within a population can be influenced by a variety of contemporary and historical factors (Hewitt and Butlin, 1997; Slatkin, 1987), which, in the absence of data from more direct ecological studies, may bias estimates of dispersal, gene flow, paternity or relatedness. In this chapter, we describe how the combined use of ecological and genetic techniques has facilitated the study of social organization and population structure in a Palaearctic bat species.

The Study Species

The brown long-eared bat (*Plecotus auritus*, Linnaeus 1758) is a small vespertilionid (5–10 g) and one of up to four *Plecotus* species found in Europe (Kiefer et al., 2002). The genus is characterized by long ears and short broad wings of low aspect ratio and low wing loading. This facilitates high maneuverability and hovering associated with a gleaning feeding strategy (Anderson and Racey, 1991) but results in slow and energetically expensive flight (Norberg, 1976).

P. auritus has a distributional range of approximately 40–64° N (Swift, 1998) and displays a seasonal cycle of behavior and reproduction typical of temperate-zone bat species (Gustafson, 1979; Oxberry, 1979). Mating and estrous are initiated in autumn; however vaginal plugs are not formed and mating continues periodically throughout winter (Racey, 1979; Strelkov, 1962). Ovulation and fertilization take place in the spring. During summer, bats congregate to form summer colonies and parturition occurs in July in the United Kingdom. Colonies are typically located in tree cavities, bird and bat boxes, and in buildings, where they often cluster in the roof apex and are easily visible. *P. auritus* is unusual among temperate-zone bat species in that adult males are also present in the summer colonies (Boyd and Stebbings, 1989; Heise and Schmidt, 1988; Park et al., 1998; Stebbings, 1966).

By October, few bats are visible at the summer roost sites and the specific whereabouts of individuals during autumn and winter, the primary mating season, is unclear. However, Strelkov (1969) classified *P. auritus* as the most stationary of all European species, with bats remaining close to (<20 km), or within, their summer roost site year-round.

Project Background

Over the past 20 years various ecological (Entwistle et al., 1996, 1997, 2000; Swift and Racey, 1983; Speakman et al., 1991), behavioral (Rydell et al., 1996), energetic (Racey et al., 1987; Speakman and Racey, 1987), physiological (Hays et al., 1992; Webb et al., 1995), and genetic studies (Burland et al., 1999, 2001) have been conducted on a population of *P. auritus* in northeast Scotland. This region lies at 57° N, close to the northern border of the species' distribution, and summer colonies have been identified only in buildings (Entwistle et al., 1997; Swift, 1998).

The overall aims of these studies related to determining the interaction of this species with its environment at high latitudes and to answering questions regarding the conservation management of this species. In the past decade, studies on this population have specifically aimed to identify the following: (1) the pattern of use of roost sites and whether *P. auritus* uses buildings within the region at random; (2) the flight capacity of this species, in terms of foraging and winter migration distances; (3) colony size and composition, focusing on levels of natal philopatry, colony relatedness, and the status of adult males; (4) mating patterns; (5) population structure, in particular the spatial distribution of individuals and extent of gene flow among colonies.

In this chapter we review and synthesize the findings of these more recent studies, and, in doing so, provide a detailed perspective on the causes and consequences of social organization and population structure in *P. auritus* at the northern border of its distribution. We suggest future avenues of research, both for the genus *Plecotus* and across the Chiroptera as a whole. The work has also led to specific conservation recommendations for *P. auritus*.

REVIEW OF STUDIES CONDUCTED ON *P. AURITUS* IN NORTHEAST SCOTLAND

Study Area

The study area was the Grampian and Highland regions of northeast Scotland, located at approximately 57° N. Within this area the primary region of study was Deeside, the valley of the River Dee, which flows east from the Cairngorm Mountains into the North Sea at the city of Aberdeen (figure 10.1). *P. auritus* roosts in buildings along the length of this river valley (excluding Aberdeen itself). Colonies in two other regions, one located along the valley of the River Spey to the west of the Cairngorm Mountains and the other around the village of Kirkmichael, located south of Glenshee, were also included in many of the studies.

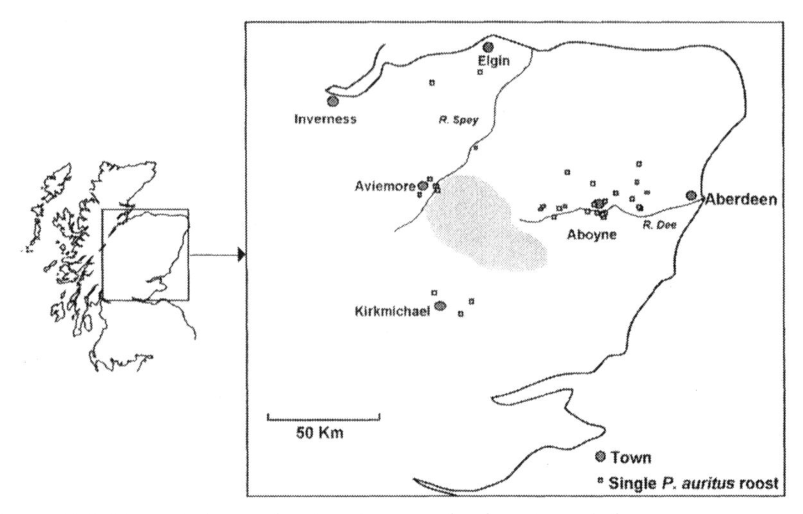

Figure 10.1 Map of the study site showing the location of the major towns and rivers, and the principal *Plecotus auritus* roosts used in the studies. (Reproduced from Burland et al., 1999.)

Roost Visits, Ringing, and Sampling Procedure

The study of *P. auritus* in northeast Scotland was initiated in 1978, since when approximately 45 *P. auritus* roost sites have been identified, with a ringing program conducted under license at approximately 30. Between 1991 and 1996, a more intensive colony-visiting schedule was employed, with individual roosts checked up to five times each summer (Burland, 1998; Entwistle et al., 2000).

Visits were conducted between April and October, and bats were captured under license by hand or static net. Sex and reproductive status were determined following Racey (1974) for females and Entwistle et al. (1998) for males, and individuals were marked using 3.0 mm aluminum bands (Mammal Society, London). During the 18 years of study, 1739 individual bats were banded and 1138 recaptures made (table 10.1).

Between 1994 and 1996, the individuals captured were sampled under license using wing biopsy (Worthington Wilmer and Barratt, 1996) to obtain tissue for genetic analysis. These individuals were genotyped using six highly polymorphic *P. auritus* microsatellite markers as described in Burland et al. (1998).

Roost and Foraging Habitat Selectivity

A detailed investigation of *P. auritus* summer roost site selectivity (Entwistle et al., 1997) demonstrated that this species is highly discriminatory. When compared with adjacent buildings, *P. auritus* roosts were older, contain more roof compartments, and were more likely to be fully lined with wood. Furthermore, the roof space was significantly warmer than in buildings not used as roosts.

Table 10.1 Summary of Bat Captures at Roosts within Northeast Scotland Between 1978 and 1996

	Total Number of Bat Captures	**Number of Bats Ringed**	**Number of Recaptures**	**Number of Bats Released Banded**
Males	1210	611	519	80
Females	1513	838	605	70
Young	402	290	14	98
Total	3125	1739	1138	248

Banded bats that were released include newborn and others sampled for the genetic study at roosts which were not included in the ringing studies.

The habitat surrounding the building also appeared to be an important factor in roost selection. Roosts were located closer to woodland and water than a random sample of buildings and had a greater total area of woodland within a radius of 0.5 km. No difference in woodland area was found at greater distances. Radiotracking demonstrated that woodland close to the roost was the primary foraging habitat, with individuals, particularly females, spending the greatest amount of time foraging within 0.5 km of the roost. The maximum recorded distances traveled during foraging were small, at 2.2 and 2.8 km for females and males, respectively (Entwistle et al., 1996). Bats did not cross-matrix habitats, but instead used field boundaries, hedgerows, tree lines, and other connective features of the landscape to move between foraging sites. The small flight distances and reluctance to fly in open air are both likely to be a result of the wing morphology of this species, which prevents efficient long-distance or fast flight.

P. auritus displays high levels of fidelity to summer roost sites, possibly as a result of this selectivity, with bats visible on 67% of roost visits (Entwistle et al., 2000). The true occupancy rate may be even higher, as bats regularly make use of cavities inaccessible to researchers (Entwistle et al., 2000). This is supported by radiotracking studies which showed that bats were frequently in the roost even when they were not visible within the roof space (Entwistle, 1994). However, males and nonreproductive females may also use cooler, alternative roost sites located close to the main summer roost, possibly to facilitate the use of torpor (Entwistle et al., 1997).

Colony Size

The size of *P. auritus* colonies was estimated both from the number of bats counted within the roost on each visit and from mark–release–recapture data. A large discrepancy was recorded between these methods. While the mean number of bats counted on roost visits was approximately 10–20 (Burland, 1998; Entwistle et al., 2000; Speakman et al., 1991), mark–recapture data revealed that actual colony size could be two to three times greater than this (30–50 bats; Entwistle et al., 2000). In addition, genetic data have suggested that, except in the most intensively studied colonies (bats captured on more than two occasions each year), estimates based on mark–recapture data may still underestimate colony numbers by 20–30 individuals (Burland et al., 2001).

Even when the possibility that some individuals remain undetected is taken into account, *P. auritus* colony size appears smaller than those for other sympatric bat species (Speakman et al., 1991). Entwistle et al. (2000) suggest that small colony size is indicative of the small distances traveled during foraging, which, in turn, limit the total foraging area available to each colony.

Colony Composition

The long-term ringing data revealed that each *P. auritus* summer roost site could be considered to house a discrete colony, composed of adults of both sex and young of the year (Entwistle et al., 2000). Across years, adults were extremely loyal to their roost, with females captured in the same site over periods of up to 13 years (Entwistle, 1994) and males up to 12 years (Burland, 1998).

There was little exchange of individuals of either sex or any age class, even among colonies located less than 1 km apart. Just six movements were detected in the 1138 recaptures (0.5% of recaptures; Burland, 1998) and no individual was caught for a second time at a new site. There was no evidence to suggest frequent immigration of unmarked adults and the proportion of marked bats in intensively studied colonies rapidly reached an asymptote close to 1 (figure 10.2).

Annual survival rates of adults were similar for the two sexes (0.651 ± 0.073 (SEM) and 0.698 ± 0.082 for females and males, respectively; Entwistle, 1994). Using these rates, Entwistle (1994) estimated average life expectancy to be 2.33 and 2.78 years for females and males, respectively. However, the annual survival figures calculated included data from years where capture effort was low, thus mortality rates may have been overestimated (Entwistle, 1994). These values may therefore be regarded as minimum estimates of survival and life expectancy.

Natal recruitment into the colony occurred in both sexes (Burland, 1998; Entwistle et al., 2000), although the overall levels of recruitment were low. Of 91 females banded during the summer of their birth, 39 were recaptured in the same roost in subsequent years. Similarly, just 21 of 97 males were subsequently caught. With little evidence for juvenile dispersal (see above), such low levels of recruitment may reflect a juvenile mortality of up to 68% in the first year (Burland, 1998). However, as the time intervals between capture of individuals can be up to 9 years (Entwistle et al., 2000), some individuals may survive and be present in the colony but evade recapture.

Mating Patterns, Colony Relatedness, and Population Structure

The ringing data have revealed that *P. auritus* forms discrete and stable colonies during summer, with individuals exhibiting strong philopatry to particular roost sites over many years. Moreover, most individuals that

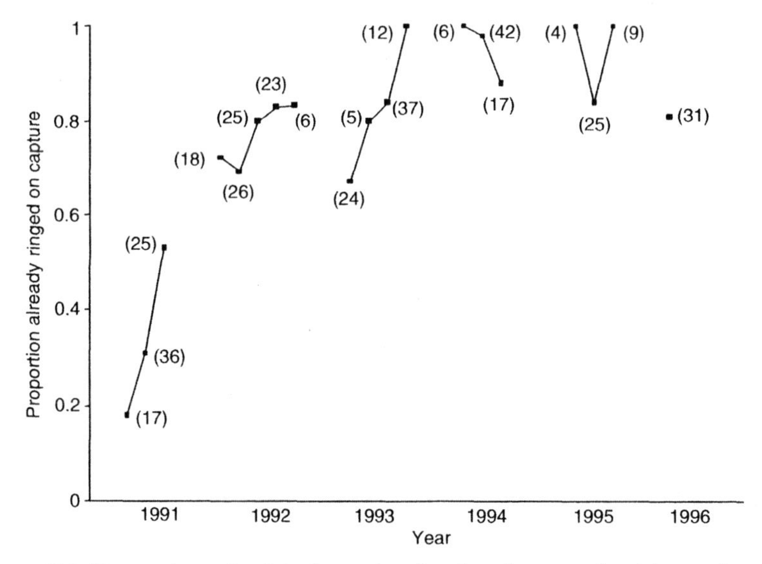

Figure 10.2 Proportion of adult bats already ringed on each visit at the most intensively studied roost between 1991 and 1996. Sample sizes for each visit are in parentheses. (Reproduced from Burland, 1998.)

survive their first year appear to remain within their natal colony. The spatial distribution of individuals can therefore be described as highly substructured (Entwistle et al., 2000) and it may be predicted that individuals from the same colony are closely related. However, as mating extends throughout winter, a time when few individuals can be located, mating patterns, colony relatedness, and genetic population structure cannot be inferred from these ringing data. Instead, genetic data generated for approximately 650 individuals were used to determine these factors (Burland et al., 1999, 2001).

Analysis of parentage determined that adult females and their offspring were always found in the same colony, supporting the inference from ringing studies that females do not move between colonies. By contrast, less than 20% of offspring were fathered by males originating from the same colony. Thus, summer colonies are not closed mating groups and individuals from different colonies appear to mix, at least during the mating season (Burland et al., 2001). The study identified few paternal half siblings within years, suggesting individual males are unable to monopolize reproduction, possibly due to the extended period of estrous and mating.

Mean colony relatedness (R) was very low (overall mean colony $R = 0.033 \pm 0.002$), and, while the distribution of pairwise relatedness estimates was extremely wide (-0.4 to 0.9), the majority of individuals within each colony were unrelated or only distantly related to one another ($R < 0.25$). Correspondingly, levels of genetic differentiation among colonies were also extremely low (mean $F_{ST} = 0.02$; Burland et al., 1999). Such low levels of relatedness and genetic differentiation are likely to result from a combination of factors (Burland et al., 2001) including high frequency of inter-colony mating, low skew in male reproductive success, high juvenile mortality rates, and low adult life expectancy. However, the mean measure of genetic differentiation among colonies was significantly greater than zero, possibly reflecting natal philopatry (Burland et al., 1999).

Despite the high levels of gene flow among colonies predicted, the identification of genetic isolation by distance within the population (Burland et al., 1999) suggests that gene flow does not equate to a model of panmixia. Instead, the direct exchange of genes appears to be restricted and is likely to occur only among neighboring colonies (Burland et al., 1999).

SOCIAL ORGANIZATION AND POPULATION STRUCTURE

The past decade has witnessed a significant accumulation of knowledge about *P. auritus*, much of which has stemmed from studies conducted in northeast Scotland. Our present understanding of social organization and population structure in this species is diagrammatically represented in figure 10.3. Possible causes and consequences are also detailed. We suggest that while many interconnected factors appear to affect the ecology and behavior of *P. auritus*, two factors, namely the temperate-zone climate and wing morphology, have a fundamental influence (figure 10.3).

During summer, *P. auritus* aggregates at traditional roost sites to form discrete mixed-sex colonies, where both sexes display natal philopatry and long-term association with the colony. The aggregation of individuals is essential to provide an appropriate thermal environment for growth and survival of offspring within the small window of time available between birth

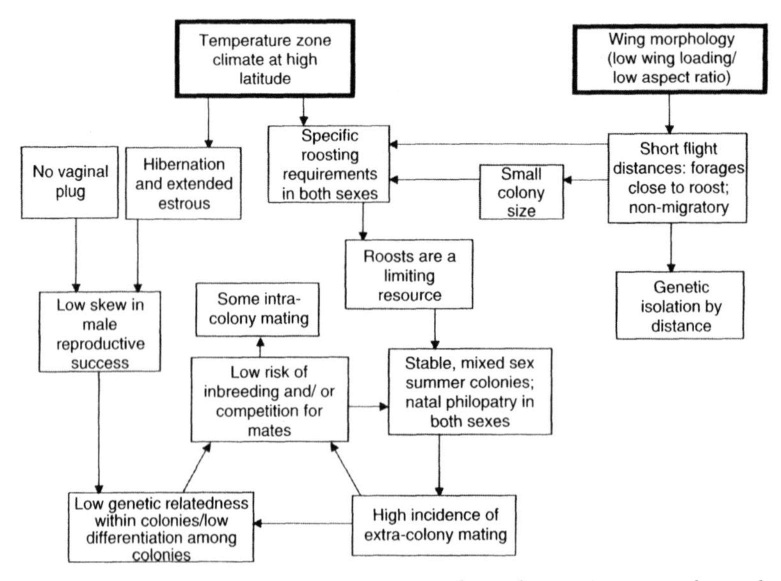

Figure 10.3 Possible causes and consequences of social organization and population structure in *Plecotus auritus* at the northern borders of its distribution.

and the decline of insect food supply in autumn (McNab, 1982). As such, the underlying influence of climate in the formation of bat colonies at temperate-zone latitudes is clear. However, these thermal benefits fail to explain the presence of adult males within the colony, the long-term loyalty of adults to a colony, or natal recruitment in both sexes. Further explanations are therefore necessary, and these may be found when ecology and behavior of this species are also considered.

P. auritus is highly selective of its summer roost site. The finding that roosts are warmer and located closer to woodland than unoccupied buildings, suggests that this selectivity is driven both by thermal constraints and by wing morphology, as follows: (1) the thermal costs of a small colony size (which may be a product of wing morphology) could be offset by the preference for warmer roost sites, especially at high latitudes, and (2) this species chooses roosting sites either close to, or within, suitable foraging habitat, possibly reflecting its inability to fly efficiently for long distances.

If, as a result of high selectivity, roost sites are a limiting resource for *P. auritus*, the pattern of natal philopatry and colony stability identified may be a result of the advantages accrued through use of traditional sites known to be previously successful. However, this does not explain the lack of movement between neighboring, well-established *P. auritus* colonies. The low levels of genetic relatedness identified within colonies suggest kin selection is not an adequate alternative explanation. Instead, Burland et al. (2001) propose that high colony stability may have facilitated the evolution of social behavior through reciprocal altruism, as previously suggested for bats by Wilkinson (1987, 1992a,b), Kerth and König (1999), and Kerth et al. (2000). However, this has not been investigated in *P. auritus*.

The presence of adult males within summer colonies, a feature unusual among temperate-zone bat species, may result from an incompatibility between torpor and spermatogenesis (Entwistle et al., 1998, 2000). This prediction is supported by the positive relationship identified between latitude and numbers of adult males within *P. auritus* colonies (Entwistle et al., 2000). By joining summer colonies, adult males may also benefit other colony members. Males exploit foraging sites that are more distant from the roost than females (Entwistle et al., 1996), and thus their presence may provide the thermal benefits of an increased colony size without incurring a cost of increased competition for resources (Entwistle et al., 2000).

It is unusual in animal species for natal recruitment and long-term association with social groups to be evident in both sexes (Greenwood, 1980). In the case of mammalian species, it is frequently the males who disperse, as a result of inbreeding avoidance and/or competition for mates (Dobson, 1982; Moore and Ali, 1984). However, the mating behavior and genetic composition of *P. auritus* colonies identified by the Scottish studies may explain why no systematic dispersal has been observed in either sex (Burland et al., 2001). First, natal dispersal as a mechanism to avoid inbreeding appears unnecessary, since the majority of fertilizations occur among individuals from different summer colonies. Second, intra-colony mating is unlikely to carry with it a high risk of inbreeding, as most individuals within a colony are unrelated, or only distantly related. Third, competition for mates among males from the same summer colony is unlikely to be high as skew in male reproductive success is low and the incidence of extra-colony mating high.

Despite the high levels of spatial organization in the summer population identified by ringing studies, the occurrence of extra-colony mating identified by genetic studies suggests mixing of individuals from different colonies must occur, at least during the mating season. With little evidence to suggest that individuals enter other summer roost sites, such contact is likely to take place at alternative transient roost sites or at hibernacula, where active males may mate with torpid females. The indication that some full siblings may be present within colonies (Burland et al., 2001) suggests that *P. auritus* may be loyal to these mating sites between years.

Identification of genetic isolation by distance within a single population strongly supports the classification of *P. auritus* as nonmigratory and suggests that dispersal distances in autumn and winter are unlikely to differ greatly from those identified during summer foraging. Alternative roosting sites, including hibernacula, may therefore be located extremely close to the summer roost site (< 1 km). The occurrence of isolation by distance, which is probably a product of limited flight capacity due to wing morphology, also reduces the likelihood that the low recruitment rates observed in juveniles, or the failure to identify high levels of movement among colonies, can be explained by long-distance dispersal.

COMPARISONS WITH OTHER SPECIES

Data available from other bat species certainly support the predicted link between wing morphology, colony size, patterns of migration, and genetic population structure described here for *P. auritus*. Enwtistle et al. (2000) found migratory species had significantly higher aspect ratio than sedentary species and demonstrated a strongly significant positive relationship between wing aspect ratio and colony size across 13 bat species. More specifically, they contrasted *P. auritus* with *Tadarida brasiliensis*, a migratory species with high aspect ratio, which uses fast-flying aerial foraging (Norberg and Rayner, 1987). This species forages extensive distances from its roost (up to 65 km), is characterized by its large colonies (≫1000 individuals, Davis et al., 1962) and shows no pattern of genetic isolation by distance across its entire sampled range (Burland and Worthington Wilmer, 2001). In fact, isolation by distance has yet to be reported in any migratory bat species (Burland and Worthington Wilmer, 2001). Whether wing morphology has a fundamental influence on social and population structure across the order Chiroptera remains to be determined and further studies are required to elucidate its role (Burland and Worthington Wilmer, 2001; Entwistle et al., 2000).

P. auritus is unusual among bat species for the amount of in-depth ecological, behavioral, and genetic data available for the same population (but see chapters 11 and 12). However, as similar data are collected on a range of bat species, it may be possible to extend figure 10.3 into a predictive model that can identify key variables that determine social organization and population structure in a range of bat species.

FUTURE RESEARCH ON *PLECOTUS*

At present, little is known about the behavior of *P. auritus* outside the summer months. It would therefore be useful to collect further data on its mating and hibernation sites. The pivotal role played by climate and wing morphology in determining social organization and population structure in *P. auritus* (figure 10.3) could be tested by conducting similar studies across the species' range. Investigations could also be conducted on the sibling species *P. austriacus*, preferably at sites shared by the two species. The extent of macrogeographic genetic structuring and response to changing landscape use should also be assessed in both *Plecotus* species to determine whether their limited flight capacity makes them vulnerable to habitat fragmentation.

CONSERVATION IMPLICATIONS

The case of *P. auritus* elegantly illustrates the need for an understanding of the ecology and behavior of a species to ensure appropriate assumptions are made regarding management decisions. Prior to these studies, little was known regarding roost site preferences, foraging habitat, patterns of colony formation and persistence, mating patterns or gene flow in this species. It was therefore unclear how *P. auritus* might respond to roost exclusion, roost/habitat destruction or habitat fragmentation. These studies have therefore markedly improved our understanding of the conservation requirements for this species.

At the initiation of this research, legislation in the United Kingdom had identified the roost as the focus for bat protection (under the Wildlife and Countryside Act 1981), although exclusion of bat colonies remained an option under advice from the appropriate government agency (Racey, 2000). The findings of these studies support a conservation strategy directed at *P. auritus* summer roost sites and suggest that roost exclusions may be highly detrimental. In particular, bats are unlikely to join adjacent colonies following exclusion, and instead may be forced to use alternative roosts of a lower "quality."

However, the research reported here, and similar studies of other species, have demonstrated a need to shift the conservation agenda to also consider factors beyond the roost site. In the case of *P. auritus*, the summer roost, the surrounding foraging habitat, and possible alternative roosting and hibernation sites, are all essential to maintain the reproductive potential of the colony and ensure its long-term survival (Entwistle, 1994; Entwistle et al., 1996, 1997). In addition, there is a clear need to consider how the increasing fragmentation and loss of connectivity in the countryside can be mitigated to avoid loss of access to foraging areas. Maintenance of habitat continuity over a wider scale is also essential to retain genetic diversity and to prevent genetic isolation of individual *P. auritus* colonies.

ACKNOWLEDGMENTS

We are extremely grateful for the help and hospitality of the householders who allowed us access to the bat roosts. Roost visits, ringing, and tissue sampling were conducted under licenses from Scottish Natural Heritage and the UK Home Office. The work described was undertaken by A.C.E. and T.M.B. (under supervision from P.A.R.) while Ph.D. students at the Zoology Department, University of Aberdeen; both were supported by NERC studentships. The genetic analysis was conducted by T.M.B. at the Institute of Zoology, Zoological Society of London, under the supervision of Elizabeth Barratt. A.C.E. was co-supervised by John Speakman. The participation of T.M.B. in the Roosting Ecology and Population Ecology Symposium was funded by the Royal Society, University of Aberdeen and Queen Mary, University of London. The manuscript was improved by helpful comments from Gary McCracken, Jerry Wilkinson, and an anonymous referee.

LITERATURE CITED

Anderson, M.E., and P.A. Racey. 1991. Feeding behavior of captive brown long-eared bats *Plecotus auritus*. Animal Behaviour, 42: 489–493.

Boyd, I.L., and R.E. Stebbings. 1989. Population changes of brown long-eared bats (*Plecotus auritus*) in bat boxes at Thetford Forest. Journal of Applied Ecology, 26: 101–112.

Begon, M., J.L. Harper, and C.R. Townsend. 1996. Ecology: Individuals, Populations, and Communities, 3rd edition. Blackwell Science, Boston.

Burland, T.M. 1998. Social Organization and Population Structure in the Brown Long-eared Bat, *Plecotus auritus*. Ph.D. Dissertation, University of Aberdeen, Aberdeen.

Burland, T.M., and J. Worthington Wilmer. 2001. Seeing in the dark: molecular approaches to the study of bat populations. Biological Reviews, 76: 389–409.

Burland, T.M., E.M. Barratt, and P.A. Racey. 1998. Isolation and characterization of microsatellite loci in the brown long-eared bat, *Plecotus auritus*, and cross species amplification within the family Vespertilionidae. Molecular Ecology, 7: 136–138.

Burland, T.M., E.M. Barratt, M.A. Beaumont, and P.A. Racey. 1999. Population genetic structure and gene flow in a gleaning bat, *Plecotus auritus*. Proceedings of the Royal Society of London, Series B, 266: 975–980.

Burland, T.M., E.M. Barratt, R.A. Nichols, and P.A. Racey. 2001. Mating patterns, relatedness and the basis of natal philopatry in the brown long-eared bat, *Plecotus auritus*. Molecular Ecology, 10: 1309–1321.

Davis, R.B., C.F. Herreid, II, and H.C. Short. 1962. Mexican free-tailed bats in Texas. Ecological Monographs, 32: 311–346.

Dobson, F.S. 1982. Competition for mates and predominant juvenile male dispersal in mammals. Animal Behaviour, 30: 1183–1192.

Entwistle, A.C. 1994. Roost Ecology of the Brown Long-eared Bat (*Plecotus auritus*, Linnaeus 1758) in North-east Scotland. Ph.D. Dissertation, University of Aberdeen, Aberdeen.

Entwistle, A.C., P.A. Racey, and J.R. Speakman. 1996. Habitat exploitation by a gleaning bat, *Plecotus auritus*. Philosophical Transactions of the Royal Society of London, Series B, 351: 921–931.

Entwistle, A.C., P.A. Racey, and J.R. Speakman. 1997. Roost selection by the brown long-eared bat *Plecotus auritus*. Journal of Applied Ecology, 34: 399–408.

Entwistle, A.C., P.A. Racey, and J.R. Speakman. 1998. The reproductive cycle and determination of sexual maturity in male brown long-eared bats, *Plecotus auritus* (Chiroptera: Vespertilionidae). Journal of Zoology (London), 244: 63–70.

Entwistle, A.C., P.A. Racey, and J.R. Speakman. 2000. Social and population structure of a gleaning bat, *Plecotus auritus*. Journal of Zoology (London), 252: 11–17.

Greenwood, P.J. 1980. Mating systems, philopatry and dispersal in birds and mammals. Animal Behavior, 28: 1140–1162.

Gustafson, A.W. 1979. Male reproductive patterns in hibernating bats. Journal of Reproduction and Fertility, 56: 317–331.

Hays, G.C., J.R. Speakman, and P.I. Webb. 1992. Why do brown long-eared bats (*Plecotus auritus*) fly in winter? Physiological Zoology, 65: 554–567.

Heise, G., and A. Schmidt. 1988. A contribution to the social organization of the long-eared bat (*Plecotus auritus*). Nyctalus, 2: 445–465.

Hewitt, G.M., and R.K. Butlin. 1997. Causes and consequences of population structure. Pp. 203–227. In: Behavioural Ecology: An Evolutionary Approach (J.R. Krebs and N.B. Davies, eds.), 4th edition. Blackwell Scientific, Oxford.

Hughes, C. 1998. Integrating molecular techniques with field methods in studies of social behavior: a revolution results. Ecology, 79: 383–399.

Kerth, G., and B. König. 1999. Fission, fusion and nonrandom associations in female Bechstein's bats (*Myotis bechsteinii*). Behaviour, 136: 1187–1202.

Kerth, G., F. Mayer, and B. König. 2000. Mitochondrial DNA (mtDNA) reveals that female Bechstein's bats live in closed societies. Molecular Ecology, 9: 793–800.

Kiefer, A., F. Mayer, J. Kosuch, O. von Helversen, and A. Veith. 2002. Conflicting molecular phylogenies of European long-eared bats (*Plecotus*) can be explained by cryptic diversity. Molecular Phylogenetics and Evolution, 25: 557–566.

Krebs, J.R. and N.B. Davies (eds.). 1987. An Introduction to Behavioural Ecology, 2nd edition. Blackwell Scientific, Oxford.

McNab, B.K. 1982. Evolutionary alternatives in the physiological ecology of bats. Pp. 151–200. In: Ecology of Bats (T.H. Kunz, ed.). Plenum Press, New York.

Moore J., and R. Ali. 1984. Are dispersal and inbreeding avoidance related? Animal Behavior, 32: 94–112.

Norberg, U.M. 1976. Aerodynamics, kinematics and energetics of horizontal flapping flight in the long-eared bat, *Plecotus auritus*. Journal of Experimental Biology, 65: 179–212.

Norberg, U.M., and J.M.V. Rayner. 1987. Ecological morphology and flight in bats (Mammalia: Chiroptera): wing adaptations, flight performance, foraging strategy and echolocation. Philosophical Transactions of the Royal Society of London, Series B, 316: 335–427.

Oxberry, B.A. 1979. Female reproductive patterns in hibernating bats. Journal of Reproduction and Fertility, 56: 359–367.

Park, K.J., E. Masters, and J.D. Altringham. 1998. Social structure of three sympatric bat species (Vespertilionidae). Proceedings of the Royal Society of London, Series B, 244: 379–389.

Racey, P.A. 1974. Ageing and assessment of reproductive status of pipistrelle bats *Pipistrellus pipistrellus* (Mammalia: Chiroptera). Journal of Zoology (London), 173: 264–271.

Racey, P.A. 1979. The prolonged storage and survival of spermatozoa in Chiroptera. Journal of Reproduction and Fertility, 56: 391–402.

Racey, P.A. 2000. Does legislation conserve and does research drive policy? The case of bats in the UK. Pp. 159–173. In: Priorities for the Conservation of Mammalian Diversity: Has the Panda had its Day? (A.C. Entwistle and N. Dunstone, eds.). Cambridge University Press, Cambridge.

Racey, P.A., J.R. Speakman, and S.M. Swift. 1987. Reproductive adaptations of heterothermic bats at the northern borders of their distribution. South African Journal of Science, 83: 635–638.

Rydell, J., A.C. Entwistle, and P.A. Racey. 1996. Timing of foraging flights in three species of bats in relation to insect activity. Oikos, 76: 243–252.

Slatkin, M. 1987. Gene flow and the geographic structure of natural populations. Science, 236: 787–792.

Speakman, J.R., and P.A. Racey. 1987. The energetics of pregnancy and lactation in the brown long-eared bat *Plecotus auritus*. PP. 367–393. In: Recent Advances in the Study of Bats (M.B. Fenton, P. Racey, and J.M.V. Rayner, eds.). Cambridge University Press, Cambridge.

Speakman, J.R., P.A. Racey, C.M. Catto, P.I. Webb, S.M. Swift, and A.M. Burnett. 1991. Minimum summer populations and densities of bats in NE Scotland, near the northern borders of their distributions. Journal of Zoology (London), 225: 327–345.

Stebbings, R.E. 1966. A population study of the Genus *Plecotus*. Journal of Zoology (London), 150: 53–75.

Strelkov, P.P. 1962. The peculiarities of reproduction in bats (Vespertilionidae) near the northern border of their distribution. Pp. 306–311. In: International Symposium on Methods in Mammalogical Investigation, Brno (J. Kratochvil, ed.). Publishing House of the Czechoslovak Academy of Sciences, Prague.

Strelkov, P.P. 1969. Migratory and stationary bats (Chiroptera) of the European part of the Soviet Union. Acta Zoologica Cracoviensia, 14: 393–440.

Swift, S.M. 1998. Long-Eared Bats. T. and A.D. Poyser, London.

Swift, S.M., and P.A. Racey. 1983. Resource partitioning in two species of bats inhabiting the same roost. Journal of Zoology (London), 200: 249–259.

Webb, P.I., J.R. Speakman, and P.A. Racey. 1995. Evaporative water loss in two sympatric species of vespertilionid bats, *Plecotus auritus* and *Myotis daubentonii*: relation to foraging mode and implications for roost site selection. Journal of Zoology (London), 235: 269–278.

Wilkinson, G.S. 1987. Altruism and co-operation in bats. Pp. 299–323. In: Recent Advances in the Study of Bats (M.B. Fenton, P. Racey, and J.M.V. Rayner, eds.). Cambridge University Press, Cambridge.

Wilkinson, G.S. 1992a. Communal nursing in the evening bat, *Nycticeius humeralis*. Behavioral Ecology and Sociobiology, 31: 225–235.

Wilkinson, G.S. 1992b. Information transfer at evening bat colonies. Animal Behaviour, 44: 501–518.

Wilson, E.O. 1975. Sociobiology: The New Synthesis. Harvard University Press, Cambridge, MA.

Worthington Wilmer, J.M., and E.M. Barratt. 1996. A non-lethal method of tissue sampling for genetic studies of chiropterans. Bat Research News, 37: 1–3.

11

Relatedness, Life History, and Social Behavior in the Long-Lived Bechstein's Bat, *Myotis bechsteinii*

Gerald Kerth

Since 1993, I have studied the social behavior and genetic population structure of the communally breeding Bechstein's bat, *Myotis bechsteinii*, to analyze the adaptive value of female sociality. Maternity colonies of *M. bechsteinii* normally consist of 15–45 females and their young; males are solitary. Individual females do not give birth every year, weaning on average *c.* 0.7 young per year. Thus, colonies are comprised of reproductive and nonreproductive bats. Low annual individual reproductive success is balanced by a high survival rate of about 80% per year for adult females. Genetic population analysis, using mitochondrial and nuclear markers, revealed strict female philopatry and male dispersal. Maternity colonies are closed societies, comprised of closely related and genetically unrelated females that probably live together for their entire lives. Within colonies, females display a fission–fusion society. Colonies regularly split into several subgroups that occupy different roosts. Individuals switch roosts frequently and subgroup composition is characterized by strong mixing of colony members. The social organization of maternity colonies suggests that the long-lived females have familiar, cooperative partners, some of which are close relatives, for raising their young. Colony members probably profit from sociality via warming by nonreproductive females, babysitting, and information transfer about roosts, whereas allonursing and information transfer about food are of little or no importance.

INTRODUCTION

Despite their tremendous diversity in ecology and natural history, most of the approximately 1100 bat species are social (Nowak, 1994). Size and composition of groups differ between species (Bradbury, 1977), but generally females form groups. This is especially the case for European bats where females always breed communally. Although sociality is widespread among female bats, relatively little is known about the adaptive value of maternity colonies. Bats have been rather neglected in studies of sociobiology, probably because their behavior and reproductive success are difficult to investigate in the field due to their cryptic lifestyle and long lives. Modern technology and molecular methods, however, now provide opportunities for studying aspects of bat biology that previously had been largely inaccessible (for a review see Burland and Worthington Wilmer, 2001). Notwithstanding, the underrepresentation of bats in studies of behavioral ecology persists. Among 560 papers published in 2000 in the journals *Animal Behaviour*, *Behaviour*, *Behavioral Ecology*, *Behavioral Ecology and Sociobiology*, and *Ethology*, 121 dealt with mammals, only four of which focused on bats. This lack of research on bats, despite their frequency among mammals (*c.* 20% of all species), hampers our understanding of sociality on a representative taxonomic background.

The existing knowledge about bat sociality points to several factors that influence group formation. However, the available data are generally inadequate to decide which factor is most important, even within a given species. Most bats depend on daily refuges against weather or predators, and with few exceptions, they cannot build roosts (Kunz, 1982; Kunz and Lumsden, 2003). Roost limitation could therefore "force" bats to aggregate ("ecological constraints": Emlen, 1994; see Burland et al., 2001, for an example). By contrast, group-living bats might gain benefits such as better thermoregulation, decreased predation risk, or benefits due to cooperation among group members (reviewed in Emlen, 1994; Krebs and Davies, 1993). Social thermoregulation is often used to explain the high frequency of sociality among female temperate bats (Neuweiler, 1993). Bats might also be social because of reduced predation risk (Fenton et al., 1994). Although social behavior has rarely been studied in bats, a variety of cooperative behaviors have been described. The striking reciprocal food sharing among adult vampire bats (*Desmodus rotundus*; Wilkinson, 1984) probably results from their unusual natural history. Other forms of cooperation, however, are probably more common. These include mutual warming of pups (*Antrozous pallidus*; Trune and Slobodchikoff, 1978), baby-sitting (*Myotis thysanodes*; O'Farrell and Studier, 1973), as well as communal nursing and information transfer about food or roosts (*Nycticeius humeralis*; Wilkinson, 1992a,b). Thus, cooperation with group members may be an important intrinsic benefit of group living in some species.

The different possible causes of sociality in bats lead to different predictions concerning behavior and population structure (Kerth, 1998; Kerth et al., 2000). If females aggregate to decrease predation risk or to improve thermoregulation, group size is important, but not necessarily group composition. This should result in a flexible individual composition within colonies and in low, if any, genetic structuring among them, as there is no a priori reason to expect that metabolic benefits or predator avoidance leads to stable groups (McCracken and Bradbury, 1981). Female natal philopatry, on the other hand, would result in related individuals staying together, generating maternally structured populations. Natal philopatry can be caused

Table 11.1 Predictions Concerning the Genetic Population Structure and the Social Behaviour of Bats Depending on Different Causes That May Lead to Sociality in Bats

Potential Causes of Sociality	**Individual Group Composition**	**Genetic Population Structure**	**Complex Social Interactions**
Ecological constraints	Unimportant	Structured	Unimportant
Social thermo-regulation	Unimportant	Nonstructured	Unimportant
Predator avoidance (dilution effect)	Unimportant	Nonstructured	Unimportant
Cooperation with colony members	Important	Structured or nonstructured	Important

by habitat limitation by, for example, limited roosts or foraging areas, benefits of staying such as familiarity with the local habitat, high dispersal costs, or combinations of these. Once females aggregate with relatives, this potentially facilitates the evolution of cooperation among colony members, resulting in even more cohesiveness and persistency of social groups, and in more complex social behaviors. By studying the genetic structure among and within colonies, and the complexity and function of social interactions including cooperative behavior, one can ascertain which factors best explain sociality in female bats (table 11.1).

Beyond its use for understanding group living in bats, a detailed sociogenetic and behavioral study of free-living bats will also provide insights into the evolution of sociality in other animals. To date, most of the available studies that have investigated the costs and benefits of living in kin-groups, examined species with one dominant breeder (nuclear families; Emlen, 1997). To broaden our understanding of family groupings, Emlen (1997) suggests that future vertebrate studies should focus on species living in kin-groups where members share in reproduction. Bats that live in kin-structured colonies fulfill these criteria. Studying the genetics and behavior of bats, therefore, promotes our understanding of the factors that shape animal societies with shared reproduction.

Since 1993, my coworkers and I have studied the behavioral ecology and sociogenetics of Bechstein's bats to understand why females are social. *Myotis bechsteinii* is monotocous, long-lived (up to 21 years) medium-sized (8–12 g), and nonmigratory. Although widespread, it is rare in Europe (Schober and Grimmberger, 1997). However, in the deciduous forests of Frankonia, Germany, where most of our field sites are located, colonies occur at high densities. During summer, Bechstein's bats roost in tree cavities, and in bird and bat boxes. Mothers breed communally in colonies that consist of 15–45 adult females; males are solitary (Kerth, 1998).

Our approach to study sociality in Bechstein's bats uses a combination of behavioral and genetic data, in concert with field experiments that manipulate social behavior. We aim to understand the evolution of complex social behavior as a function of relatedness among several breeding females. We hope that our data on Bechstein's bats will improve the general understanding of extended families, and allow us to evaluate models of reproductive sharing in animal societies (e.g., Clutton-Brock, 1998). Finally, our data on habitat use, individual survival, reproductive success, gene flow, and group dynamics provide a basis for designing and implementing conservation strategies for the endangered Bechstein's bat.

Our current knowledge of the behavior of Bechstein's bat is based on field observations of individually marked bats. One colony has been studied since 1993, and three others since 1996. All colony members carry implanted PIT-tags that allow us to identify individuals in a roost (mostly bat boxes) without handling. For behavioral observations, all bat boxes in the area of our main study colony were modified for automatic infrared video monitoring. Females of this colony are also marked with colored plastic rings. Every year, between the end of April, when the bats arrive from unknown hibernacula, until October, when they leave again, we monitored all boxes at least three times a week for the presence of bats. This resulted in 6283 individual "recaptures" in four colonies over 5 years (1996–2000). Furthermore, we repeatedly captured bats (2–4 times per year) to take body measures, check reproductive status, and mark juveniles (Kerth, 1998; Kerth and König, 1999).

In addition, we analyzed the genetic structure of 10 colonies that ranged from 1 to about 60 km apart. We collected individual wing-tissue samples from which we extracted DNA. To evaluate maternal relatedness we used four markers within the mitochondrial DNA. Additionally, we used up to 11 nuclear microsatellites to estimate individual relatedness within and differentiation among colonies (Kerth et al., 2000, 2002a, 2002b).

LIFE HISTORY

I determined survival and reproductive success of adult females present in four colonies in 1996–2000. The mean number of adult females per colony was 19.6. Colonies fluctuated in size (see Kerth et al., 2002a, for details on yearly fluctuations and the age structure in the four colonies). Two colonies decreased: one by 20% (from 20 to 16 females) and the other by 43% (from 28 to 16 females). One colony remained constant (21 females) and one colony increased by 25% (from 28 to 35 females). The total number of bats declined from 97 to 88. Mean annual survival rate, based on recaptures, was 81.6% (range 57–100%). Forty of the females present in 1996 were still alive in spring 2001. Survival rate over 5 years was considerably higher for 18 females born in 1995 compared with 79 females that were already older than 1 year in 1996 (61% versus 37%). For 28 females, I determined the reproductive status every year between 1996 and 1999.

They weaned 2.8 ± 0.9 (mean $\pm$ SD) young in 4 years (range 1–4). This result corresponds closely to the mean annual number of young (0.67 ± 0.23) determined for 51 females over 3–5 years between 1996 and 2000.

GENETIC POPULATION STRUCTURE

Microgeographic Differentiation among Colonies

The population structure of *M. bechsteinii* differs strikingly for mitochondrial (mtDNA) and nuclear DNA (nDNA). Based on mtDNA, colonies consist of one, maximally two matrilines, most colonies being clearly distinguished by unique haplotypes (Kerth et al., 2000). Low intra-colony variability and strong haplotype segregation among colonies are reflected by extremely high F_{ST} values (up to 0.96). This suggests that females are strictly philopatric, a conclusion that is supported by colony pedigrees (see below) and field data (Kerth et al., 2000; 2002a; 2002b). Haplotype distribution among 18 solitary males living close to one colony showed that males disperse between colony sites, revealing the absence of dispersal barriers (Kerth et al., 2000). In contrast, nuclear microsatellites revealed a very weak, though significant, population differentiation ($F_{ST} = 0.015$; Kerth et al., 2002b).

Individual Relatedness within Colonies

We determined individual genealogy in four complete maternity colonies living in close proximity, based on genetic maternity exclusions in combination with knowledge about age and colony membership of marked bats (Kerth et al., 2002a). Although colonies fluctuated in size, no immigration occurred over 5 years (6283 individual readings on 530 control days), and 75% of the 88 colony members present in 2000 lived together with close relatives ($r \geq 0.25$). Nevertheless, mean colony relatedness was nearly zero (0.02).

Genetic Colony Structure and Sex-Specific Dispersal

Mean colony relatedness in Bechstein's bats is low despite strict female philopatry. Colonies consist of closely related females living together with individuals that, based on their nDNA, are genetically unrelated, although they are born in the same colony and often belong to the same matriline (according to their mtDNA type). Such a population structure indicates that considerable gene flow among colonies is mediated via males. The noticeable different population structures observed for mtDNA ($F_{ST} = 0.66$–0.96) and nDNA ($F_{ST} = 0.015$) microsatellites confirm sex-biased dispersal and that females mate with males born in foreign colonies. Computer simulations suggest that when male dispersal is strong (possibly complete), effective colony size on average exceeds 21 females, and the sex ratio is even (as observed), the documented weak nDNA population differentiation is possible even for absolute female philopatry (Kerth et al., 2002b).

SOCIAL BEHAVIOR

Roosting Associations

Bechstein's bat colonies frequently split into subgroups that occupy different day roosts. We monitored the roost occupation of individually marked females (PIT-tags) belonging to one colony during 3 years (Kerth and König, 1999). Over 399 control days, the colony split in two or three subgroups on average. Despite marked mixing, individual subgroup composition was nonrandom. Bats associated according to their reproductive status, with lactating females preferentially roosting together. Genetic similarity, based on nuclear and mitochondrial markers, had no consistent influence on the degree of association. Combining three association indices suggested that nonrandom associations occur even in the absence of shared roost and group size preferences (Kerth and König, 1999).

High associations among reproducing females may reflect the importance of cooperation among mothers. Significantly lower associations among females of different reproductive status, which still roost together frequently, suggest that nonreproductive bats may follow a mixed strategy. They may compromise between cooperating with mothers, for example, by increasing subgroup size and thereby roost temperature, and optimizing their own energetic needs, by temporarily preferring lower roost temperatures in smaller subgroups or different roosts to become torpid during the day. Differences in individual energetic costs and benefits of communal roosting for reproductive and nonreproductive females, however, remain to be demonstrated. If they exist, communal roosting of nonreproductive and reproductive females could indicate "altruistic warming" if such behavior benefits mothers and babies but involves costs to nonreproducing bats. Because the reproductive status of individuals changes between years and females live together for many years, possible "altruistic warming" by nonreproductive females could be stabilized via reciprocity.

Social Interactions

Nightly infrared video monitoring of occupied bat boxes revealed that females regularly allogroom each other and often have face to face contact (Kerth, 1998). We quantified social interactions among eight lactating females during 15 nights (each female was observed for 150–360 minutes; Kerth et al., 2003). One hundred and sixty-four social interactions could not be clearly categorized, but we observed 136 face contacts (mean $\pm$ SD duration: 4.9 ± 2.8 seconds) and 44 allogrooming events (20.2 ± 19.2 seconds). Eighty-seven percent of the face contacts occurred between an incoming and a resident female within 3 minutes of arrival. In addition to the potential role of allogrooming in parasite load reduction (Wilkinson, 1985), both allogrooming and face contacts may also function in colony mate recognition.

Behavioral confrontation tests showed that female Bechstein's bats could detect conspecifics belonging to foreign colonies, and behaved aggressively toward them (Kerth et al., 2002b).

Allonursing

To evaluate whether females nurse foreign young, I ringed thirty-one 20-day-old juveniles that had reached a forearm length of 32 mm. At this age, juveniles take their first flights, on average being nursed for another 2 weeks (Kerth, 1998). Using 78 nightly nursing events, I could assign 17 marked juveniles to lactating females. All observed mother–offspring pairs were genetically confirmed using 9–11 nuclear microsatellites. From these results, there was no indication of allonursing for the marked juveniles. However, I sometimes observed juveniles that seemed to attempt to steal milk and, on one occasion, two different females consecutively nursed one unmarked juvenile. In conclusion, allonursing occurs rarely in *M. bechsteinii* and is therefore unlikely to be of crucial benefit for communal breeding in this species.

Information Transfer about Food

Sociality in bats has further been explained by the argument that colonies serve as information centers where females transfer knowledge about food (Wilkinson, 1992b). To test whether information transfer occurs in Bechstein's bats, we used radiotelemetry to study the nightly habitat use of 10 adult females belonging to one colony, during 2 years (Kerth et al., 2001b). We expected that if colony members regularly exchange information about food, then they should often forage together and move in pairs or groups among different feeding areas. However, all bats revisited their own foraging area over several nights. Five females repeatedly radiotracked at different seasons maintained their individual hunting areas even between years. Because females were very loyal to individual foraging areas that were typically substantial distances from each other, information transfer about food is unlikely to be the crucial factor promoting sociality in *M. bechsteinii*.

Information Transfer about Roosts

In contrast to most other mammals that breed in shelters, female Bechstein's bats, like many other bat species (Lewis, 1995), frequently switch day roosts during one breeding season (Kerth, 1998; Kerth and König, 1999). Thus, they must often decide where to spend the day. Selecting the right roost is important, because roost quality, especially microclimatic condition, influences survival and reproduction in bats (reviewed in Altringham, 1996). In a field experiment, we allowed marked females to choose between relatively warm versus cold bat boxes, while controlling for site preferences. Roost occupancy suggested that roost selection was based on roost temperature directly (Kerth et al., 2001a). Frequent switching between roosts that differ in quality, and expected costs of finding and testing new roosts, suggest that information transfer among colony members could facilitate knowledge about suitable roosts.

In order to test this hypothesis, I conducted a second experiment, providing bats with pairs of suitable and unsuitable roosts. Our bat boxes have two entrances in sequence; blocking the interior entrance creates an "unsuitable" roost, because a bat can still enter the box but cannot roost inside. Nine pairs of boxes with one box randomly manipulated as "unsuitable" were placed in the home range of one colony. Each box was equipped with an automatic reading device allowing for continuous monitoring. Subsequently, during a 10-week period following the lactation period, all pairs of boxes were monitored for the arrival of bats marked with PIT-tags. In two box pairs, only males were recorded. The remaining seven pairs were discovered by females within 1–4 weeks. Because the blocked box within a pair was equally likely to be entered by the first bat that found the respective pair as an unblocked box, our roost manipulation (blocking) proved not to be detectable from the outside. Subsequently, however, new bats arrived in unblocked boxes significantly more often than in blocked ones (figure 11.1). These preliminary experimental data support the hypothesis that colony members exchange information about suitable roosts. If bats located roosts independently, the arrival rate of "new" individuals should have been equal at both types of boxes.

This experiment is ongoing. We expect that a larger sample size will provide us with a detailed insight into the arrival pattern of individuals

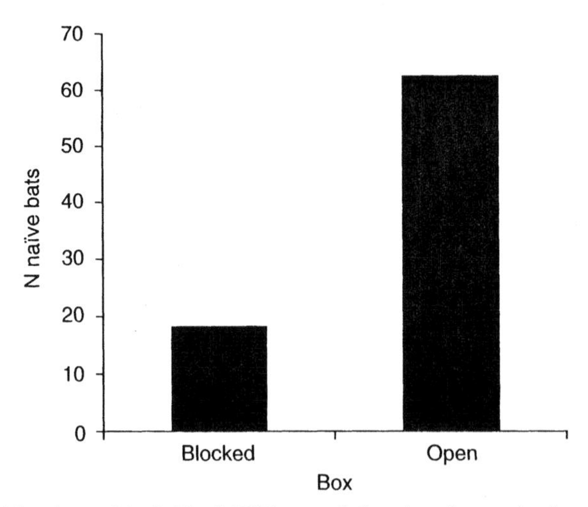

Figure 11.1 Number of individual PIT-tagged females that arrived at blocked and unblocked (open) bat boxes after their initial discovery by a colony member (see text for details). Within six box pairs at which "new" females were recorded after boxes had been discovered by a first bat, unblocked boxes were visited by a significantly higher number of females than blocked boxes (Wilcoxon matched pair test: $n=6$, $Z=-1.99$, $P=0.023$, one-tailed).

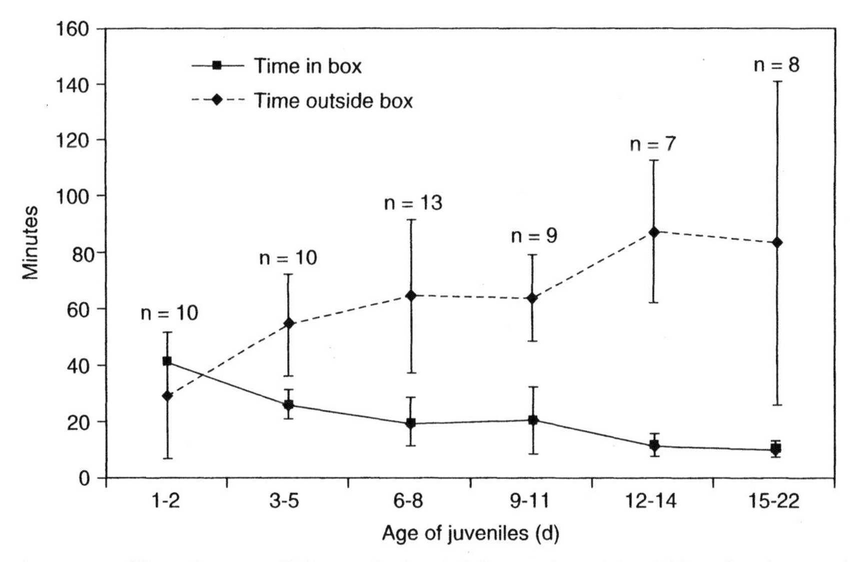

Figure 11.2 Time (mean ± SD) per single nightly nursing visit within a bat box, and per single foraging trip, as a function of the age of the juveniles. A total of 19 females were observed during the lactation period in 1994. Because it was impossible to observe all 19 individuals each night, numbers of females observed (indicated by *n*) differ between age classes.

in new roosts, allowing us to evaluate whether colony members reciprocate in exchanging information, or whether certain individuals serve as scouts. Furthermore, we hope to understand the mechanism by which individuals transfer information about roosts (for example because of local enhancement).

Babysitting

Lactating females return to a roost several times at night to nurse their offspring. When pups were less than 3 days old, mothers spent about 45 minutes per stay, foraging for about 30 minutes in between nursing bouts. As juveniles grew, mothers decreased time per single visit (figure 11.2). Although an individual mother spends plenty of time together with her newborn offspring, a pup of a solitary female would be alone for a significant proportion of the night. However, female Bechstein's bats breed communally and there was a negative correlation between time alone (pups without any adult) and group size (figure 11.3). This correlation, however, was only significant when the young were less than 10 days old, that is, when they were, presumably, still heterothermic. Moreover, when I removed all cases with only one pup present in a box, there was no longer a significant correlation between time alone and group size (Spearman $R = -0.03$, $n = 21$, $P = 0.91$). If females warm juveniles, the offspring may benefit from the nearly permanent attendance of adults, even if they do not get milk from

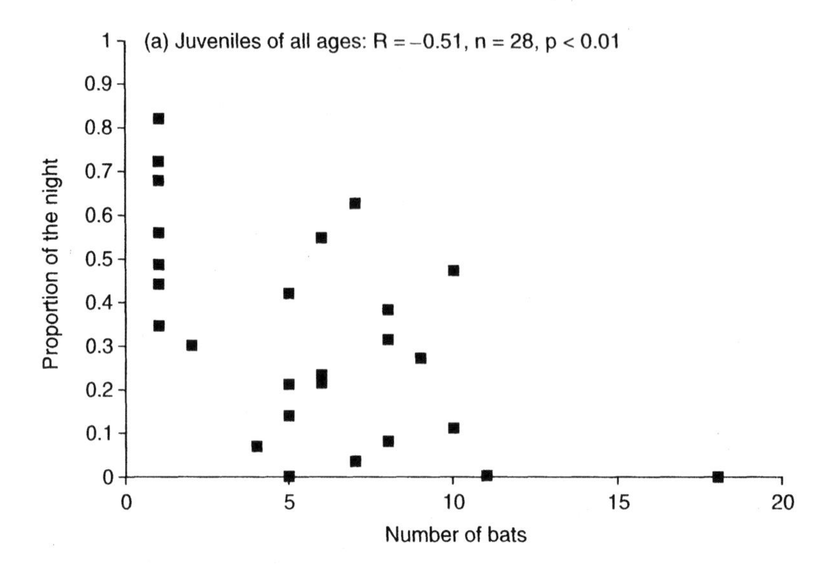

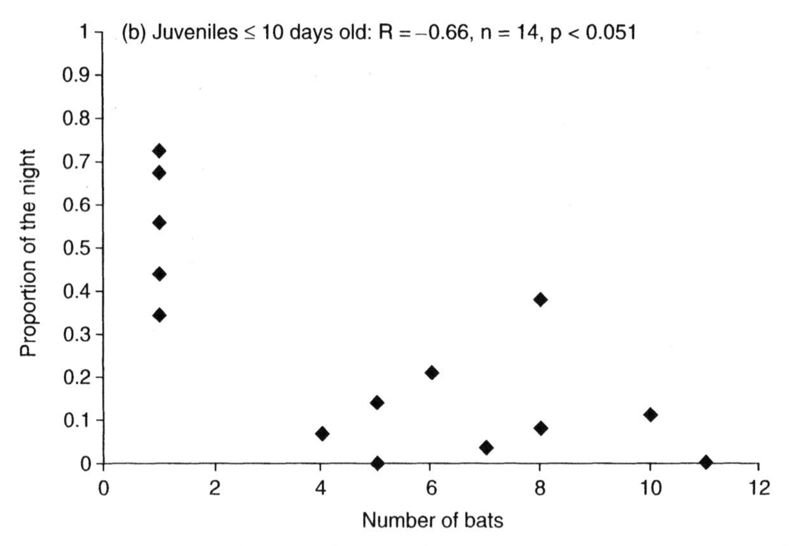

Figure 11.3 Proportion of the night juveniles were left alone (without an adult female) as a function of the number of young present in a box. Spearman's correlation coefficients are given (a) for a total of 28 observation nights in 1994 and 1995, and (b) and (c) for 14 nights each, after dividing the dataset on the basis of age of young.

foreign mothers. It is well documented for bats that juvenile growth depends on the temperature inside a roost (reviewed in Altringham, 1996). In conclusion, pups may benefit energetically from not being alone. Above a threshold of a few bats, increasing group sizes, however, seem to

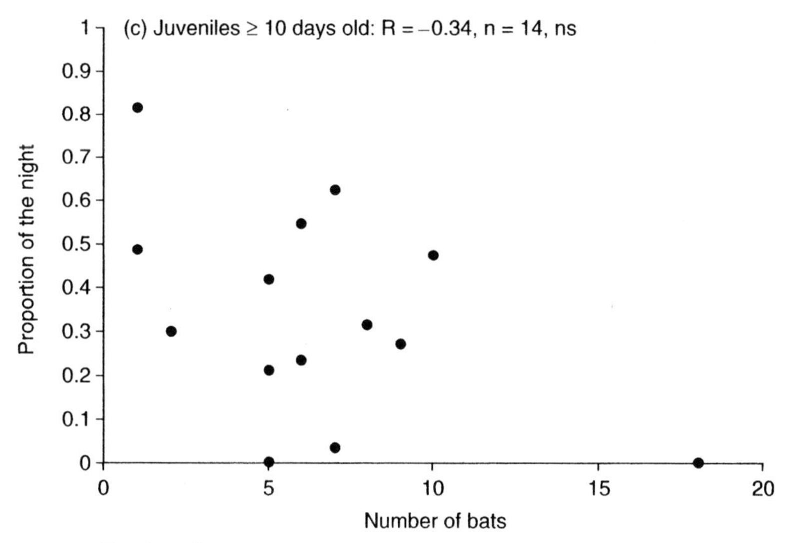

Figure 11.3 Continued.

have no further effect. Babysitting among reproductive females that provide heating during the night might also explain the nonrandom association pattern found in day roosts.

CONCLUSIONS

Our genetic and behavioral data both revealed that female Bechstein's bats do not move between colonies, even where colonies are in close proximity to each other, and no barriers to dispersal exist (Kerth et al., 2000). Such strong colony fidelity might reflect limited roost availability. However, if individual roosts were a limited resource, one would expect high roost fidelity. Instead females often switch roost boxes (Kerth and König, 1999), suggesting that they are adapted to an environment where natural roosts are not very rare. Roost limitation, therefore, cannot sufficiently explain strict female natal philopatry and, as a consequence, group living in Bechstein's bats. However, natal philopatry could be a side effect of the importance of profound habitat or roost knowledge for successful reproduction in females (Kerth et al., 2001a,b). Because colonies comprise long-lived females, they could serve as centers for communal knowledge about local resources (e.g., roosts). Knowledge of a communal colony, therefore, might be an important benefit of sociality in long-lived, philopatric bat species. The absolute social seclusion of Bechstein's bat colonies also suggests that individual group composition is more important than group size (Kerth et al., 2000). Thus, it seems unlikely that colonies mainly serve thermoregulatory or antipredatory functions that benefit all colony members equally at the same time. Only if social thermo-regulation or antipredatory behaviors are costly for some individuals and

Table 11.2 Predictions Concerning Group Composition and Social Behavior of Bats, as a Function of Different Causes of Sociality, and the Results Found for Female Bechstein's Bats (*Myotis bechsteinii*)

Potential Causes of Sociality	Group Composition		Social Interactions		Result
	Expected	Observed	Expected	Observed	
Ecological constraints	Stable	Stable	Not complex	Complex	Not supported
Social thermo-regulation	Flexible	Stable	Not complex	Complex	Not supported
Predator avoidance (dilution effect)	Flexible	Stable	Not complex	Complex	Not supported
Cooperation with colony members	Stable	Stable	Complex	Complex	Supported

require altruism could they explain the existence of closed societies, in which kin selection or reciprocity among long-term associated colony members could stabilize altruistic behavior.

The existence of complex social behavior (allogrooming, face contacts, and individual roosting associations), as well as assumed benefits of three cooperative behaviors (information transfer about roosts, babysitting, and probably altruistic warming), support the idea that females form stable colonies to live with familiar, cooperative partners (table 11.2). Since colony members vary strongly in their degree of relatedness (Kerth et al., 2002a), and because relatives do not preferentially associate, kin selection per se is unlikely to explain the social system in Bechstein's bats (compare Burland et al., 2001; McCracken and Bradbury, 1981; Wilkinson, 1985). However, if Bechstein's bats can discriminate among colony members depending on their relatedness, cooperation can, in part, still be favored by kin selection even in the absence of significant levels of mean colony relatedness (Kerth et al., 2002a).

Our data suggest that *M. bechsteinii* is adapted to a stable forest habitat. Because of strict female philopatry, colonies are demographically independent (Kerth et al., 2000). Thus, conservation plans should consider maternity colonies as separate reproductive units and protect them by maintaining deciduous forest that provides many roosts and suitable foraging areas. To preserve this endangered species, forest management should avoid large clear cuttings, and conserve nonfragmented mature deciduous forests that allow females to breed successfully in a predictable habitat, and which also enable males to disperse freely between colony sites.

ACKNOWLEDGMENTS

Cooperation by numerous coworkers was crucial for me to succeed during 9 years of studying Bechstein's bats. I cordially thank all of them for their essential help in gathering and analyzing the data presented. To David Hosken, Barbara König, Gary McCracken, and two anonymous referees I am grateful for helpful comments on the manuscript, and additionally to Barbara König for her constant and invaluable support. Our work during the past years has been supported by the German Science Foundation, the German Federal Agency of Nature Conservation, and the Swiss National Science Foundation.

LITERATURE CITED

Altringham, J.D. 1996. Bats: Biology and Behaviour. Oxford University Press, Oxford.

Bradbury, J.W. 1977. Social organization and communication. Pp. 1–72. In: Biology of Bats, vol. 3 (W. A. Wimsatt, ed.). Academic Press, New York.

Burland, T.M., and J. Worthington Wilmer. 2001. Seeing in the dark: molecular approaches to the study of bat populations. Biological Reviews, 76: 389–409.

Burland, T.M., E.M. Barratt, R.A. Nichols, and P.A. Racey. 2001. Mating pattern, relatedness and the basis of natal philopatry in the brown long-eared bat *Plecotus auritus*. Molecular Ecology, 10: 1309–1321.

Clutton-Brock, T.H. 1998. Reproductive skew, concessions and incomplete control. Trends in Ecology and Evolution, 7: 282–285.

Emlen, S.T. 1994. Benefits, constraints and the evolution of the family. Trends in Ecology and Evolution, 9: 282–285.

Emlen, S.T. 1997. Predicting family dynamics in social vertebrates. Pp. 228–253. In: Behavioural Ecology. An Evolutionary Approach (J.R. Krebs and N.B. Davies, eds.), 4th edition. Blackwell Science, Oxford.

Fenton, M.B., I.L Rautenbach, S.E. Smith, C.M. Swanepoel, J. Grosell, and J. van Jaarsveld. 1994. Raptors and bats: threats and opportunities. Animal Behaviour, 48: 9–18.

Kerth, G. 1998. Sozialverhalten und genetische Populations-struktur bei der Bechsteinfledermaus (*Myotis bechsteinii*). Wissenschaft und Technik, Berlin.

Kerth, G., and B. König. 1996. Transponders and an infrared videocamera as methods in a field study on the social behaviour of Bechstein's bats (*Myotis bechsteinii*). Myotis, 34: 27–34.

Kerth, G., and B. König. 1999. Fission, fusion and nonrandom associations in female Bechstein's bats (*Myotis bechsteinii*). Behaviour, 136: 1187–1202.

Kerth, G., F. Mayer, and B. König. 2000. Mitochondrial DNA (mtDNA) reveals that female Bechstein's bats live in closed societies. Molecular Ecology, 9: 793–800.

Kerth, G., K. Weissmann, and B. König. 2001a. Day roost selection in female Bechstein's bats (*Myotis bechsteinii*): a field experiment to determine the influence of roost temperature. Oecologia, 126: 1–9.

Kerth, G., M. Wagner, and B. König. 2001b. Roosting together, foraging apart: information transfer about food is unlikely to explain sociality in female Bechstein's bats (*Myotis bechsteinii*). Behavioral Ecology and Sociobiology, 50: 283–291.

Kerth, G., K. Safi, and B. König. 2002a. Mean colony relatedness is a poor predictor of colony structure and female philopatry in the communally breeding Bechstein's bat (*Myotis bechsteinii*). Behavioral Ecology and Sociobiology, 52: 203–210.

Kerth, G., F. Mayer, and E. Petit. 2002b. Extreme sex-biased dispersal and inbreeding avoidance in the communally breeding, nonmigratory Bechstein's bat (*Myotis bechsteinii*). Molecular Ecology, 11: 1491–1498.

Kerth, G., B. Almasi, N. Ribi, D. Thiel, and S. Lüpold. 2003. Social interactions among wild female Bechstein's bats (*Myotis bechsteinii*) living in a maternity colony. Acta Ethologica, 5: 107–114.

Krebs, J.R., and N.B. Davies. 1993. An Introduction to Behavioural Ecology, 3rd edition. Blackwell Scientific, Oxford.

Kunz, T.H. 1982. Roosting ecology of bats. Pp. 1–55. In: Ecology of Bats (T.H. Kunz, ed.). Plenum Press, New York.

Kunz, T.H., and L.F. Lumsden. 2003. Ecology of cavity and foliage roosting bats. Pp. 3–89. In: Bat Ecology (T.H. Kunz and M.B. Fenton, eds.). University of Chicago Press, Chicago.

Lewis, S.E. 1995. Roost fidelity of bats: a review. Journal of Mammalogy, 76: 481–496.

McCracken, G.F., and J.W. Bradbury. 1981. Social organization and kinship in the polygynous bat *Phylostomus hastatus*. Behavioral Ecology and Sociobiology, 8: 1–34.

Neuweiler, G. 1993. Biologie der Fledermäuse. Georg Thieme, Stuttgart.

Nowak, R.M. 1994. Walker's Bats of the World. Johns Hopkins University Press, Baltimore.

O'Farrell, M.J., and E.H. Studier. 1973. Reproduction, growth, and development in *Myotis thysanodes* and *M. lucifugus* (Chiroptera: Vespertilionidae). Ecology, 54: 18–30.

Schober, W., and E. Grimmberger. 1997. The Bats of Europe and North America. T.H.F. Publications, Neptune, NJ.

Trune, D.R., and C.N. Slobodchikoff. 1978. Position of immatures in pallid bat clusters: a case of reciprocal altruism. Journal of Mammalogy, 58: 469–478.

Wilkinson, G.S. 1984. Reciprocal food sharing in vampire bats. Nature, 308: 181–184.

Wilkinson, G.S. 1985. The social organization of the common vampire bat (*Desmodus rotundus*). I. Pattern and cause of association. Behavioral Ecology and Sociobiology, 17: 111–122.

Wilkinson, G.S. 1992a. Information transfer at evening bat colonies. Animal Behaviour, 44: 501–518.

Wilkinson, G.S. 1992b. Communal nursing in the evening bat, *Nycticeius humeralis*. Behavioral Ecology and Sociobiology, 31: 225–335.

12

Causes and Consequences of Genetic Structure in the Greater Horseshoe Bat, *Rhinolophus ferrumequinum*

Stephen J. Rossiter, Gareth Jones, Roger D. Ransome, & Elizabeth M. Barratt

Social organization plays a fundamental role in determining the apportionment of genetic variation in animal populations. Genetic structure may, in turn, have far-reaching consequences for individual fitness and behavior. A review of long-term studies of the greater horseshoe bat (*Rhinolophus ferrumequinum*) shows that population structure is shaped by a combination of breeding behavior, dispersal, and population history. Strong female philopatry and polygyny might be expected to produce genetic drift among colonies. The absence of strong differentiation among most colonies is probably explained by common ancestry, as well as contemporary gene exchange via male dispersal, extra-colony copulation, and occasional inter-colony transfer of females. In contrast, where they do occur, genetic differentiation and differences in genetic variability between isolated populations probably reflect the species' colonization history, and also point to reduced gene flow over greater distances. Although females breed with males originating from both within and outside their own colony, relatedness and Hardy–Weinberg analyses indicate that neither active inbreeding nor outbreeding occurs. Nevertheless, offspring with higher mean d^2 values, a measure of outbreeding based on microsatellite allele divergence within the individual, were more likely to survive to adulthood, indicating that outbreeding is important for individual fitness. Within the colony, significant relatedness structure, coupled with elevated relatedness among matrilineal kin, may provide suitable conditions for the evolution of kin-selected behavior. Indeed, radiotracking showed that female kin preferentially share foraging sites and night-roosts. Such associations may result from long-term maternal tutoring, rather than short-term kin-directed cooperation.

INTRODUCTION

The horseshoe bats (family Rhinolophidae) comprise 69 species, of which most are confined to the Old World tropics (Nowak, 1994). The medium-sized (13–34 g) greater horseshoe bat (*Rhinolophus ferrumequinum*) occurs throughout the southern Palaearctic, from northwest Europe to Japan (Schober, 1998). In Europe, this species has undergone population decline and range contraction during the last century, and is now rare over most of its distribution (Stebbings and Griffith, 1986). *R. ferrumequinum* has been the subject of extensive research over many years, and much of its biology is well characterized. Females form summer maternity colonies to give birth to and raise single young and return annually to their natal roost (Ransome, 1991). Adult males, in contrast, are less social, roosting at underground sites throughout much of the year (Ransome, 1991). In Britain, maternity colonies begin to form in May, and reach peak numbers around mid-June to July, by which time most surviving females return to breed. Males of up to 14 years old may join the colony during early summer, but, with the exception of yearlings, usually leave when parturition begins. Colonies in Britain usually contain 50 to 200 breeding females. Females begin to breed in their second or third year, and, although they may miss some years, may breed up to 29 years (Ransome, 1995a).

In Britain, several investigations have focused on a maternity colony of about 40 breeding females, which forms annually in the attic of Woodchester Mansion, Gloucestershire (51°2′ N, 2°90′ W). An autecological study, initiated in the mid-1950s, continues today. Since 1982 all bats have been ringed, and, from 1993, bats have been tissue-sampled as neonates (e.g., Ransome, 1968, 1989, 1995a). The long-term study at Woodchester presents an outstanding opportunity to relate roosting ecology to population genetic structure in bats. Patterns of recruitment, dispersal, turnover, and breeding may all affect the extent of genetic differentiation among social groups, as well as determining group size, composition, and relatedness. Here we draw upon recent studies of social organization and population differentiation in *R. ferrumequinum*, to evaluate the relationship between breeding behavior and gene flow, and to examine the potentially far-reaching consequences of genetic structure for individual fitness and behavior.

CAUSES AND CONSEQUENCES OF GENETIC STRUCTURE AMONG COLONIES

Breeding Behavior

The accumulation of gene correlations within, and thus genetic variance between, animal populations, is affected by the breeding and dispersal behavior of the individuals within them (e.g., Chesser et al., 1993). Breeding behavior, in addition to affecting social organization, may determine levels of outbreeding and inbreeding, as well as individuals' reproductive success, which can further influence the distribution of genetic variation both within and among populations. In particular, breeding systems directly influence the effective population size (N_e), a measure of the number of individuals that contribute gametes to the next generation. Polygyny, characterized by unequal reproductive success among males, may reduce N_e, potentially accelerating genetic drift (Hartl, 1987). If paternal input differs among populations, polygyny can promote genetic divergence.

Mating behavior in bats, as in other mammals, is largely determined by the distribution, density, and home range of females (Clutton-Brock, 1989). In most bat species the female range is not readily defendable, and males guard either a resource that is visited by females (resource defense polygyny) or, where they occur, small stable groups of females (harem polygyny) (reviewed by McCracken and Wilkinson, 2000).

In Britain, mating in *R. ferrumequinum* begins in late August, coinciding with the weaning of offspring. Radiotracking and capture surveys show that females are increasingly less faithful toward the maternity roost during this period, and regularly visit mating territories of males scattered throughout the area (authors' unpublished data). Solitary males, aged 3 years upwards, establish seasonal mating territories from mid-summer to November, and again in the following spring (Ransome, 1991). Territories may be small caves, mines, cellars, or a specific part of a larger underground system (Ransome, 1991). During the mating period, up to eight adult females have been observed to occupy territories with a single male (Ransome, 1991). Individual males may reoccupy the same territory for up to 16 years, and are often revisited by the same females (R.D. Ransome, unpublished data). Following insemination, a vaginal plug forms from the coagulation of the accessory gland secretions of males, which blocks the female reproductive tract (Mann, 1964). Plugs may thus prevent sperm competition, and so promote reproductive skew. Sperm are stored until fertilization occurs in spring (Matthews, 1937). Based on observation, therefore, mating appears to be polygynous, with some repeated partnerships between years.

Genetic analysis of the Woodchester population reveals some congruence between actual parentage, and predictions based on observation. Parentage inference, based on seven microsatellite loci (Rossiter et al., 1999), was used to examine paternity of 131 offspring of known maternity born between 1993 and 1997 (Rossiter, 2000; Rossiter et al., 2000a). Candidate fathers were sampled at underground sites throughout the area, and comprised Woodchester-born individuals and immigrants from neighboring colonies. Paternity was assigned to 58 (44.3%) young born over 5 years. For each cohort analyzed, between 5 (21%) and 10 (39%) of the 28 genotyped candidate males were identified as fathers. Therefore, paternity of each cohort is shared, and, based on the data available, individual annual success did not exceed three young (figure 12.1).

Annual reproductive skew recorded among *R. ferrumequinum* males, although low, may be attributable to the formation of temporary mating groups, coupled with the occurrence of vaginal plugs. Conversely, other aspects of this species' biology probably preclude higher levels of skew, such as female dispersal during the mating season, and possible switching between the territories of different males (authors' unpublished data). Woodchester females, for example, have been recorded in caves near to neighboring maternity colonies (D.J. Priddis and R.D. Ransome, pers. obs). Moreover, regular bouts of arousal recorded in hibernating *R. ferrumequinum* (Park et al., 2000) may allow males to copulate with torpid females, as observed in *Nyctalus noctula* (Gebhard, 1995), thereby providing opportunities for sneaking and thus preventing the monopolization of mates. The occurrence of mating in winter is evident from the rising proportion of females with plugs, detectable by palpation of the abdomen (R.D. Ransome, pers. obs.), although it is not known whether these matings lead to pregnancy. Therefore, while autumn and spring breeding pairs appear to involve active female choice for males or their territories (see Rossiter et al., 2000b), winter copulations may be unsolicited.

Despite annual variation in male breeding success, 82% of sampled candidate fathers were awarded offspring in 5 years, and therefore most *R. ferrumequinum* males probably achieve some paternities during their

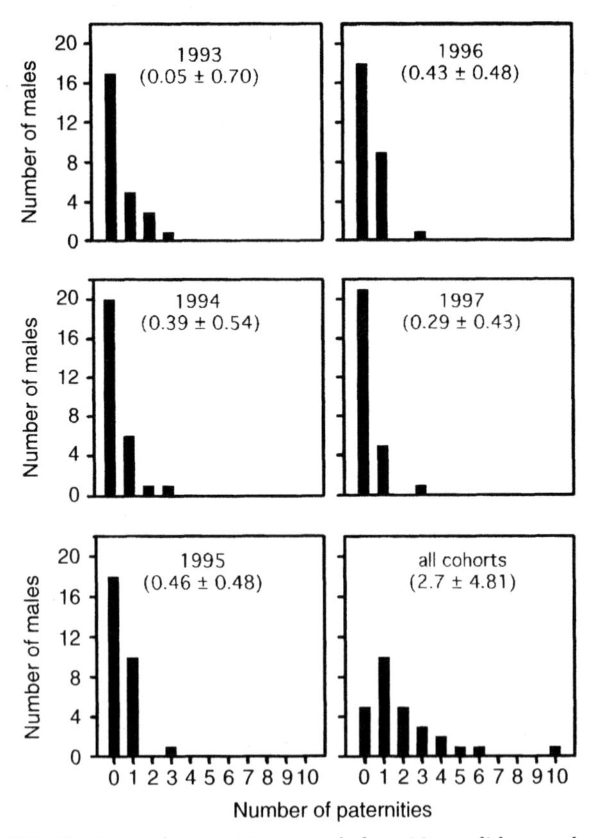

Figure 12.1 Distributions of paternities awarded to 23 candidate males for separate cohorts from 1993 to 1997, and for all cohorts combined. Mean and variance values are given in parentheses. (From Rossiter et al., 2000a.)

reproductive lifetime. Thirteen individuals, including several known territory-holders, sired offspring in more than one year. This led to an increase in overall variance in paternity success over 5 years (figure 12.1). As males may mate with females from multiple colonies, more work is needed to confirm whether the reproductive success of Woodchester males remains skewed on a wider geographical scale. Also, the possible basis of female choice in this species remains unresolved: Do females choose males per se, or do they select specific sites and copulate with the occupant? Females are known to breed with males originating from both within and outside their natal colony. For example, of 23 males assigned paternity of Woodchester offspring, 19 (83%) were also born at Woodchester. Only four originated from other colonies. Consequently, outbreeding does not appear to occur routinely. This is supported by relatedness analysis (Queller and Goodnight, 1989), in which the average pairwise relatedness between breeding pairs (0.02 ± 0.04, $n = 37$) was not less than the average calculated among all male–female pairs within the colony (0.002 ± 0.006, $n = 840$). Our finding that the most prolific male occupied a territory just 2 km from Woodchester, suggests that males possibly compete for sites close to the maternity roost. This needs to be confirmed by further study. In accordance with observations of stable mating group composition between years, repeated breeding pairs occurred more than by chance, with six recorded cases of females re-breeding with the same male. Repeated pairings may be adaptive if females reselect high-quality mates, or if the resulting co-ancestry among the offspring facilitates kin-directed cooperation. Alternatively, full-sibs may result from sperm competition if some males produce consistently superior sperm or larger sized ejaculates (e.g., Hosken, 1997). However, relatedness analysis shows that, generally, full-sibs do not occur in high numbers; the average relatedness statistics for known half-sibs (0.27 ± 0.03, $n = 104$) did not differ from that of maternal half-sibs of unknown paternity (0.28 ± 0.02, $n = 157$). This apparent discrepancy between results may indicate that females and/or males adopt more than one mating strategy. For example, long-term territory holders may sire more full-sibs, while less successful males (e.g., non-territory holders or sneaks) may sire greater numbers of maternal half-sibs.

Dispersal and Population History

As well as breeding behavior, genetic structure can also be influenced by the extent to which animals move (Slatkin, 1985). Offspring natal dispersal, for example, may homogenize gene pools, counteracting genetic drift (e.g., McCracken and Bradbury, 1977). Similarly, temporary seasonal movements, when coinciding with successful mating, may also lead to gene flow, potentially over large distances (e.g., Webb and Tidemann, 1996).

Surveys of *R. ferrumequinum* hibernacula in Britain show that seasonal dispersal behavior is variable. While the majority of Woodchester-born individuals, of both sexes, hibernate within 15 km of the maternity roost, others travel up to 40 km each year to hibernacula (R.D. Ransome, pers. obs.). Longer movements have also been observed elsewhere (e.g., Hooper and Hooper, 1956), and a 2-year-old male was recently discovered hibernating over 160 km from its birth site (D.J. Priddis, pers. comm.). The dispersal distances of females to hibernacula may be more limited, due to strong summer natal philopatry. Indeed, in 11 years just 1 of 119 individuals born at Woodchester, and known to be alive by subsequent recapture, has missed a summer there (Ransome, 1995a). The successful permanent transfer of females between colonies is also rare, with just two immigrations into Woodchester recorded in 8 years (Ransome, 1989). The behavioral traits of *R. ferrumequinum* that may affect gene flow are thus well understood. Female philopatry, multiple paternity, and little inter-colony female exchange, suggest that genetic mixing among colonies occurs mainly via extra-colony mating, rather than the physical transfer of bats. Thus, commuting distances of females to male territories, natal dispersal distances of males, and the distribution of territories will all affect genetic structure (e.g., Worthington Wilmer et al., 1999).

Just as dispersal counters genetic drift, factors limiting dispersal promote differentiation. Geographic barriers, habitat fragmentation, and tendencies toward site fidelity may all contribute toward population genetic structure. These genetic influences will shape existing patterns of differentiation, laid down by past processes such as population bottlenecks, and range expansion and contraction during climatic fluctuations. In *R. ferrumequinum*, population declines and contractions over Europe have resulted in the loss and isolation of many maternity colonies. In Britain, at the northernmost limit of its distribution, the highly fragmented remnant population of between 4000 and 6600 bats, representing approximately 20 known maternity colonies, is confined to parts of southwest England and southern Wales (Harris et al., 1995).

Microsatellite analysis of seven British maternity colonies and two samples of *R. ferrumequinum* from continental Europe (figure 12.2) suggests that population structure reflects both the social organization and population history of this species (Rossiter, 2000; Rossiter et al., 2000b). Most neighboring colonies (<20 km) are not genetically differentiated ($F_{ST} = 0.00$ to 0.04), and thus gene flow over short distances, predominantly via extra-colony mating, is sufficient to negate the possible effects of female philopatry and moderate polygyny. By comparison, higher pairwise genetic distances between English and Welsh colonies ($F_{ST} = 0.08$–0.14), together with markedly lower genetic diversity in the Welsh population (mean allelic diversity of 3.67 compared with 6.67 for the English colonies), suggests that these two population fragments are genetically isolated, and thus gene flow is restricted over greater distances. Indeed, reduced genetic diversity within the Welsh colonies may point to a population split pre-dating documented declines, or may even result from more ancient colonization events. Yet despite population isolation, and the incidence of intra-colony mating recorded at Woodchester, no colony deviated from Hardy–Weinberg

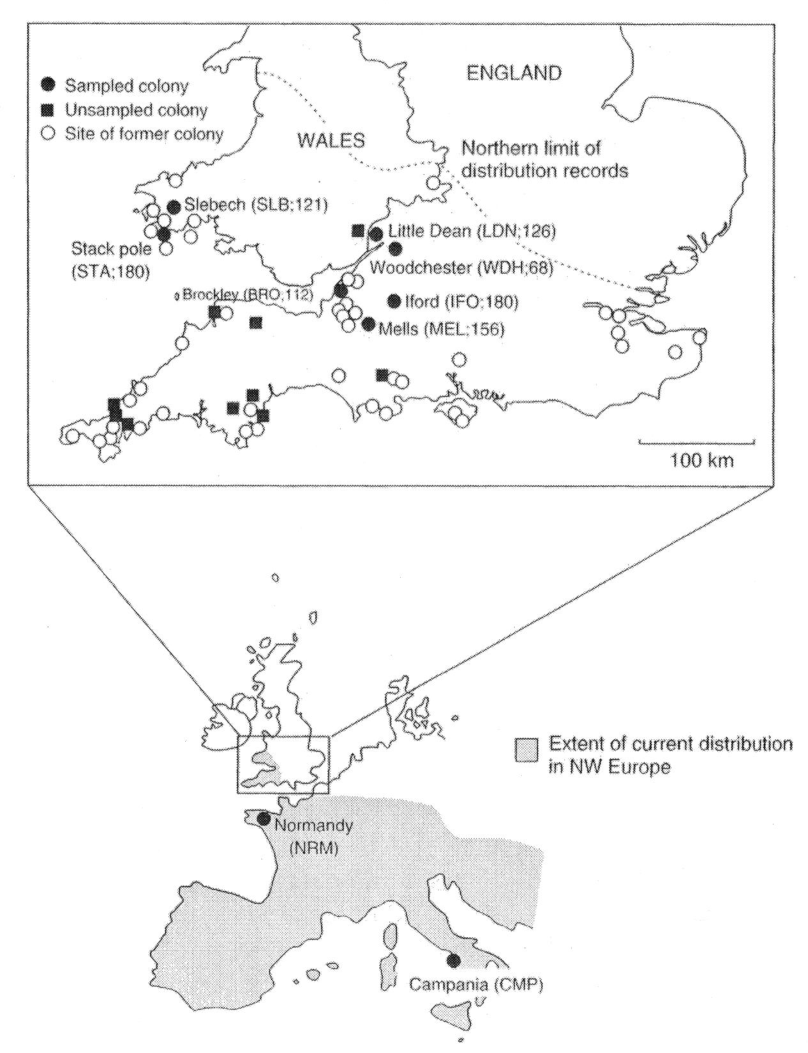

Figure 12.2 The location of British colonies and European populations sampled for population genetic analysis. Colony sizes are given in parentheses. Also indicated are unsampled colonies, past colonies, the northern distribution limit in Britain, and the species range across northwest Europe. (From Rossiter et al., 2000b.)

equilibrium, and thus breeding in *R. ferrumequinum* colonies is effectively random.

GENETIC VARIABILITY AND STRUCTURE WITHIN ANIMAL POPULATIONS

Consequences for Fitness

Behavioral subdivision of populations may impose genetic costs. Consanguineous breeding, from either random or assortative mating, increases levels of homozygosity, and therefore the expression of deleterious recessive alleles. The potential adverse effect of homozygosity on the fitness traits of progeny, termed "inbreeding depression," is well established, having been reported in a range of taxa (reviewed by Avise, 1994). In contrast, heterozygosity resulting from outbreeding can have potential fitness benefits (Shull, 1948). Indeed, in species and populations where outbreeding and associated heterosis occur regularly, negative effects of inbreeding may arise through a loss of heterosis, as opposed to inbreeding depression (see Amos and Hoelzel, 1992).

Although it is a generally held view that genetic isolation and associated drift pose threats to natural populations, the relationship between genetic variability and fitness is equivocal. Inbreeding, for example, may be an adaptive strategy to raise co-ancestry, whereas outbreeding does not always lead to heterozygote advantage (see examples in Rossiter, 2000). Indeed, gene flow between populations that are adapted to different local conditions may reduce fitness ("outbreeding depression"), possibly through the breakdown of co-adapted gene complexes (Templeton, 1986). To date, surprisingly few investigations have documented trends between genetic variability and fitness in wild animal populations. A recent study of overwinter survival in *R. ferrumequinum* offspring (Rossiter et al., 2001), described below, suggests that genetic factors may have an important influence on individual fitness in wild bat populations.

In Britain, offspring of *R. ferrumequinum* are usually born around mid-July, and survival is affected by weather conditions (Ransome, 1990). Warm springs promote earlier births (Ransome and McOwat, 1994), leading to greater numbers of offspring reaching hibernacula (Ransome, 1989). Conversely, prolonged cold winters result in above-average mortality rates (Ransome, 1991), with very poor conditions causing population crashes (Ransome, 1989; Ransome and McOwat, 1994). Thus, depending on weather conditions, the number of bats alive 1 year after birth can vary between 0 and 52% of the original cohort (Ransome, 1990, 1998).

To expand on these findings, a more detailed examination of the potential factors affecting mortality in 154 Woodchester offspring was recently undertaken (Rossiter, 2000; Rossiter et al., 2001). Of 13 variables that were considered, including mother and offspring phenotypic characters, offspring genotype, and life-history traits, only the genetic variable mean d^2 (Coulson et al., 1998)—a measure of outbreeding based on microsatellite allele divergence within the individual—was significantly positively associated with male survival to breeding age (figure 12.3). Contrary to earlier studies, weather effects were not identified, possibly due to relatively stable, mild weather conditions prevailing during the study period. Although the basis of enhanced survival of outbred male offspring has not been resolved, the detection of heterosis, even in the absence of obvious active outbreeding, highlights the importance of maintaining opportunities for gene flow in this species. Indeed, facilitation of genetic mixing among colonies, via the

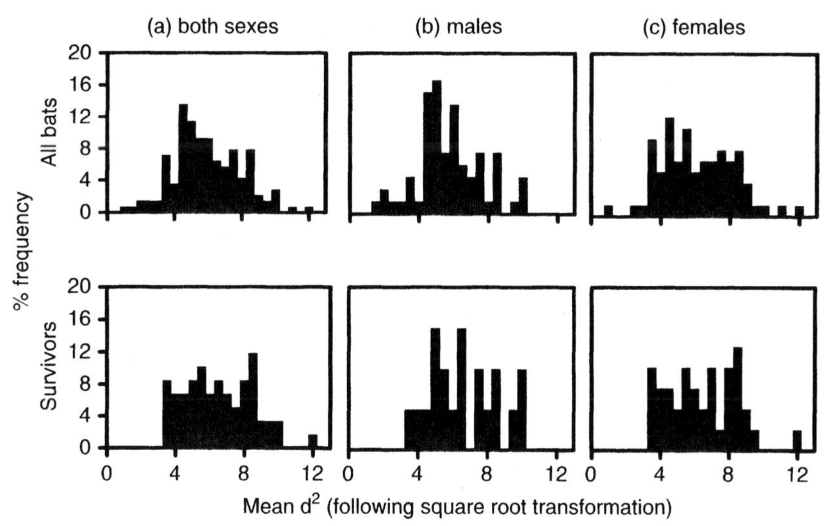

Figure 12.3 Percentage frequency distributions for survivors, versus all bats, to breeding age, for (a) both sexes combined, (b) males, and (c) females. Logistic regression models showed significant differences between survivors versus all bats in males ($P < 0.05$) and both sexes ($P < 0.01$). (From Rossiter et al., 2001.)

protection of mating sites and habitat, may contribute to offspring survival and thus population growth.

Consequences for the Evolution of Social Behavior

Genetic structure can have important implications for social behavior. Genetic similarity may increase the scope for kin selection, while altruism may also be favored where costs to the donor are outweighed by benefits to the recipient, and opportunities occur for reciprocation (e.g., Wilkinson, 1987). Several aspects of *R. ferrumequinum* behavior may be expected to facilitate the evolution of altruism. Coloniality, coupled with strong female philopatry over many years, leads to the spatial aggregation of matrilineal relatives. Furthermore, parentage inference shows that high numbers of half-sibs, and a few full-sibs, exist within the colony. Stable colonies are composed of highly gregarious members, which roost in bodily contact with each other. However, such social organization does not appear to translate into high background colony relatedness. In Woodchester, for example, average relatedness among females approximates to zero ($R = 0.03 \pm 0.03$) (Rossiter et al., 2002), while similar low values reported for colonies of other bat species have raised questions regarding the general importance of close genetic ties in the evolution and maintenance of sociality in bats (Burland and Worthington Wilmer, 2001). By comparison, greater opportunities for inclusive fitness benefits probably exist within individual matrilines (Rossiter et al., 2002). Relatedness estimates recorded within 15 multi-generation matrilineal pedigrees at Woodchester, comprising

3–12 females, ranged from 0.64 ± 0.12 to 0.17 ± 0.09. Moreover, significant structure detected among these matrilines indicates that high intra-matriline relatedness is not homogenized by common paternity.

To date, two studies, both focusing on the Woodchester colony, have sought to determine whether the theoretical potential for altruism is realized in natural colonies of *R. ferrumequinum*. Examination of roosting behavior suggests that cooperative interactions such as food sharing and allogrooming—kin-biased or otherwise—probably occur either very rarely or not at all (P. Wohlund, unpublished data). However, radiotracking of 14 adult females from two Woodchester matrilines (figure 12.4) has revealed that matrilineal relatives share foraging and night-roosting sites to a greater extent than nonkin (Rossiter et al., 2002). Estimates of pairwise home-range overlap, based on radio-fixes collected over three successive summers, correlated with Hamilton's coefficient of relatedness, despite individual variation. Greatest overlap occurred between females and their adult daughters, which sometimes shared both core areas and night roosts over several years (figure 12.5).

Previous research on range expansion in *R. ferrumequinum* has shown that mother–daughter pairs do not share foraging areas during the first summer, and that the foraging distances of offspring only approach those of adults at around 60 days (Jones et al., 1995), perhaps due to constraints imposed by incomplete skeletal growth before this time (see Rossiter et al., 2002).

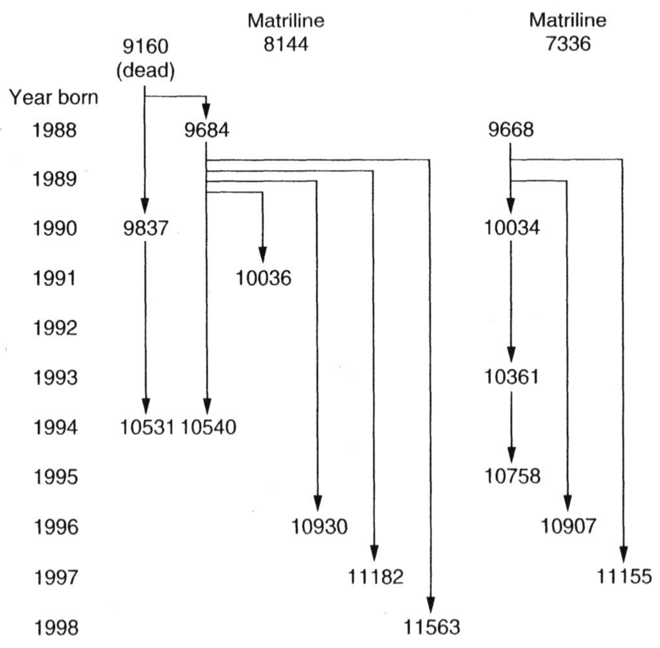

Figure 12.4 Fourteen females, from two matrilines, radiotracked to determine the extent of foraging associations among relatives. (From Rossiter et al., 2002.)

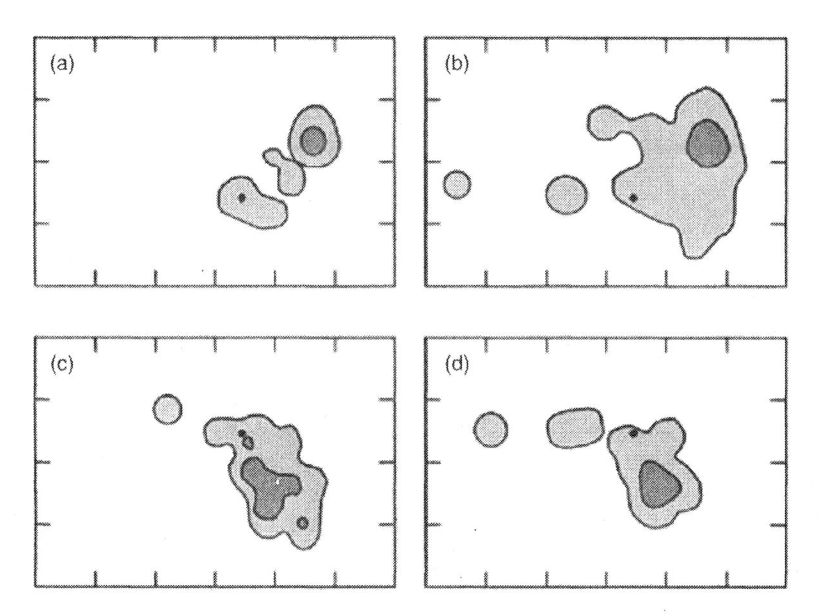

Figure 12.5 Utilization distribution (UD) for two mother–young pairs. Parts (a) and (b) represent the UDs of individuals 9684 and 10930 respectively, and parts (c) and (d) represent individuals 10034 and 9668. Fifty percent and 80% UDs are shown by light and dark shading, respectively. Woodchester Mansion is represented by a filled black circle for each grid; tick marks represent 2 km increments. (From Rossiter et al., 2002.)

Thus, mother–daughter foraging associations appear to develop after weaning. The adaptive significance of such interactions, if any, is not clear. However, the low recorded incidence of spatiotemporal associations, coupled with long-term fidelity for feeding grounds, indicates that range overlap cannot be ascribed to short-term information transfer. Instead, shared home ranges may result from the vertical transmission of information about favorable feeding areas from mothers to their weaned daughters, and may therefore be better explained by extended maternal care rather than kin-selected cooperation. A mechanism of maternal tutoring could also explain why weaned juveniles are less likely to reach adulthood if their mothers die during their first year (Ransome, 1995b).

CONCLUSIONS

Newly emerging results from studies of wild *R. ferrumequinum* have contributed greatly to our understanding of how this species' social organization and population history may influence genetic structure, and, in turn, the behavior and fitness of individuals. Recent evidence of heterosis from long-term life-history data has shown that genetic variability and structure may have important consequences for offspring survival. Meanwhile, combined relatedness and behavioral analyses have highlighted the potential impact that long-term female natal philopatry may have on both genetic and social structure within the colony. Social architecture based on matrilines can produce favorable conditions for individuals to accrue inclusive fitness benefits. The potential advantage of living in a matrilineal society composed of relatives of multiple generations is illustrated by the possibility that descendant kin acquire knowledge of favorable feeding sites through a system of vertical information transfer. Such sharing of social information within matrilines of bats could theoretically underpin natal philopatry itself, and offers an exciting avenue for future study.

ACKNOWLEDGMENTS

We thank the Woodchester Mansion Trust for access to the maternity roost, and the many people who have assisted with the collection of data and samples over the years. Bats were captured and sampled under licensees from English Nature, the Countryside Council for Wales, and the Home Office. This work was funded by the Natural Environment Research Council, Bat Conservation International, and English Nature.

LITERATURE CITED

Amos, B., and R. Hoelzel. 1992. Applications of molecular genetic techniques to the conservation of small populations. Biological Conservation, 61: 133–144.

Avise, J.C. 1994. Molecular Markers, Natural History and Evolution. Chapman and Hall, London.

Burland, T.M., and J. Worthington Wilmer. 2001. Seeing in the dark: molecular approaches to the study of bat populations. Biological Reviews, 76: 389–409.

Chesser, R.K., D.W. Sugg, O.E. Rhodes, J.N. Novak, and M.H. Smith. 1993. Evolution of mammalian social structure. Acta Theriologica, 38: 163–174.

Clutton-Brock, T.H. 1989. Mammalian mating systems. Proceedings of the Royal Society of London, Series B, 236: 339–372.

Coulson, T.N., J.M. Pemberton, S.D. Albon, M. Beaumont, T.C. Marshall, J. Slate, F.E. Guinness, and T.H. Clutton-Brock. 1998. Microsatellites reveal heterosis in red deer. Proceedings of the Royal Society of London, Series B, 265: 489–495.

Gebhard, J. 1995. Observations on the mating behavior of *Nyctalus noctula* (Schreber, 1774) in the hibernaculum. Myotis, 32–33: 123–129.

Harris, S., P. Morris, S. Wray, and D. Yalden. 1995. A Review of British Mammals: Population Estimates and Conservation Status of British Mammals other than Cetaceans. Joint Nature Conservation Committee, Peterborough, UK.

Hartl, D. 1987. A Primer of Population Genetics, 2nd edition. Sinauer Associates, Sunderland, MA.

Hooper, J.H.D., and W.M. Hooper. 1956. Habits and movements of cave dwelling bats in Devonshire. Proceedings of the Zoological Society of London, 127: 1–26.

Hosken, D.J. 1997. Sperm competition in bats. Proceedings of the Royal Society of London, Series B, 264: 385–392.

Jones, G. 2000. The ontogeny of behavior in bats: a functional perspective. Pp. 362–392. In: Ontogeny, Functional Ecology and Evolution of Bats (R.A. Adams and S.C. Pedersen, eds.). Cambridge University Press, Cambridge.

Jones, G., P.L. Duvergè, and R.D. Ransome. 1995. Conservation biology of an endangered species: field studies of greater horseshoe bat. Symposia of the Zoological Society of London, 67: 309–324.

Mann, T. 1964. The Biochemistry of Semen and of the Male Reproductive Tract. Methuen, London.

Matthews, L.H. 1937. The female sexual cycle of the British horseshoe bats, *Rhinolophus ferrumequinum insulanus* Barrett-Hamilton *and R. hipposideros minutus* Montagu. Transactions of the Zoological Society of London, 23: 224–267.

McCracken, G.F., and J.W. Bradbury. 1977. Paternity and genetic heterogeneity in the polygynous bat *Phyllostomus hastatus*. Science, 198: 303–306.

McCracken, G.F., and G.S. Wilkinson. 2000. Bat mating systems. Pp. 321–362. In: Reproductive Biology of Bats (E.G. Crichton and P.H. Krutzsch, eds.). Academic Press, London.

Nowak, R.M. 1994. Walker's Bats of the World. Johns Hopkins University Press, London.

Park, K.J., G. Jones, and R.D. Ransome. 2000. Torpor, arousal and activity of hibernating Greater Horseshoe Bats (*Rhinolophus ferrumequinum*). Functional Ecology, 14: 580–588.

Queller, D.C., and K.F. Goodnight. 1989. Estimating relatedness using genetic markers. Evolution, 43: 258–275.

Ransome, R.D. 1968. The distribution of the greater horseshoe bat, *Rhinolophus ferrumequinum*, during hibernation, in relation to environmental factors. Journal of Zoology (London), 154: 77–112.

Ransome, R.D. 1989. Population changes of greater horseshoe bats studied near Bristol over the past twenty-six years. Biological Journal of the Linnean Society, 38: 71–82.

Ransome, R.D. 1990. The Natural History of Hibernating Bats. Christopher Helm, London.

Ransome, R.D. 1991. Greater horseshoe bat *Rhinolophus ferrumequinum*. Pp. 88–94. In: The Handbook of British Mammals (G.B. Corbet and S. Harris, eds.), 3rd edition. Blackwell Scientific, Oxford.

Ransome, R.D. 1995a. Earlier breeding shortens life in female greater horseshoe bats. Philosophical Transactions of the Royal Society of London, Series B, 350: 153–161.

Ransome, R.D. 1995b. Does significant maternal care continue beyond weaning in greater horseshoe bats? Bat Research News, 36: 102–103.

Ransome, R.D. 1998. The impact of maternity roost conditions on populations of greater horseshoe bats. English Nature Research Report 292. English Nature, Peterborough, UK.

Ransome, R.D., and McOwat, T.P. 1994. Birth timing and population changes in greater horseshoe bat (*Rhinolophus ferrumequinum*) colonies are synchronised by climatic temperature. Zoological Journal of the Linnean Society, 112: 337–351.

Rossiter, S.J. 2000. Causes and Consequences of Genetic Structure in the Greater Horseshoe Bat (*Rhinolophus ferrumequinum*). Ph.D. Dissertation, University of Bristol, Bristol.

Rossiter, S.J., G. Jones, R.D. Ransome, and E.M. Barratt. 1999. Characterization of microsatellite loci in the greater horseshoe bat *Rhinolophus ferrumequinum*. Molecular Ecology, 8: 1959–1961.

Rossiter, S.J., G. Jones, R.D. Ransome, and E.M. Barratt. 2000a. Parentage, reproductive success and breeding behaviour in the greater horseshoe bat

(*Rhinolophus ferrumequinum*). Proceedings of the Royal Society of London, Series B, 267: 545–551.

Rossiter, S.J., G. Jones, R.D. Ransome, and E.M. Barratt. 2000b. Genetic variation and population structure in the endangered greater horseshoe bat *Rhinolophus ferrumequinum*. Molecular Ecology, 9: 1131–1135.

Rossiter, S.J., G. Jones, R.D. Ransome, and E.M. Barratt. 2001. Outbreeding increases offspring survival in wild greater horseshoe bats (*Rhinolophus ferrumequinum*). Proceedings of the Royal Society of London, Series B, 268: 1055–1061.

Rossiter, S.J., G. Jones, R.D. Ransome, and E.M. Barratt. 2002. Relatedness structure and kin-biased foraging in the greater horseshoe bat (*Rhinolophus ferrumequinum*). Behavioral Ecology and Sociobiology, 51: 510–518.

Schober, W. 1998. Die Hufeisennasen Europus. Westarp Wissenschaften, Hohenwarsleben, Germany.

Shull, G.H. 1948. What is heterosis? Genetics, 33: 439–446.

Slatkin, M. 1985. Gene flow in natural populations. Annual Review of Ecology and Systematics, 16: 393–430.

Stebbings, R.E., and F. Griffith. 1986. Distribution and Status of Bats in Europe. Institute of Terrestrial Ecology, NERC, Huntington, UK.

Templeton, A.R. 1986. Coadaptation and outbreeding depression. Pp. 105–116. In: Conservation Biology. The Science of Scarcity and Diversity (M.E. Soulé, ed.). Sinauer Associates, Sunderland, MA.

Webb, N.J., and C.R. Tidemann. 1996. Mobility of Australian flying-foxes, *Pteropus* spp. (Megachiroptera): evidence from genetic variation. Proceedings of the Royal Society of London, Series B, 263: 497–502.

Wilkinson, G.S. 1987. Altruism and cooperation in bats. Pp. 299–323. In: Recent Advances in the Study of Bats (M.B. Fenton, P. Racey, and J.M.V. Rayner, eds.). Cambridge University Press, Cambridge.

Worthington Wilmer, J., L. Hall, E.M. Barratt, and C. Moritz. 1999. Genetic structure and male mediated gene flow in the ghost bat (*Macroderma gigas*). Evolution, 53: 1582–1591.

13

Population Genetic Structure of Very Large Populations: The Brazilian Free-Tailed Bat, *Tadarida brasiliensis*

Amy L. Russell & Gary F. McCracken

Species with restricted distributions or low potential for movement often are distinguished by population structures that can be readily defined by geographical distribution. By contrast, highly vagile or migratory species such as many birds, bats, insects, and marine organisms, are often characterized by population structures that are much harder to predict. Using published data from allozymes and from mitochondrial DNA (mtDNA) sequences, we present an analysis of genetic structuring in the Brazilian free-tailed bat, *Tadarida brasiliensis*, a highly vagile mammal that is characterized by extremely large population sizes. We compare molecular genetic analyses between and within subspecies with structure that has been inferred from morphological, behavioral, and ecological data. Mitochondrial DNA sequence data indicate that significant differences exist between populations in North America and those in South America. A cladistic analysis of gene flow suggests genetic structuring among females in North America that is not consistent with the current subspecific taxonomy. Neither allozyme nor mtDNA data support distinction between the North American subspecies, nor do they support the genetic uniqueness of Cockrum's (1969) hypothesized migratory groups of *T. b. mexicana*. Data indicate that this subspecies has evolved as a single genetic unit, with an extremely large effective population size. Effective population sizes of the magnitude suspected for these bats are extremely rare for mammalian species.

INTRODUCTION

The population structure of highly vagile or migratory species can be difficult to determine. Species that migrate may exhibit different population structures at different times of the year and in different places (summer range, winter range, transitional range(s)). These movement patterns can make the species hard to track, and can result in informational biases toward one part of their life history. Additionally, males and females may exhibit different migration patterns, and population structure may be inaccurately assessed if such variations in behavior are not recognized.

Tadarida brasiliensis is one of the most abundant bats in the Western Hemisphere, occurring in colonies numbering up to tens of millions of individuals (table 13.1). It is found throughout Central and most of South America and in North America from the Atlantic to the Pacific

Table 13.1 Estimates of Historical Numbers of Bats in Major Colonies of *T. b. mexicana* in the Southwestern United States

Colony	Estimated Size	Year of Estimate
Texas		
Bracken Cave	20×10^6	1957
Goodrich Cave	$14–18 \times 10^6$	1957
Rucker Cave	$12–14 \times 10^6$	1957
Frio Cave	10×10^6	1957
Ney Cave	10×10^6	1957
Fern Cave	$8–12 \times 10^6$	1957
Devil's Sink Hole	$6–10 \times 10^6$	1957
Davis Cave	6×10^6	1957
Valdina Sink	4×10^6	1957
	Abandoned	1987
Quarry Colony	4×10^6	1989
Webb Cave	$<0.6 \times 10^6$	1957
Wilson	$<0.6 \times 10^6$	1957
Y-O Ranch Cave	$<0.6 \times 10^6$	1957
New Mexico		
Carlsbad Caverns	8.7×10^6	1936
	4×10^6	1957
	218,000	1973
Arizona		
Eagle Creek Cave	$25–50 \times 10^6$	1963
	30,000	1969
Oklahoma		
Vickery Cave	1×10^6	1969
Four caves in Western Oklahoma (includes Vickery)	$>3 \times 10^6$	1952

Modified from McCracken (2003).

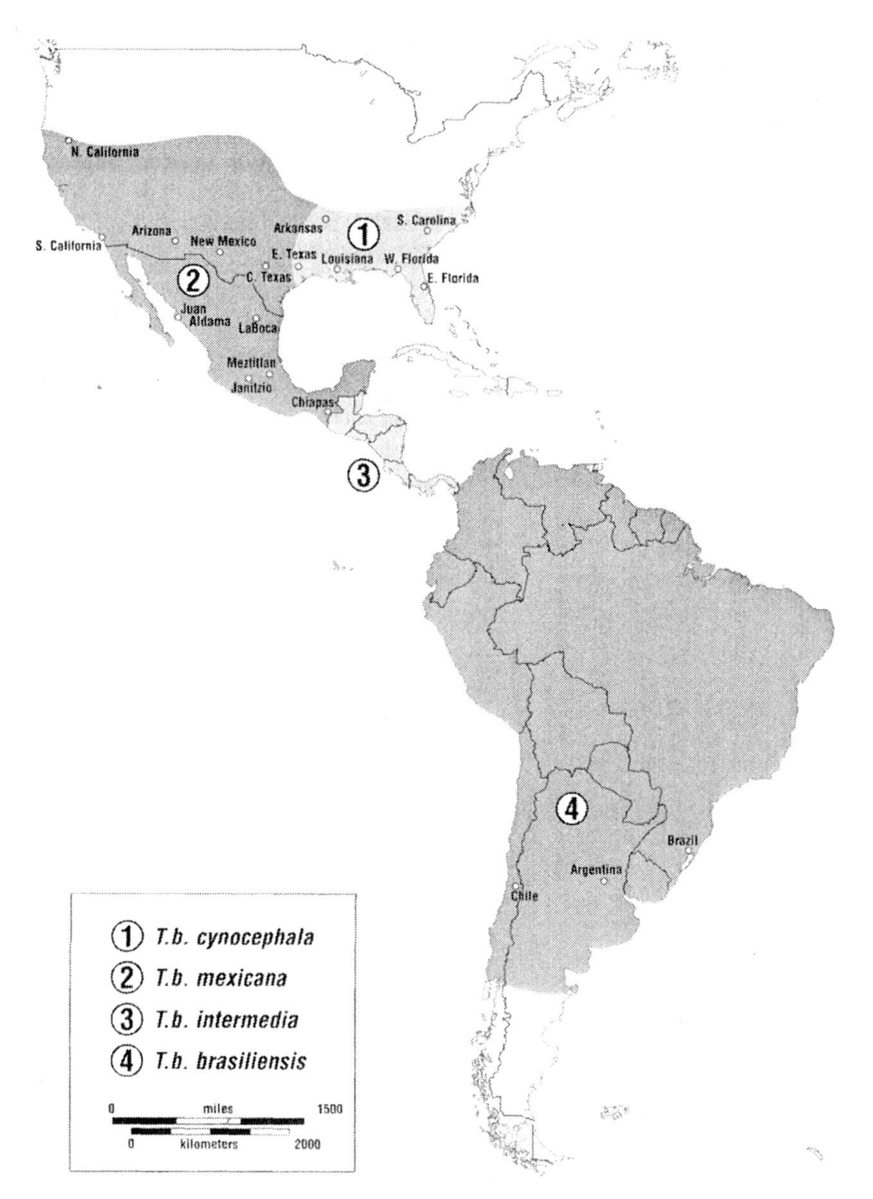

Figure 13.1 Geographical range of recognized subspecies of *Tadarida brasiliensis* throughout North, Central, and South America (from Wilkins, 1989). The locations of populations sampled by Russell and McCracken (unpublished results) are indicated.

coast, north to 40° N latitude (Hall, 1981; figure 13.1). Three subspecies of *T. brasiliensis* are currently recognized within North and Central America, with *T. b. cynocephala* in the southeastern United States, *T. b. mexicana* in the southwestern and western United States and most of Mexico, and *T. b. intermedia* from southern Mexico to Panama. A fourth subspecies, *T. b. brasiliensis*, occupies the entire species range in South America.

Although it is one of the most well studied species of bats in North America, most conclusions about population structuring and taxonomic subdivisions of *T. brasiliensis* are based on inferences from behavioral studies and banding data rather than from genetic analyses. However, inferences from behavior can be easily confounded by many characteristics of these bats. *Tadarida brasiliensis* is capable of flying very long distances. Individuals are known to migrate annually up to 1500 km (Villa-R. and Cockrum, 1962), and to fly up to 50 km from their roost on single foraging flights (Davis et al., 1962). Males and females display different patterns of movement throughout the year, with migratory females typically moving long distances to maternity colonies while many males appear to engage in local movements in the vicinity of their winter roosts (Davis et al., 1962; Villa-R. and Cockrum, 1962). Their capacity for extended flights makes the species extremely difficult to monitor and their sex-specific differences in movement patterns can confuse attempts to characterize gene flow. Here we evaluate patterns of genetic structuring considering data from previous studies of behavior, banding recoveries, and morphology and from published and new molecular genetic data.

DIFFERENTIATION BETWEEN SUBSPECIES

Evidence from Behavioral Studies

The majority of behavioral and life-history studies have concerned the North American subspecies *T. b. cynocephala* and *T. b. mexicana*. Relatively little information is available for populations in the Central and South American portions of the species' range, and taxonomic recognition of *T. b. intermedia* and *T. b. brasiliensis* is based largely on geography. Like *T. b. mexicana*, *T. b. brasiliensis* is known to migrate in much of its range (Marques, 1991), but population subdivision, as proposed by Cockrum (1969) for *T. b. mexicana*, has not been investigated in the Southern Hemisphere.

Recognition of *T. b. cynocephala* and *T. b. mexicana* is based substantially on behavioral differences in migration, hibernation, and roosting habits (Barbour and Davis, 1969; Carter, 1962). *Tadarida brasiliensis mexicana* roosts in caves and man-made structures in colonies of several million individuals and generally migrates to winter roosts in Mexico. Mating has generally been thought to occur in transitional roosts in Mexico during the northward migration (Cockrum, 1969; McCracken et al., 1994), but has also been observed in Texas shortly after the bats arrive in March (Annika Keeley, pers. comm.). The vast majority of the *T. b. mexicana* that migrate are pregnant females (Cockrum, 1969; Davis et al., 1962; Villa-R., 1956), with large numbers of adult males remaining in Mexico. Males that migrate typically occupy smaller colonies separate from the large maternity colonies. Young are born in maternity colonies in late May to mid-June, and become volant at about 6 weeks of age. Adult females then begin to leave the maternity caves, often moving further north before the southward migration begins in August and September (Glass, 1982).

Tadarida brasiliensis cynocephala roosts primarily in man-made structures in colonies typically not exceeding several thousand individuals (Bain, 1981). This subspecies undergoes only local seasonal movements, spending the winter months in hibernation or torpor (LaVal, 1973; Sherman, 1937). However, close examination of winter colonies of *T. b. cynocephala* in Florida shows that these bats are often active in winter (Kiser, 1996). While monitoring a large bat house at the University of Florida, Kiser and Glover (1997) found marked fluctuations in population size throughout the year, indicating substantial local movement during the winter months. While these bats maintain some activity throughout the year, there is no evidence that *T. b. cynocephala* undertake long-distance migration similar to that of *T. b. mexicana*.

While these behavioral characteristics have been cited as distinguishing *T. b. cynocephala* and *T. b. mexicana* (Barbour and Davis, 1969; Carter, 1962), there is substantial plasticity in these same traits and evidence suggests that the behavioral differences between the two North American subspecies may be overemphasized. Western populations of *T. b. mexicana* in Utah, Nevada, California, and Oregon are remarkable in their behavioral similarity to *T. b. cynocephala*. Roosting in buildings and rock crevices in populations of fewer than several thousand, these bats remain in the United States through the winter (Grinnell, 1918; Jewett, 1955; Krutzsch, 1955; Perkins et al., 1990). It has also been shown that individuals in migratory populations of *T. b. mexicana* can engage in extended periods of torpor. Individuals of a migratory population of *T. b. mexicana* maintained at 5 °C followed the typical pattern of lowering their body temperature and metabolic rate to levels observed in hibernating species of bats (Herreid, 1963; Orr, 1958). Putatively migratory populations of *T. b. mexicana* also occur during the winter in eastern Texas where a colony of *T. b. mexicana* remained active throughout the winter months in College Station, only 137 km from colonies of *T. b. cynocephala* (Spenrath and LaVal, 1974). Bennett (2000) also described a colony in Huntsville, Texas, as simultaneously containing both *T. b. mexicana* and *T. b. cynocephala*, and a similarly mixed colony has been reported from under a bridge in Houston, Texas (Barbara French, Bat Conservation International, pers. comm.). These populations may present opportunities for interbreeding between *T. b. mexicana* and *T. b. cynocephala*. Specimens that are intermediate between *T. b. cynocephala* and *T. b. mexicana* in cranial measurements have been described from a population in southeastern Texas, and may result from interbreeding between the two subspecies (Schmidley et al., 1977).

Evidence from Morphological Studies

Measurements of the skull (i.e., greatest length of skull and zygomatic breadth) and forearm length have been cited as distinguishing *T. b. mexicana* and *T. b. cynocephala* (Carter, 1962; Owen et al., 1990). The subspecies are

Table 13.2 Morphological Measurements for Two Defining Characters of Three Subspecies of *T. brasiliensis*

Subspecies	**Sex**	**Mean**	**Range**
Skull length (mm)			
cynocephala	Male	17.48	16.7–18.4
	Female	17.16	16.1–18.2
mexicana	Male	16.94	16.2–17.7
	Female	16.72	16.1–17.4
brasiliensis	Both	17.28	16.1–17.8
Zygomatic breadth (mm)			
cynocephala	Male	10.27	9.6–10.8
	Female	10.10	9.6–10.6
mexicana	Male	9.78	9.3–10.4
	Female	9.70	9.3–10.2
brasiliensis	Both	9.95	9.1–10.5

Data from Owen et al. (1990).

typified by differences in the means, but there is broad overlap in the ranges of these morphological characters (see table 13.2 for length of skull and zygomatic breadth). Because of this overlap, it has been argued that these characters poorly differentiate the taxa (McCracken and Gassel, 1997; Schmidley et al., 1977; Schwartz, 1955), and most arguments for the separation of the taxa have relied more on behavioral and geographical differences than on morphology.

Evidence from Molecular Genetic Analyses

The taxonomic status of the subspecies of *T. brasiliensis* has been a point of some debate, especially regarding *T. b. cynocephala* and *T. b. mexicana*. The taxa comprising the *brasiliensis* group of the genus *Tadarida* were initially differentiated at the species level (Shamel, 1931), but have since been lowered to the subspecies level (Schwartz, 1955). Subsequent publications have questioned this change in taxonomy (Carter, 1962; Owen et al., 1990). Owen et al. (1990) reported a nearly fixed difference between *T. b. cynocephala* and *T. b. mexicana* at a single allozyme locus. From this evidence, the morphometric data, and the differences in behavior and geographical distribution, they concluded that *T. b. cynocephala* and *T. b. mexicana* warrant species status even though a small amount of gene flow might be occurring between the subspecies.

Other tests of the genetic divergence of *T. b. mexicana* and *T. b. cynocephala* have concluded that any taxonomic recognition is questionable (McCracken and Gassel, 1997; Russell and McCracken, unpublished results). Using data from 22 loci, McCracken and Gassel (1997) found that levels of genetic similarity and genetic structuring between the subspecies were in the range of that typically seen between populations of the same subspecies. Several lines of evidence indicated that samples from near the western limits of putative *T. b. cynocephala* in Arkansas represented an area of genetic introgression, and the authors concluded that there was evidence of substantial gene flow between the subspecies.

The genetic structuring of populations can be affected by a combination of current and historical processes, including restricted gene flow, range fragmentation or expansion, and long-distance colonization (Templeton, 1998). Because allozyme data are typically analyzed as differences in allele frequencies, inferences regarding the phylogenetic relationships of alleles are not considered. Phylogenetic analyses from DNA sequence data, on the other hand, illuminate historical relationships between haplotypes and can distinguish the impacts of current and historical processes on the species. We have examined mtDNA control region sequence variation among the four mainland subspecies of *T. brasiliensis*: *T. b. intermedia*, *T. b. brasiliensis*, *T. b. cynocephala*, and *T. b. mexicana* (figure 13.1). Sequence data were obtained from a 407 bp segment of the mitochondrial D-loop, a rapidly evolving, noncoding region used often in studies of intraspecific population structuring (Kerth et al., 2000; Petit et al., 1999; Wilkinson and Fleming, 1996).

An analysis of molecular variance (AMOVA; Excoffier et al., 1992) revealed that over half of the total sequence variation is attributable to differences between subspecies, with *T. b. brasiliensis* being clearly different from the other three subspecies in the analysis (table 13.3). A phylogenetic analysis of the four subspecies using a Bayesian likelihood criterion also reveals a clear differentiation of the South American from North American subspecies (figure 13.2). Analyses including only *T. b. cynocephala*, *T. b. mexicana*, and *T. b. intermedia* showed little differentiation between subspecies ($P = 0.064$) and a comparatively higher level of population-level structuring within the subspecies (10.4%).

While AMOVA analyses could not detect highly significant variation among the three North American subspecies, these analyses are limited in that they test for genetic structuring among groups that were defined a priori. Significant but unanticipated groupings might go undetected. To further analyze relationships between the North American populations that may not conform to a priori subspecific designations, we performed a cladistic analysis of gene flow (Slatkin and Maddison, 1989) using the mtDNA sequence data. By mapping taxonomic assignments onto a cladogram as an independent character (figure 13.3), we tested whether the number of character-state changes on the observed cladogram is significantly less than a null distribution derived from a number of randomly joining trees. Thus, character-state

Table 13.3 AMOVA Analyses of mtDNA Sequence Data of Recognized Subspecies, Indicating the Proportion of Total Variance Attributed to Each Hierarchical Level

	All Subspecies	**Excludes *T.b. brasiliensis***	**Excludes *T.b. cynocephala***	**Excludes *T.b. mexicana***	**Excludes *T.b. intermedia***
Among subspecies	54.26%	6.93%	63.15%	68.21%	57.19%
Among population, within subspecies	6.13%	10.41%	5.42%	4.69%	5.81%
Within population	39.61%	82.66%	31.43%	27.10%	37.00%
Θ_{CT}	0.5426*	0.0693	0.6315*	0.6821*	0.5719*
P value	<0.0001	0.0642	0.0025	0.0051	<0.0001

Data from Russell and McCracken (unpublished).
Five separate analyses were conducted: one including data from all subspecies sampled, and four excluding data from a different subspecies.
*Significant structuring among the subspecies analyzed.

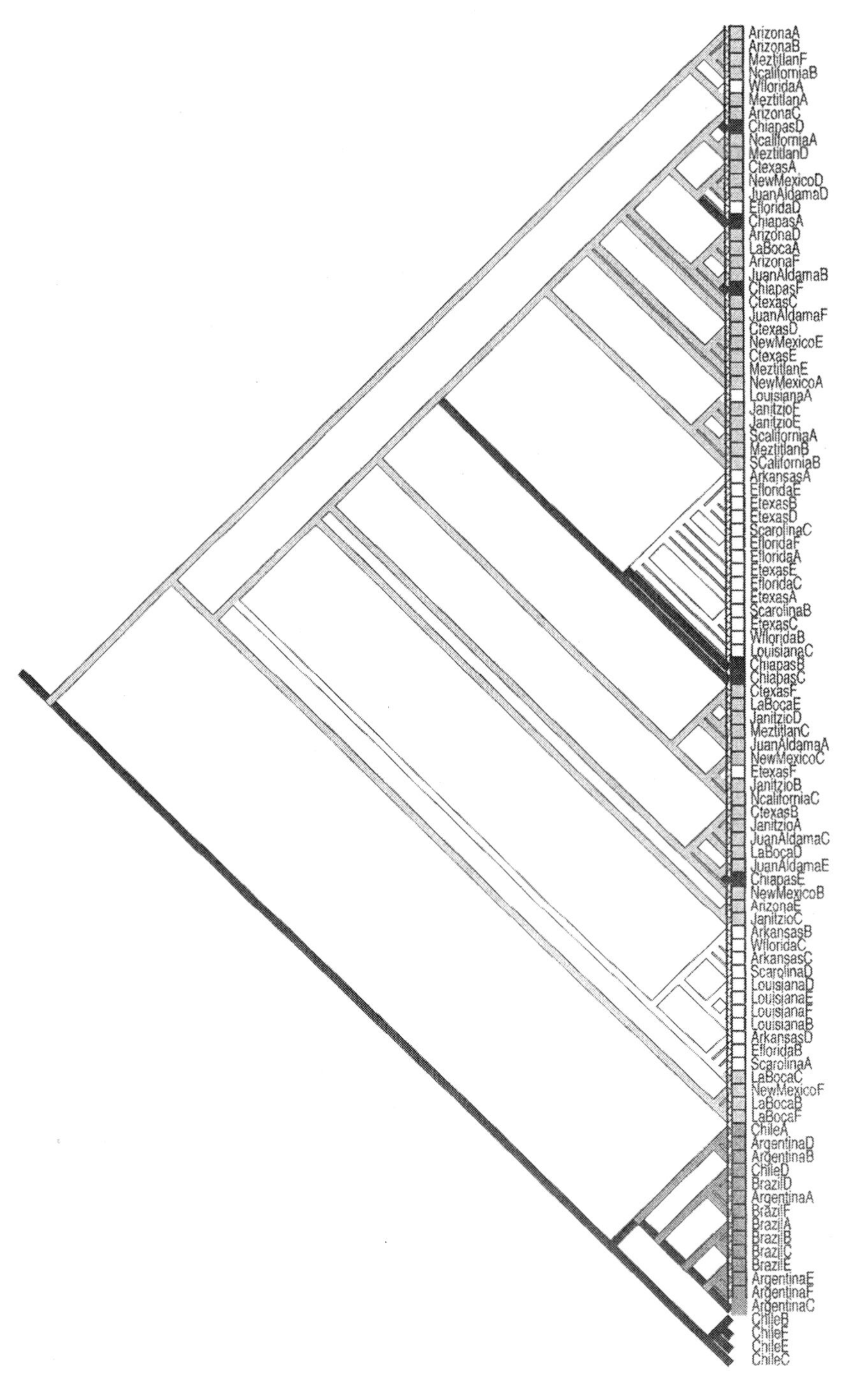

Figure 13.2 A likelihood-based Bayesian phylogeny of samples of the subspecies *T. b. cynocephala* (white), *T. b. mexicana* (light gray), *T. b. intermedia* (black), and *T. b. brasiliensis* (dark gray).

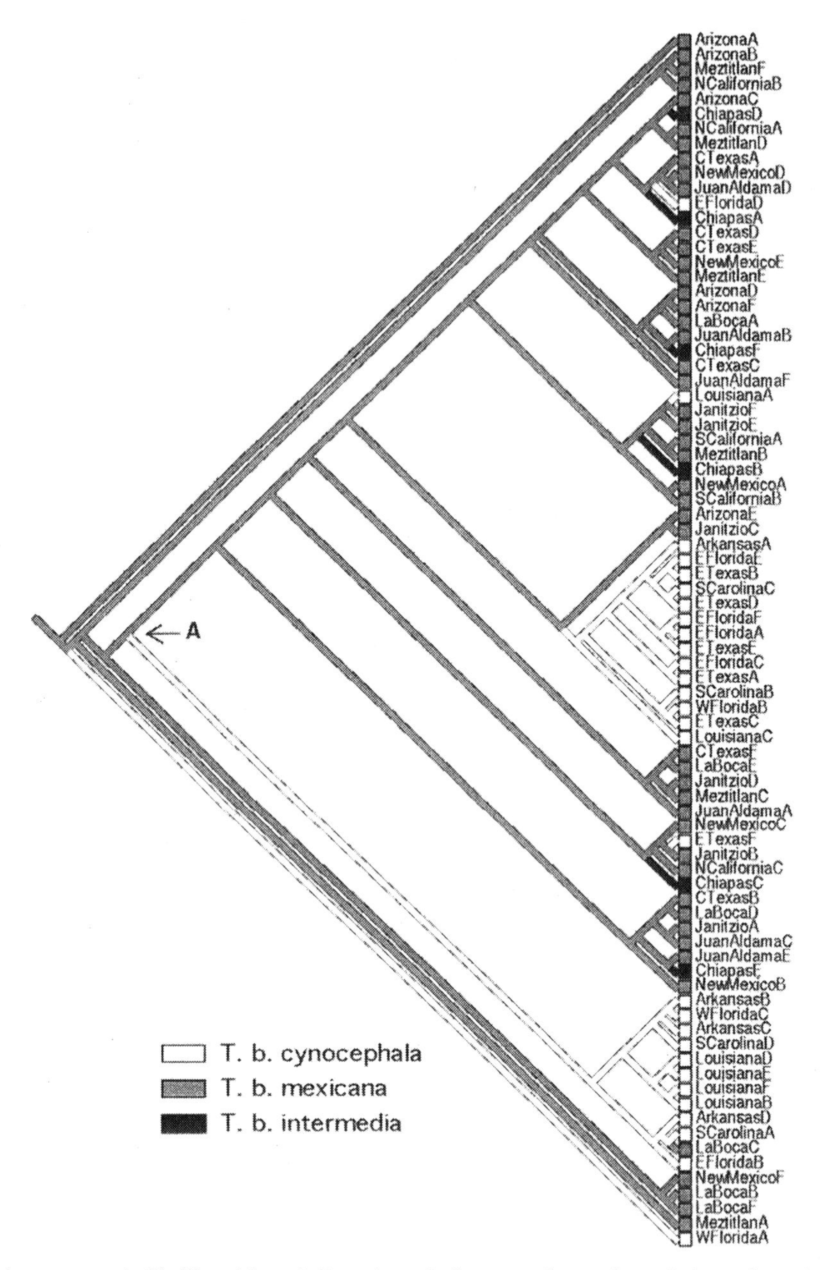

Figure 13.3 A likelihood-based Bayesian phylogeny of samples of the subspecies *T. b. cynocephala* (white), *T. b. mexicana* (gray), and *T. b. intermedia* (black). Subspecies assignment is mapped onto the tree as a character; putative migration events between subspecies occur at the inferred character-state changes. For example, a migration event between a subset of *T. b. mexicana* and *T. b. cynocephala* populations is inferred to occur at node A as indicated on the phylogeny. We refer to this tree as the "observed phylogeny." Using MacClade (Maddison and Maddison, 2000), we constructed 1000 random-joining trees using the same taxa as the observed phylogeny.

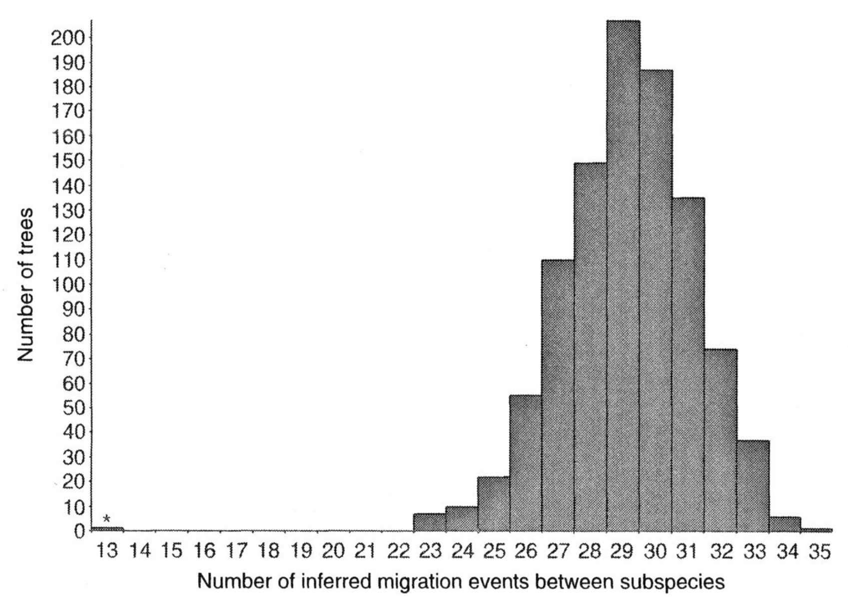

Figure 13.4 The total number of migration events on the observed phylogeny (*) compared with the distribution of total migration events on each of 1000 random-joining trees. This analysis shows that there are significantly fewer migration events required to explain the observed phylogeny than the random-joining trees, indicating that the clades on the observed phylogeny are structured nonrandomly. See text for further explanation.

changes represent inferred gene flow events between the assigned groups, that is, between the subspecies. In this analysis, a likelihood-based Bayesian phylogeny (Huelsenbeck and Ronquist, 2001) required significantly fewer gene flow events than a null distribution from 1000 randomly joining trees (figure 13.4). This indicates that there have been barriers to gene flow between North American populations. Examination of the phylogeny reveals that this genetic structuring among subspecies is due to the segregation of *T. b. cynocephala* into two major nonsister clades, both of which are embedded within a larger clade of sequences from *T. b. mexicana* (figure 13.3). However, sequences from *T. b. cynocephala* are not structured geographically among those two clades and the reason for this genetic structure is presently unclear. These genetic analyses indicate that *T. b. cynocephala* as a subspecies is not genetically coherent, and that substantial gene flow exists between *T. b. cynocephala* and *T. b. mexicana* (McCracken and Gassel, 1997; Owen et al., 1990; Russell and McCracken, unpublished data). The observed lack of differentiation between *T. b. intermedia* and *T. b. mexicana* is not surprising, given that the single population of *T. b. intermedia* included in the analysis is from the extreme northern end of the subspecies' range, and thus might be expected to undergo significant amounts of genetic exchange with *T. b. mexicana*.

POPULATION STRUCTURING WITHIN *TADARIDA BRASILIENSIS MEXICANA*

Evidence from Banding Studies

Extensive banding studies in the 1950s and 1960s involved over 430,000 banded bats with approximately 1.5% recovery (reviewed in Cockrum, 1969; Glass, 1982; McCracken et al., 1994) and led Cockrum (1969) to conclude that *T. b. mexicana* exists as four separate groups of subpopulations that may be genetically isolated due to different migratory corridors and different migratory behaviors (figure 13.5).

While the majority of banding recoveries are consistent with the movements of bats within and not between the population subgroups defined by Cockrum, there also are numerous recoveries demonstrating movement

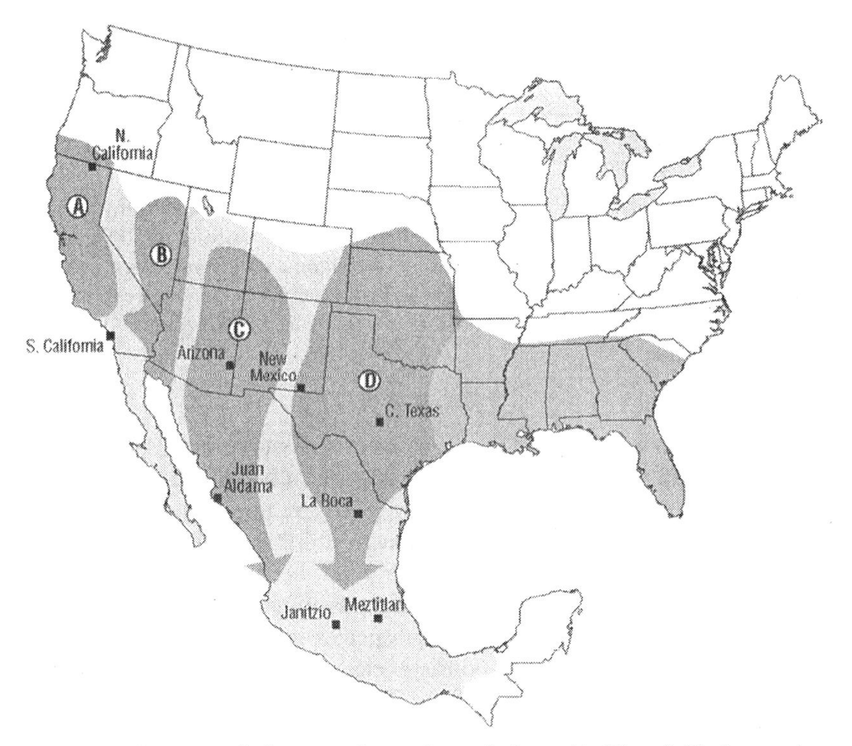

Figure 13.5 Ranges of the putative subpopulations (A–D) of *T. b. mexicana* (from Cockrum, 1969). Populations sampled by Russell et al. (2005) are indicated. Group A is a nonmigratory coastal group, experiencing only local seasonal movements. Group B is located in southeastern California and the Great Basin. Cockrum (1969) hypothesized that this group migrates relatively short distances into southern California and the Baja Peninsula, but this has not been substantiated. Group C was defined as migrating long distances from Arizona and New Mexico into western Mexico, separated from the other migratory group by the Sierra Madre Occidental mountain range. Group D was defined as migrating between Texas, New Mexico, and Oklahoma and Mexico on the eastern side of the Sierra Madre Occidental.

between groups. In studies at Carlsbad Caverns in western New Mexico, Constantine (1967) documented recoveries of 64 banded bats in places other than where they were banded. Of these, 13 recoveries (20%) involved individuals that moved between the eastern and western migratory groups (groups C and D; figure 13.5). Notably, the longest documented movement was from Carlsbad Caverns, New Mexico, in the eastern migratory group to Las Garrochas Cave, Jalisco, in the western migratory group (Villa-R. and Cockrum, 1962). Furthermore, the presence of large colonies in locations between the migratory groups, such as Ojuela Cave in the Sierra Madre Occidental (McCracken et al., 1994) or the San Luis Valley of Colorado (Svoboda et al., 1985), suggests that population subgroups are not as distinct as suggested by Cockrum (1969), and that there may be gene flow between these putative migratory groups of *T. b. mexicana*.

Evidence from Molecular Studies

Beginning in the mid-1980s, several studies examined molecular markers to quantify levels of divergence between Cockrum's (1969) proposed migratory groups of *T. b. mexicana* (McCracken and Gassel, 1997; McCracken et al., 1994; Russell et al., 2005; Svoboda et al., 1985).

Svoboda et al. (1985) examined six polymorphic allozyme loci in samples from summer maternity colonies within the two migratory groups (groups C and D; figure 13.5) and a large bachelor colony located in a region between the same two groups. *F*-statistics (Wright, 1978) provided evidence for genetic structure (av. $F_{ST} = 0.052$). However, Svoboda et al. (1985) concluded that this was due to differences among the sampled populations rather than genetic differentiation between migratory groups. McCracken et al. (1994) questioned the evidence of Svoboda et al. (1985), noting that their inter-populational differences were due to a single esterase locus and that the remaining five loci they examined showed no evidence of differences between populations (average $F_{ST} = 0.019$).

Hypothesizing that individuals from different populations mix during migration, McCracken et al. (1994) investigated levels of population structuring between the two putative migratory groups (groups C and D; figure 13.5) using samples from both summer maternity colonies and winter colonies. Data from 38 allozyme loci showed that neither the summer nor winter colonies were genetically structured into distinct geographic units. Analyses of allele frequencies in maternity colonies indicated that any single large colony contained the genetic diversity present in all migratory *T. b. mexicana*.

McCracken and Gassel (1997) extended these genetic analyses to include nonmigratory populations from California and *T. b. cynocephala* from Florida and Arkansas. Their analysis revealed no measurable differentiation among nonmigratory and migratory *T. b. mexicana* and detected no significant genetic structuring between *T. b. mexicana* and *T. b. cynocephala*.

Using more extensive sampling than these previous studies, we analyzed DNA sequence variation in the D-loop of the mitochondrial control region

Table 13.4 AMOVA Analysis of mtDNA Sequence Data of Putative Subpopulations of *T. b. mexicana*, Indicating the Proportion of Total Variance Attributed to Each Hierarchical Level

Among groups	1.66%
Among population, within groups	10.82%
Within population	87.52%
Φ_{CT}	0.0166
P value	0.3077

Data from Russell et al. (2005).
Φ_{CT} is equivalent to Wright's (1978) F_{ST} for allele frequency data. The significance of the Φ_{CT} value was tested by comparing the observed value to a distribution obtained from 10,000 random permutations of populations among the migratory groups.

from populations from both the United States and Mexico, including the coastal nonmigratory group and the two migratory groups (groups A, C, and D; figure 13.5). An AMOVA analysis indicated that differences between these putative subpopulations were insignificant relative to the variation within them (table 13.4). A cladistic analysis of gene flow also failed to detect any significant geographic structuring among haplotypes (figure 13.6). The number of migration events inferred from the Bayesian phylogeny fell within the null distribution expected from randomly joining trees (figure 13.7). Thus, all molecular genetic analyses have failed to find any support for genetic structuring among populations of *T. b. mexicana*. Because the mitochondrial genome is inherited only from the mother, conclusions from this analysis are limited to the movements of females.

ESTIMATES OF EFFECTIVE POPULATION SIZE (N_e)

A total of over 150 million *T. b. mexicana* are estimated to inhabit major maternity colonies in the southwestern United States (table 13.1). Banding data have shown that many of these colonies are linked by frequent exchange of individuals (Cockrum, 1969; Constantine, 1967; Glass, 1982), and the genetic studies reviewed above indicate that gene flow among colonies is sufficient to prevent geographic structuring of populations (Russell et al., 2005). Thus, all information indicates that these bats are characterized by an extraordinarily large effective population size.

To estimate N_e from our molecular data, we used two methods to estimate Θ; where $\Theta = 2N_{e(f)}u$ for haplotype data, u is the mutation rate per nucleotide, and $N_{e(f)}$ is the effective population size of females (Russell et al., 2005). A phylogenetically based method uses the maximum likelihood criterion to estimate Θ, and thus $N_{e(f)}$, for historical populations; we call this estimate Θ_H (Kuhner et al., 1998). Tajima (1983) showed that Θ can also be estimated from the mean number of pairwise differences. This estimate of Θ applies to

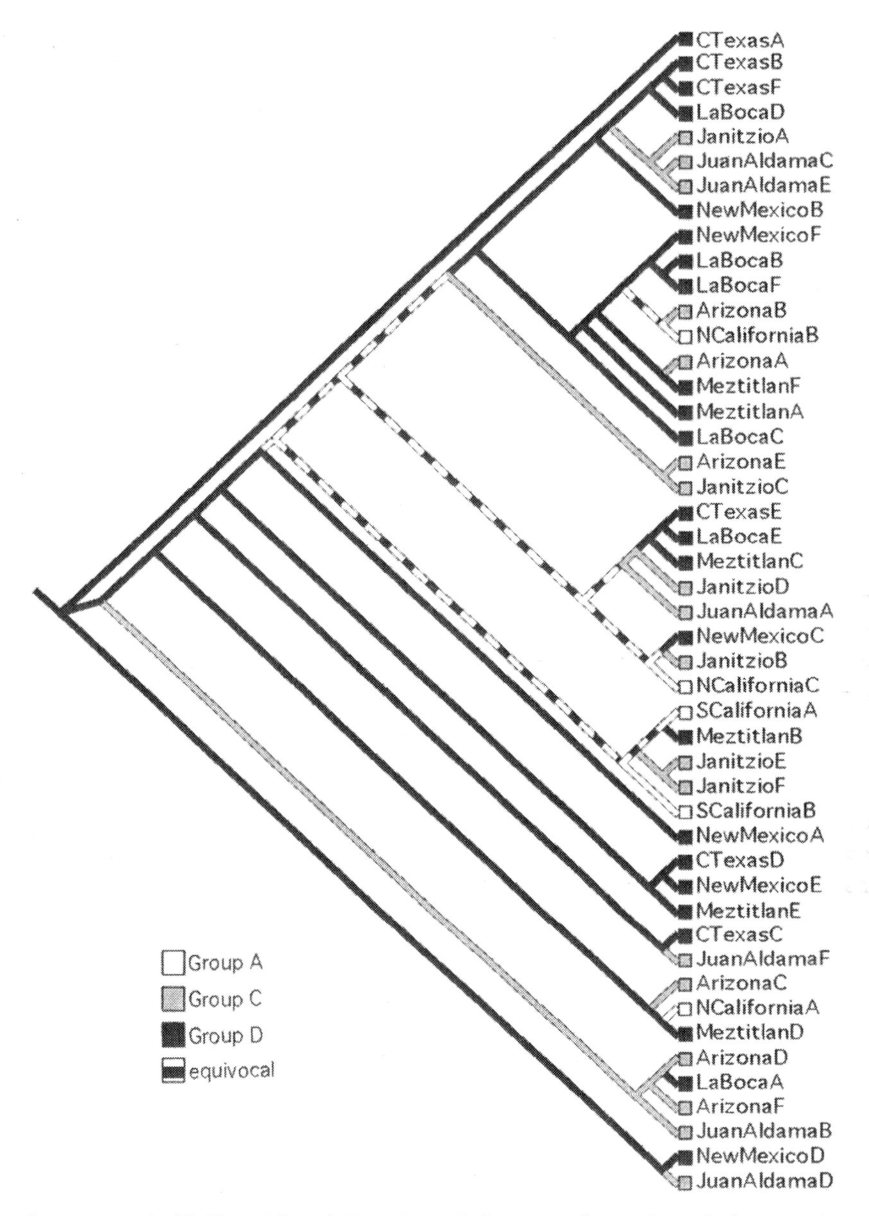

Figure 13.6 A likelihood-based Bayesian phylogeny of samples of the putative subpopulations of *T. b. mexicana*: easternmost long-distance migratory group (group D, black), western long-distance migratory group (group C, gray), and western coastal nonmigratory group (group A, white). Migratory group assignment is mapped onto the tree as a character; putative migration events between groups occur at inferred character-state changes.

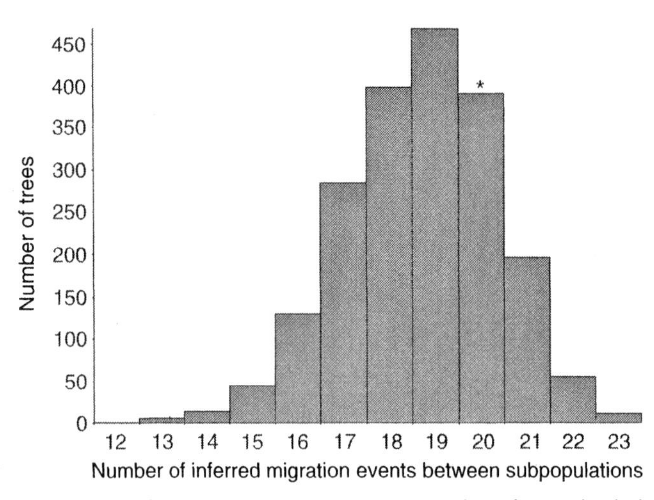

Figure 13.7 The number of migration events on the observed phylogeny (*) compared with the distribution of total migration events on each of 1000 random-joining trees.

current populations, and we refer to it as Θ_C. We have conducted a literature search for estimates of both Θ_H and Θ_C, emphasizing species expected to have high population sizes and/or high levels of gene flow (table 13.5). Estimates of $N_{e(f)}$ of historical populations have ranged from 5 million for widely ranging grey wolves (Vilà et al., 1999) to 500 million for *Anopheles* mosquitoes (Walton et al., 2000). For *T. b. mexicana* we estimate $\Theta_H = 1.668$, indicating a historical effective population size of approximately 11 million females—more than twice that of historical populations of grey wolves. Our estimate of Θ_C (= 41.781) indicates that *T. b. mexicana* has undergone a recent expansion to a current effective population size of 418 million females. This estimate is much larger than the highest we have found for bird ($N_{e(f)} = 224{,}347$; Griswold and Baker, 2002), bat ($N_{e(f)} = 159{,}000$; Wilkinson and Fleming, 1996), marine ($N_{e(f)} = 800{,}000$; Avise, 1992), or insect species ($N_{e(f)} = 4{,}226{,}667$; Walton et al., 2000). All estimates of $N_{e(f)}$ from Θ values, of course, depend on accurate estimates of u, the mutation rate per nucleotide. For our comparisons of $N_{e(f)}$ across taxa, we have used published estimates of u consistent in both general taxonomic group (i.e., mammal, bird, fish) and locus. Available estimates of $N_{e(f)}$ for other mammals include an historic $N_{e(f)}$ of 13,000 for the endangered Morro Bay kangaroo rat (Matocq and Villablanca, 2001) and a current $N_{e(f)} = 460{,}000$ for the coyote (Vilà et al., 1999). For the highly endangered Morro Bay kangaroo rat, current census population size is doubtless much less than 13,000 (Matocq and Villablanca, 2001), and for coyotes current census population size is estimated at 7 million (Vilà et al., 1999). These examples serve to illustrate that N_e is influenced by much more than the census population size. Comparisons of the N_e parameter among taxa with different life histories are still informative.

Table 13.5 Estimates of Θ (= $2N_{e(f)}u$) and $N_{e(f)}$ from Selected Species

Species	Common Name	Historical Estimates		Current Estimates		References
		Θ_H	$N_{e(f)}$	Θ_C	$N_{e(f)}$	
Fringilla coelebs	Common chaffinch	–	–	0.0103[a]	224,347[d]	Griswold and Baker, 2002
Leptonycteris curasoae yerbabuenae	Lesser long-nosed bat	–	–	0.0159[a]	159,000[e]	Wilkinson and Fleming, 1996
Brevoortia tyrannus	Atlantic menhaden	–	–	0.0319[b]	800,000[g]	Avise, 1992
Dipodomys heermanni morroensis	Morro Bay kangaroo rat	0.0013[a]	13,000[e]	–	–	Matocq and Villablanca, 2001
Canis latrans	Coyote	0.373[a]	3,730,000[e]	0.046[a]	460,000[e]	Vilà et al., 1999
Canis lupus	Grey wolf	0.744[a]	5,000,000[g]	0.026	173,000[g]	Vilà et al., (1999)
Anopheles dirus sp. D	Asian mosquito	0.757[c]	500,000,000[g]	0.00634	4,226,667[f]	Walton et al., 2000
Tadarida brasiliensis mexicana	Mexican free-tailed bat	1.668[a]	16,680,000[e]	41.781[a]	417,810,000[e]	Russell et al., (2005)

Estimates of Θ_H were derived using a phylogenetically based analysis (Kuhner et al., 1998), yielding parameter estimates for historical populations. Estimates of Θ_C (also denoted in the literature as Θ_π or π) were derived from analyses using haplotype frequencies (Tajima, 1983), yielding estimates for current populations. Estimates of $N_{e(f)}$ were either taken directly from the literature or, where noted, derived from Θ values, using values of u available in the literature. Estimates of u were specific to the taxonomic group (birds, mammals, arthopods, etc.) and locus. All loci are on the mitochondrial genome.

[a]Estimated from control region sequence data.

[b]Estimated from restriction fragment length polymorphism (RFLP) data of the entire mtDNA genome.

[c]Estimated from cytochrome oxidase 1 sequence data.

[d]Calculated from Θ, with $u = 2.3 \times 10^{-8}$ (u from Kvist et al., 1998).

[e]Calculated from Θ, with $u = 5 \times 10^{-8}$ (u from Vilà et al., 1999).

[f]Calculated from Θ, with $u = 7.6 \times 10^{-10}$ (u from Walton et al., 2000).

[g]Estimate of $N_{e(f)}$ taken directly from reference.

The conclusion that the effective number of female *T. b. mexicana* is extraordinarily large is also supported by the observed diversity in mtDNA haplotypes. In a dataset representing 44 individuals from nine colonies throughout the subspecies' range, we identified 43 unique haplotypes. Similarly high levels of haplotype diversity have been found in other species with large population sizes (70 unique haplotypes from 84 individuals of *Anopheles dirus*; Walton et al., 2000). Such high haplotype diversity is an expected characteristic of an extremely large population and we interpret this diversity as the consequence of unusually large current effective population sizes.

CONCLUSIONS

Behavioral, ecological, and morphological studies have suggested species-level or subspecific distinctions among North American populations of *T. brasiliensis*. However, close examination reveals substantial plasticity in these behavioral and ecological traits, broad overlap in morphology, and little genetic differentiation between the subspecies. While extensive banding studies led to the conclusion that *T. b. mexicana* exists as four behaviorally distinct migratory groups, genetic studies do not support the purportive population-level structuring.

Despite the traditionally emphasized differences in the behavior of the subspecies, the most rigorous genetic studies to date suggest genetic exchange. Using mtDNA sequence data, we document a more complex phylogeographic pattern. Significant differences exist between the South American subspecies *T. b. brasiliensis* and the North American subspecies *T. b. intermedia*, *T. b. mexicana*, and *T. b. cynocephala*. These differences may be due to a barrier to gene flow or to isolation-by-distance. These alternative hypotheses remain to be tested with more complete sampling of populations in Central and northern South America. Mitochondrial DNA sequence data do not support structuring of *T. b. cynocephala*, *T. b. mexicana*, and *T. b. intermedia* along established taxonomic designations, but a cladistic analysis of gene flow shows a structuring of haplotypes of *T. b. cynocephala* into two clades that do not correspond with geographic distributions. One potential explanation for this pattern is that these clades represent two separate invasions into the southeastern United States from *T. b. mexicana* populations, followed by convergence to cynocephala-type behavior and morphology. This hypothesis could be examined by testing for concordant patterns at other genetic loci, particularly nuclear loci.

Previous genetic studies provide no support for population-level structuing within *T. b. mexicana*. Using mtDNA sequence data, we also find no genetic evidence of structuring among colonies of *T. b. mexicana*. From these analyses, we conclude that populations of *T. b. mexicana* are characterized by high levels of gene flow, to the extent that the entire subspecies may be evolving as a single genetic unit. Estimates of Θ_H and Θ_C ($= 2N_{e(f)}u$), where Θ_H concerns the historical population and Θ_C concerns the current population, reveal that the historical size of the *T. b. mexicana* population was comparable to that of other wide-ranging species, and suggests a recent and substantial increase in population size. This conclusion is supported by the large census population sizes documented for colonies in the southwestern United States and Mexico, as well as the large amount of genetic diversity maintained in the subspecies.

ACKNOWLEDGMENTS

The authors thank Tom Kunz and Akbar Zubaid for the invitation to participate in the Roosting Ecology and Population Biology Symposium. Lisa Bailey, Tom Near, Leslie Saidak, Randy Small, Nancy Simmons, and Sunitha Vege provided invaluable advice regarding data collection and analysis. Maps were prepared by the University of Tennessee Cartography laboratory. We thank Randy Small, Carl-Gustaf Thulin, Maarten Vonhof, and an anonymous reviewer for comments on the manuscript. This research was supported by the American Museum of Natural History, Bat Conservation International, Sigma Xi, and the University of Tennessee.

LITERATURE CITED

Avise, J.C. 1992. Molecular population structure and the biogeographic history of a regional fauna: a case history with lessons for conservation biology. Oikos, 63: 62–76.

Bain, J.R. 1981. Roosting ecology of three Florida bats: *Nycticeius humeralis*, *Myotis austroriparius*, and *Tadarida brasiliensis*. M.S. Thesis, University of Florida, Gainesville, FL.

Barbour, R.W., and W.H. Davis. 1969. Bats of America. University Press of Kentucky, Lexington, KY.

Bennett, B. 2000. Pesticide accumulation and DNA content variation in two subspecies of Brazilian free-tailed bats. M.S. Thesis, Sam Houston State University, Huntsville, TX.

Carter, D.C. 1962. The systematic status of the bat *Tadarida brasiliensis* (I. Geoffroy) and its related mainland forms. Ph.D. Dissertation, Agricultural and Mechanical College of Texas, College Station, TX.

Cockrum, E.L. 1969. Migration in the guano bat, *Tadarida brasiliensis*. Miscellaneous Publications, The University of Kansas Museum of Natural History, 51: 303–336.

Constantine, D.G. 1967. Activity patterns of the Mexican free-tailed bat. University of New Mexico Publications in Biology, 7: 1–79.

Davis, R. 1966. Homing performance and homing ability in bats. Ecological Monographs, 36: 201–237.

Davis, R.B., C.F. Herreid II, and H.L. Short. 1962. Mexican free-tailed bats in Texas. Ecological Monographs, 32: 311–346.

Excoffier, L., P.E. Smouse, and J.M. Quattro. 1992. Analysis of molecular variance inferred from metric distances among DNA haplotypes: application to human mitochondrial DNA restriction data. Genetics, 131: 479–491.

Genoways, H.H., P.W. Freeman, and C. Grell. 2000. Extralimital records of the Mexican free-tailed bat (*Tadarida brasiliensis mexicana*) in the central United States and their biological significance. Transactions of the Nebraska Academy of Sciences, 26: 85–96.

Glass, B.P. 1982. Seasonal movements of Mexican free-tail bats *Tadarida brasiliensis mexicana* banded in the great plains. Southwestern Naturalist, 27: 127–133.

Grinnell, H.W. 1918. A synopsis of the bats of California. University of California Publications in Zoology, 17: 223–404.

Griswold, C.K., and A.J. Baker. 2002. Time to the most recent common ancestor and divergence times of populations of common chaffinches (*Fringilla coelebs*) in Europe and North Africa: insights into Pleistocene refugia and current levels of migration. Evolution, 56: 143–153.

Hall, E.R. 1981. The Mammals of North America, 2nd edition. Wiley, New York.

Herreid, C.F., II. 1963. Survival of a migratory bat at different temperatures. Journal of Mammalogy, 44: 431–433.

Huelsenbeck J.P., and F.R. Ronquist. 2001. MRBAYES: Bayesian inference of phylogeny. Bioinformatics, 17: 754–755.

Jewett, S.G. 1955. Free-tailed bats, and melanistic mice in Oregon. Journal of Mammalogy, 36: 458–459.

Kerth, G., F. Mayer, and B. König. 2000. Mitochondrial DNA (mtDNA) reveals that female Bechstein's bats live in closed societies. Molecular Ecology, 9: 793–800.

Kiser, W.M. 1996. Observations of a winter colony of LeConte's free-tailed bats *Tadarida brasiliensis cynocephala* in Lee Co., Alabama. Bat Research News, 37: 116.

Kiser, W.M., and K.V. Glover. 1997. Acceptance of an artificial roost by LeConte's free-tailed bats, *Tadarida brasiliensis cynocephala*. Bat Research News, 38: 1–4.

Krutzsch, P.H. 1955. Observations on the Mexican free-tailed bat, *Tadarida mexicana*. Journal of Mammalogy, 36: 236–242.

Kuhner, M.K., J. Yamato, and J. Felsenstein. 1998. Maximum likelihood estimation of population growth rates based on the coalescent. Genetics, 149: 429–434.

Kvist, L., M. Ruokonen, A. Thessing, J. Lumme, and M. Orell. 1998. Mitochondrial control region polymorphism reveals high amount of gene flow in Fennoscandian willow tit (*Parus montanus borealis*). Hereditas, 128: 133–143.

LaVal, R.K. 1973. Observations on the biology of *Tadarida brasiliensis cynocephala* in southeastern Louisiana. American Midland Naturalist, 89: 112–120.

Maddison, D.R., and W.P. Maddison. 2000. MacClade 4: Analysis of Phylogeny and Character Evolution, version 4.0. Sinauer Associates, Sunderland, MA.

Marques, R.V. 1991. Ciclo reprodutivo e aspectos do comportamento de *Tadarida brasiliensis brasiliensis* (I. Geoffroy, 1824)—Chiroptera, Molossidae—em ambiente urbano na região de Porto Alegre, RS, Brasil. M.S. Thesis, Pontifícia Universidade Católica do RS, Porto Alegre.

Matocq, M.D., and F.X. Villablanca. 2001. Low genetic diversity in an endangered species: recent or historic pattern? Biological Conservation, 98: 61–68.

McCracken, G.F. 2003. Estimates of population sizes in summer colonies of Brazilian free-tailed bats. Pp. 21–30. In: Monitoring Trends in Bat Populations of the United States and Territories: Problems and Prospects (T.J. O'Shea and M. Bogan, eds.). U.S. Geological Survey, Biological Resources Discipline, Information and Technology Report. USGS/BRB/ITR-2003-0003.

McCracken, G.F., and M.F. Gassel. 1997. Genetic structure in migratory and nonmigratory populations of Brazilian free-tailed bats. Journal of Mammalogy, 78: 348–357.

McCracken, G.F., M.K. McCracken, and A.T. Vawter. 1994. Genetic structure in migratory populations of the bat *Tadarida brasiliensis mexicana*. Journal of Mammalogy, 75: 500–514.

Orr, R.T. 1958. Keeping bats in captivity. Journal of Mammalogy, 39: 339–344.

Owen, R.D., R.K. Chesser, and D.C. Carter. 1990. The systematic status of *Tadarida brasiliensis cynocephala* and Antillean members of the *Tadarida brasiliensis* group, with comments on the generic name *Rhizomops* Legendre. Occasional Papers, The Museum, Texas Tech University, 133: 1–18.

Perkins, J.M., J.M. Barss, and J. Peterson. 1990. Winter records of bats in Oregon and Washington. Northwestern Naturalist, 71: 59–62.

Petit E., L. Excoffier, and F. Mayer. 1999. No evidence of bottleneck in the post-glacial recolonization of Europe by the noctule bat (*Nyctalus noctula*). Evolution, 53: 1247–1258.

Russell, A.L., R.A. Medellin, and G.F. McCracken, 2005. Genetic variation and migration in the Mexican free-tailed bat (*Tadarida brasiliensis mexicana*), Molecular Ecology, 14: 2207–2222.

Schmidley D.J., K.T. Wilkins, R.L. Honeycutt, and B.C. Weynand. 1977. The bats of Texas. Texas Journal of Science, 28: 127–143.

Schwartz, A. 1955. The status of the species of the *brasiliensis* group of the genus *Tadarida*. Journal of Mammalogy, 36: 106–109.

Shamel, H.H. 1931. Notes on the American bats of the genus *Tadarida*. Proceedings of the U.S. National Museum, 78:1–27.

Sherman, H.B. 1937. Breeding habits of the free-tailed bats. Journal of Mammalogy, 18: 176–187.

Slatkin, M., and W.P. Maddison. 1989. A cladistic measure of gene flow inferred from the phylogenies of alleles. Genetics, 123: 603–613.

Spenrath, C.A., and R.K. LaVal. 1974. An ecological study of a resident population of *Tadarida brasiliensis* in eastern Texas. Occasional Papers, The Museum, Texas Tech University, 21: 1–14.

Svoboda, P.L., J.R. Choate, and R.K. Chesser. 1985. Genetic relationships among southwestern populations of the Brazilian free-tailed bat. Journal of Mammalogy, 66: 444–450.

Tajima, F. 1983. Evolutionary relationship of DNA sequences in finite populations. Genetics, 105: 437–460.

Templeton, A.R. 1998. Nested clade analyses of phylogeographic data: testing hypotheses about gene flow and population history. Molecular Ecology, 7: 381–397.

Vilà, C., I.R. Amorim, J.A. Leonard, D. Posada, J. Castroviejo, F. Petrucci-Fonseca, K.A. Crandall, H. Ellegren, and R.K. Wayne. 1999. Mitochondrial DNA phylogeography and population history of the grey wolf *Canis lupus*. Molecular Ecology, 8: 2089–2103.

Villa-R., B. 1956. *Tadarida brasiliensis mexicana* (Saussure), el murcielago guanero, es una subspecie migratoria. Acta Zoologica Mexicana, 1: 1–11.

Villa-R., B., and E.L. Cockrum. 1962. Migration in the guano bat *Tadarida brasiliensis mexicana* (Saussure). Journal of Mammalogy, 43: 43–64.

Walton, C., J.M. Handley, W. Tun-Lin, F.H. Collins, R.E. Harbach, V. Baimai, and R.K. Butlin. 2000. Population structure and population history of *Anopheles dirus* mosquitoes in southeast Asia. Molecular Biology and Evolution, 17: 962–974.

Wilkins, K.T. 1989. *Tadarida brasiliensis*. Mammalian Species, 331: 1–10.

Wilkinson, G.S., and T.H. Fleming. 1996. Migration and evolution of lesser long-nosed bats *Leptonycteris curasoae*, inferred from mitochondrial DNA. Molecular Ecology, 5: 329–339.

Wright, S. 1978. Evolution and the Genetics of Populations: Variability Within and Among Natural Populations. University of Chicago Press, Chicago.

14

Evolutionary Dynamics of the Short-Nosed Fruit Bat, *Cynopterus sphinx* (Pteropodidae): Inferences from the Spatial Scale of Genetic and Phenotypic Differentiation

Jay F. Storz, Hari R. Bhat, Johnson Balasingh, P. Thiruchenthil Nathan, & Thomas H. Kunz

We report the results of a population-genetic study of the short-nosed fruit bat, *Cynopterus sphinx* (Pteropodidae). The purpose of our study was to assess the relative importance of drift, gene flow, and spatially varying selection in shaping patterns of genetic and phenotypic variation across a latitudinal climatic gradient in peninsular India. At a microgeographic scale, polygynous mating resulted in a substantial reduction of effective population size. However, at a macrogeographic scale, rates of migration were sufficiently high to prevent a pronounced degree of stochastic differentiation via drift. Spatial analysis of genetic and phenotypic differentiation revealed that clinal variation in body size of *C. sphinx* cannot be explained by a neutral model of isolation by distance. The geographic patterning of morphometric variation is most likely attributable to spatially varying selection and/or the direct influence of latitudinally ordered environmental effects. The combined analysis of genetic and phenotypic variation indicates that recognized subspecies of *C. sphinx* in peninsular India represent arbitrary subdivisions of a continuous spectrum of clinal size variation.

INTRODUCTION

What is the relative importance of drift versus gene flow in driving the genetic differentiation of partially isolated populations? What is the relative importance of spatially varying selection versus stochastic processes in maintaining clinal variation in phenotypic traits? These questions are of central importance to our understanding of local adaptation and the determinants of geographic variation (Endler, 1977; García-Ramos and Kirkpatrick, 1997; Haldane, 1948; Slatkin, 1973, 1978). Many key inferences about the role of selection in maintaining clinal variation have been obtained by relating inferred dispersal distances to cline widths (Barton and Gale, 1993). Highly vagile animals such as bats and birds are of particular interest in this regard because the persistence of clinal variation requires recurrent selection to counterbalance the homogenizing effect of gene flow. For example, cline widths of approximately 50 km have been documented for Robertsonian fusions between populations of the phyllostomid bat *Uroderma bilobatum* in Central America (Baker, 1981). The standard deviation of single-generation dispersal distances was estimated to be 11 km, indicating that stable maintenance of the observed cline would require selection coefficients of >0.37 against heterozygotes in the contact zone (Barton, 1982).

Relative to the effects of selection, it is considerably more difficult to quantify the importance of drift as a cause of spatial differentiation. In a review of empirical data on cline widths in mammalian populations, Barton (1990) suggested that spatial patterns of differentiation may be greatly affected by nonrandom breeding and drift at the local population level. Similarly, Patton and Smith (1990) argued that the extreme levels of genetic and phenotypic differentiation observed in North American pocket gophers (genus *Thomomys*) are largely attributable to the localized effects of sampling drift caused by polygynous mating and extinction–recolonization dynamics. Since pocket gophers are characterized by highly restricted dispersal capabilities, stochastic fluctuations in allelic frequencies at a microspatial scale may be an important determinant of broad-scale population genetic structure (Daly and Patton, 1990; Patton and Feder, 1981; Patton and Yang, 1977).

Measuring rates of drift requires estimates of effective population size (N_e). It is therefore important to determine whether estimates of N_e for local populations can be used to predict macrogeographic patterns of genetic differentiation. This is relevant to the broader issue of whether local population processes can be extrapolated to broad-scale patterns of genetic differentiation (Lidicker and Patton, 1987).

Here we summarize an analysis of genetic and phenotypic variation in the short-nosed fruit bat, *Cynopterus sphinx*, across a range of spatial scales in peninsular India. We start by examining the influence of polygynous mating and overlapping generations on N_e. We then assess the extent to which estimates of N_e at the local population level can inform the analysis of genetic differentiation at a broader geographic scale. After characterizing the geographic patterning of neutral genetic variation, we compare relative levels of genetic and phenotypic differentiation to assess the possible role of spatially varying selection in maintaining a latitudinal cline in body size. Specifically, we compare relative levels of between-population divergence in body size and neutral DNA markers to assess whether the observed pattern of clinal size variation can be explained by a neutral model of isolation by distance.

NATURAL HISTORY

Cynopterus sphinx is a harem-forming frugivorous bat that is widely distributed across the Indomalayan Region (Storz and Kunz, 1999). In one of our main study sites in western India (Pune, Maharashtra, 18°32′ N, 73°51′ E), harem breeding groups of *C. sphinx* roost in tents constructed within flower/fruit clusters of the kitul palm (*Caryota urens*; Storz et al., 2000b). The breeding population of *C. sphinx* in Pune is subdivided into diurnal roosting colonies, each containing one to five harems and often one or more satellite males in adjacent roosts. Colonies comprise all bats occupying flower/fruit cluster tents in a single tree, or cluster of two or three adjacent trees (figure 14.1). In many cases, there is only one tent-roosting harem per colony. In other

Figure 14.1 (a) A kitul palm tree *Caryota urens* that housed a colony of *Cynopterus sphinx* in Pune, India. Three flower/fruit clusters that were modified as tents (T3, T4, and T5) and occupied by bats in the 1997 wet season are indicated. Harems of *C. sphinx* roosting in the altered crowns of mature fruit clusters (tent T4 and tent T5) in the 1998 dry season, are shown in (b) and (c), respectively. The harem males in each tent are indicated by arrows.

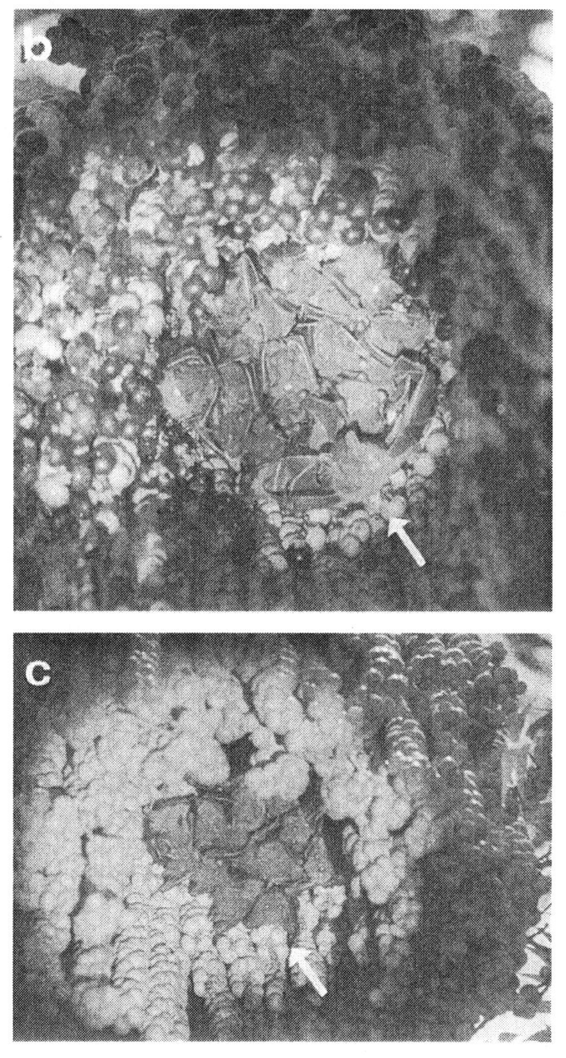

Figure 14.1 Continued.

colonies, several harems occupy different tents in the same or adjacent trees. Harems consist of a single adult male roosting in association with one to 37 reproductive females and their dependent young. Harem size averages 6.1 adults in the wet season (SD = 3.5) and 13.6 adults in the dry season (SD = 8.5). The same harem social configuration is maintained year-round, despite a high degree of synchrony and seasonality in the timing of reproduction. Adult females often remain associated as roost-mates from one parturition period to the next, and group cohesion is unaffected by turnover of harem males. Juveniles of both sexes disperse after weaning and sexually immature bats are never present in harems at the time of parturition (Storz et al., 2000a,b).

THE LOCAL POPULATION LEVEL: GENETIC CONSEQUENCES OF POLYGYNY

In animal taxa characterized by polygynous mating systems, variance in male reproductive success is a primary determinant of N_e (Nunney, 1993; Wright, 1938) and may thus exert a powerful influence on the likely course of microevolutionary events. In populations characterized by polygynous mating and overlapping generations, N_e increases as a positive function of generation interval (Nunney, 1993, 1996). When generations overlap, the ratio of N_e to adult census number (N) is predicted to fall within the range of 0.25–0.75 under most demographic circumstances (Nunney, 1993, 1996; Nunney and Elam, 1994). In species characterized by a relatively rapid maturation period scaled to adult lifespan, N_e/N exhibits an asymptotic convergence to 0.5 with increasing generation time (Waite and Parker, 1996). According to theory, extreme circumstances are required to reduce $N_e/N < 0.25$ (Nunney, 1993, 1996; Nunney and Elam, 1994). It remains to be determined whether the high variance in male reproductive success thought to characterize populations of harem-forming mammals and lek-mating birds is generally capable of producing such circumstances.

One of the objectives of our study of *C. sphinx* was to test the hypothesis that polygynous mating results in a significantly reduced N_e in a population with overlapping generations. This hypothesis was tested in a natural population of *C. sphinx* in Pune, India. Using 10-locus microsatellite genotypes of adults and progeny from consecutive breeding periods (Storz et al., 2001b,c), variance in male mating success was inferred from the relative proportion of successfully reproducing males and the size distribution of paternal sibships comprising each offspring cohort. The influence of the mating system on N_e was then assessed using a model designed for age-structured populations that incorporated demographic and genetic data obtained from individually marked bats.

Same-age offspring were assigned to sibships using a likelihood-based analysis of paternal relatedness (Storz et al., 2001c). Within-season variance in male mating success was then estimated from the size distribution of paternal sibships comprising each offspring cohort. Accordingly,

$$\sigma^2 = \frac{N_m \sum_i p_i^2 - \left(\sum_i p_i\right)^2}{N_m^2}, \tag{1}$$

where N_m is the total number of sexually mature males in the population (including nonbreeding males), and p is the number of pups in the ith paternal sibship. The standardized variance in male mating success (I_{bm}) was calculated as

$$I_{bm} = \frac{N_m \sum_i p_i^2 - \left(\sum_i p_i\right)^2}{n^2}, \tag{2}$$

where n is the total number of pups in the offspring cohort. In any given breeding period, there is variance in progeny number among successful males in addition to the variance between successful and unsuccessful males that results when $(N_m - s)$ of the males do not mate (where s is the number of paternal sibships).

The ratio of effective size to adult census number (N_e/N) for the *C. sphinx* study population was estimated using the method of Nunney (1993: equation A2):

$$N_e/N = [4r(1-r)T] \div [A_m(1-r) + A_f r] + [I_{bm}(1-r) + I_{bf} r] + [A_m I_{Am}(1-r) + A_f I_{Af} r], \quad (3)$$

where r is the operational sex ratio (expressed as the proportional number of sexually mature males), T is the mean generation interval $[=(T_m + T_f)/2$, where T_i is the generation interval of sex $i]$, A_i is the average adult lifespan of both sexes, I_{Ai} is the standardized variance in adult lifespan of both sexes, I_{bf} is the standardized variance in female fecundity per breeding period, and I_{bm} is the standardized variance in male mating success per breeding period (for details of parameter estimation, see Storz et al., 2001c).

Results of our analysis indicated that the study population of *C. sphinx* was characterized by an extremely high variance in male mating success (figure 14.2), as expected from its harem-forming mode of social organization (Storz et al., 2000a,b). The distribution of paternity was more highly skewed in the 1997 (wet season) cohort of offspring than in the 1998 (dry season) cohort. Differences in the degree of polygyny between the two offspring cohorts were primarily attributable to seasonal variation in the dispersion of females. Tight clustering of females in diurnal roosts facilitates a male mating strategy of resource-defense polygyny (Storz et al., 2000a,b).

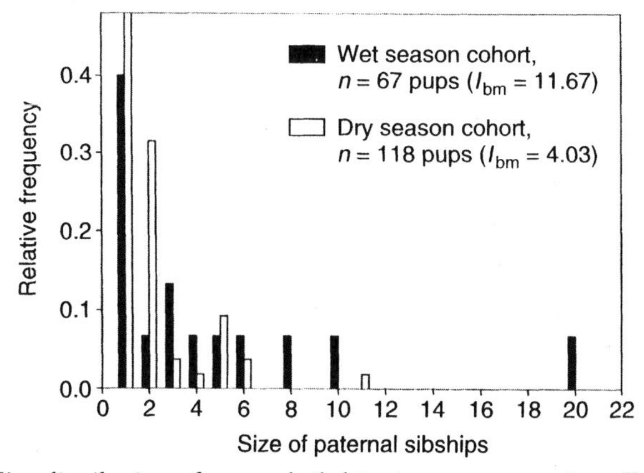

Figure 14.2 Size distribution of paternal sibships in two consecutive offspring cohorts (wet season and dry season) of *C. sphinx* in Pune, India. A "sibship" of size 1 represents a pup with no shared paternity in the same age cohort.

When aggregations of reproductive females are distributed among a limited number of roosts that males can defend as territories, a small fraction of the adult male population will likely succeed in monopolizing opportunities for mating. In the dry season, female dispersion is highly clumped and average harem size is 2.3-fold higher than in the wet season. Pups born in the wet season are conceived 115–135 days previously, during the post-partum estrus period (midway through the dry season) when the potential for polygyny is greatest (Storz et al., 2000a,b). The proportionally greater degree of polygyny reflected in the wet-season offspring cohort was therefore consistent with seasonal differences in average size of harems (Storz et al., 2001b).

In conjunction with estimates of parameters describing reproduction and demography of the *C. sphinx* study population, substitution of the average within-season I_{bm} into equation (3) resulted in an N_e/N estimate of 0.42. By contrast, substitution of the Poisson-expected value of I_{bm} resulted in an N_e/N estimate of 0.51. Thus, as a result of polygynous mating, the predicted rate of drift ($1/2N_e$ per generation) in the *C. sphinx* study population was 17.6% higher than expected from a Poisson distribution of male mating success. The estimated N_e/N for the *C. sphinx* study population was substantially lower than would be expected if a more egalitarian mating system prevailed. However, despite the high within-season variance in male mating success, the estimated N_e/N was well within the 0.25–0.75 range expected for age-structured populations under otherwise unexceptional demographic conditions (Nunney, 1993, 1996; Nunney and Elam, 1994).

The life-history schedule of *C. sphinx* (and that of bats in general) is characterized by a disproportionately short sexual maturation period scaled to adult lifespan. Consequently, the influence of polygynous mating on N_e/N is mitigated by the extensive overlap of generations (Nunney, 1993; Waite and Parker, 1996). In *C. sphinx*, as in other long-lived, polygynous mammals, continual turnover of breeding males ultimately ensures a broader sampling of the adult male gamete pool than indicated by the disproportionate posterity of top-ranking individuals within a single season.

Our estimate of N_e/N for the *C. sphinx* study population may be generally applicable to a large number of phyllostomid and pteropodid bat species, most of which are characterized by polygynous mating systems and overlapping generations (McCracken and Wilkinson, 2000; Wilkinson, 1987). Our estimate of N_e/N for *C. sphinx* is considerably lower than similar estimates obtained for 10 other mammalian species (median = 0.66, range = 0.51–1.27; table 14.1). Consideration of lineage-specific life-history schedules suggests that bats may be characterized by generally low N_e/N ratios relative to other mammalian taxa (Storz et al., 2001c). However, compared with other orders of mammals, bats do not exhibit levels of genetic heterozygosity or karyotypic diversity consistent with long-term small N_e values (Bush et al., 1977; Coyne, 1984). Among mammalian taxa characterized by an extensive overlap of generations, variation in long-term N_e may have little to do with differences in mating systems. Instead, rates of drift over evolutionary time-scales

Table 14.1 Demographic Estimates of the Ratio of Variance Effective Size to Adult Census Number (N_e/N) for Populations of 11 Mammalian Species

Species	N_e/N	T_m, T_f	**References**
White-toothed shrew, *Crocidura russula*	0.60	1.0, 1.0	Bouteiller and Perrin, 2000
Short-nosed fruit bat, *Cynopterus sphinx*	0.42	8.79, 7.88	Storz et al., 2001c
Human (Gainj), *Homo sapiens*	1.27	38.3, 31.3	Storz et al., 2001d
Savannah baboon, *Papio cynocephalus*	0.51	10.1, 10.2	Storz et al., 2002
Grizzly bear, *Ursus arctos*	0.82	10.1, 10.1	Nunney and Elam, 1994
Old World rabbit, *Oryctolagus cuniculus*	0.56	1.8, 1.8	Nunney and Elam, 1994
Grey squirrel, *Sciurus carolinensis*	0.57	1.79, 2.44	Nunney and Elam, 1994
Banner-tailed kangaroo rat, *Dipodomys spectabilis*	0.56	1.7, 1.7	Nunney and Elam, 1994
Wild horse, *Equus caballus*	0.82	7.23, 7.86	Nunney and Elam, 1994
Moose, *Alces alces*	0.84	4.8, 4.8	Nunney and Elam, 1994
American bison, *Bison bison*	0.72	7.0, 7.0	Nunney and Elam, 1994

Estimates of N_e/N for *C. russula*, *H. sapiens*, and *P. cynocephalus* were obtained using the Hill (1972, 1979) equation for age-structured populations. The remaining estimates were obtained using a simplified version of the Hill equation based on the assumption of age-independent survival and fertility (Nunney, 1993; Nunney and Elam, 1994). T_m and T_f are the mean generation times (in years) for males and females, respectively.

are likely highest in lineages characterized by stochastic variation in population numbers.

GEOGRAPHIC PATTERNING OF GENETIC AND PHENOTYPIC VARIATION

Under equilibrium conditions in an island model of population structure, and assuming that the rate of mutation is negligibly small relative to the rate of migration, Wright (1969: 291) demonstrated that $F_{ST} = (1 + 4N_e m)^{-1}$ (where F_{ST} is the standardized variance in allelic frequencies and $N_e m$ is the migration rate scaled by local effective size). Kimura and Weiss (1964) showed that, as a general rule of thumb, the homogenizing effect of gene flow will be sufficient to counteract stochastic divergence via drift when $N_e m > 4$. In contrast, the diversifying effect of drift will be sufficient to overpower gene flow when $N_e m << 1$. Results of our analysis reported above indicate that, in the absence of outside emigration, the Pune study population of *C. sphinx* would be characterized by an instantaneous effective size of 108.2 ($N_e/N = 0.42$, average $N = 257.5$ adults; Storz et al., 2001c). Assuming that this estimate is generally applicable to local populations of *C. sphinx* across the region surveyed, a per generation migration rate of > 0.037 would be required to meet the criterion that $N_e m > 4$.

To assess the scale and magnitude of spatial differentiation in *C. sphinx*, a total of 251 adults were sampled for a joint analysis of genetic and morphometric variation. Bats were sampled from eight localities along a latitudinal transect that spanned a linear distance of 1080 km across peninsular India (from 18° N to 9° N; figure 14.3). Morphometric variation of *C. sphinx* was assessed by examining seven external characters that jointly summarize overall body dimensions and wing area (Storz, 2002; Storz et al., 2001a). Principal components analysis on the variance–covariance matrix of the seven morphometric traits was used to extract a multivariate index of overall body size. Genetic analysis was based on a total of six polymorphic microsatellite loci and partitioning of genetic variance within and among populations was assessed using Weir and Cockerham's (1984) estimators of *F*-statistics.

Inferring the Relative Importance of Drift and Gene Flow in Causing Genetic Differentiation

The geographic patterning of microsatellite variation revealed a relatively low degree of genetic subdivision across peninsular India ($F_{ST} = 0.030$, 95% CI $= 0.013$–0.045). However, simple summary statistics provide no insight into the relative importance of drift and gene flow in causing the observed level of genetic differentiation. Two separate criteria can be used to determine whether a particular population has attained migration–drift equilibrium: (1) a significant association between pairwise F_{ST} versus and distance, and/or (2) a scatterplot of pairwise F_{ST} versus distance that reveals a positive and monotonic relationship over the full range of distance values (Hutchison and Templeton, 1999).

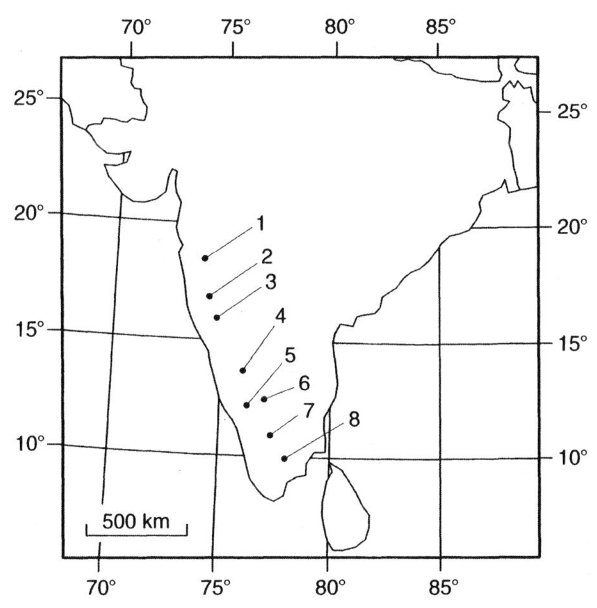

Figure 14.3 Map of peninsular India showing localities where *Cynopterus sphinx* was sampled for the analysis of morphometric and genetic variation. Names of sampling localities, geographic coordinates, and elevation (recorded to the nearest 10 m) are as follows: (1) Pune, Maharashtra (18°32′ N 73°51′ E, 600 m), (2) Kolhapur, Maharashtra (16°42′ N 74°13′ E, 560 m), (3) Belgaum, Karnataka (15°54′ N 74°36′ E, 900 m), (4) Shimoga, Karnataka (13°56′ N 75°35′ E, 650 m), (5) Thithimathi, Karnataka (12°05′ N 76°00′ E, 860 m), (6) Mysore, Karnataka (12°18′ N 76°37′ E, 780 m), (7) Metupalayam, Tamil Nadu (11°18′ N 76°59′ E, 450 m), and (8) Othakadai, Tamil Nadu (9°56′ N 78°07′ E, 150 m).

Pairwise estimates of F_{ST} exhibited a monotonic increase as a positive function of separation distance (figure 14.4). Matrix randomization tests revealed a statistically significant relationship between arcsin $\sqrt{F_{ST}}$ and ln-distance ($r = 0.533$, $P = 0.008$). Absolute standardized residuals from a regression of arcsin $\sqrt{F_{ST}}$ against ln-distance also exhibited a significant positive correlation with ln-distance ($r = 0.401$, $P = 0.041$). Thus, the spatial patterning of pairwise F_{ST} clearly indicates that *C. sphinx* has attained migration–drift equilibrium under extremely high levels of gene flow across peninsular India. Moreover, the microsatellite data revealed no evidence of major phyletic breaks across the region surveyed (figure 14.5). Having verified migration–drift equilibrium among the sampled populations, the overall F_{ST} value translates into an estimate of $N_e m = 8.08$ using Wright's (1969) infinite alleles/island model approximation. As might be expected for a species with such strong dispersal capabilities, this $N_e m$ estimate far exceeds the level required to counteract stochastic divergence via drift.

Genetic evidence suggests that *C. sphinx* has undergone a historical population contraction in the Indian subcontinent, possibly as a result of climatically induced range shifts during the late Quaternary (Storz and Beaumont, 2002).

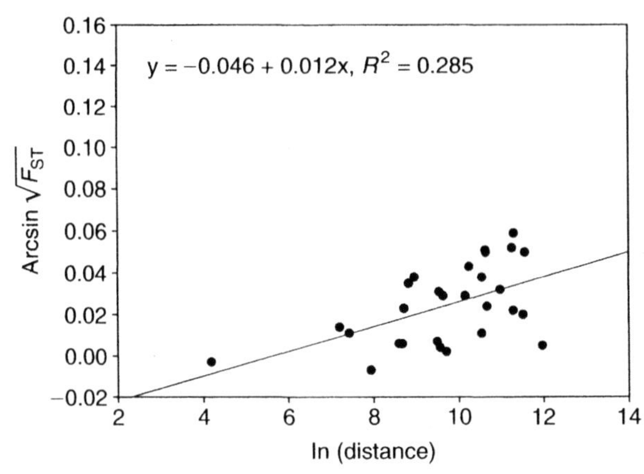

Figure 14.4 Least-squares linear regression of arcsin $\sqrt{F_{ST}}$ against ln-distance for each pairwise combination of populations. Statistical significance of regression coefficients was assessed using a matrix randomization test.

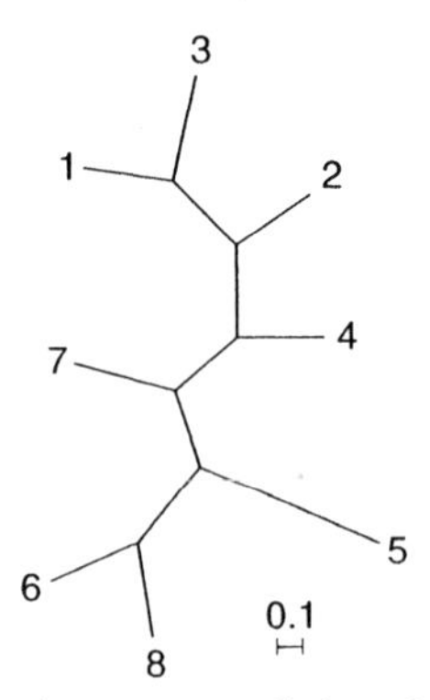

Figure 14.5 Neighbor-joining phenogram of *C. sphinx* populations in peninsular India, based on a matrix of the co-ancestry distance of Reynolds et al. (1983; eqn. 5.12 in Weir, 1996). The consensus phenogram was based on 1000 bootstrap replicates of the microsatellite data set. Numbers refer to populations listed in the legend for figure 14.3.

Thus, we might expect *C. sphinx* to be characterized by a nonequilibrium mode of population structure that reflects the predominant role of drift relative to gene flow. This mode of population structure would be implicated by a random association between pairwise F_{ST} and separation distance in conjunction with a wide degree of scatter between plotted points (Hutchison and Templeton, 1999). However, since the observed pattern of microsatellite variation in *C. sphinx* is clearly consistent with an isolation-by-distance relationship, levels of gene flow across peninsular India must have been high enough to maintain (or re-establish) migration–drift equilibrium following the historical reduction in effective population size.

Since F_{ST} is based on the infinite alleles model of mutation (Kimura and Crow, 1964), its suitability for the analysis of microsatellite variation depends on the spatial and/or temporal scale of divergence under consideration. Slatkin (1991, 1993) derived expressions for inbreeding coefficients in terms of allelic genealogies and demonstrated that F_{ST} measures the difference in within- and between-population coalescence times scaled by the average coalescence time. As a measure of genetic divergence, F_{ST} is therefore independent of mutation rate (μ), provided that the average coalescence time is less than $1/\mu$. Simulation results of Slatkin (1993) indicate that high mutation rates characteristic of microsatellite loci could potentially mask a pattern of isolation by distance. However, for the spatial scale considered in this study, rates of migration are so high relative to possible rates of mutation that Weir and Cockerham's (1984) unbiased estimator of F_{ST} ($= \theta$) can be expected to provide a more accurate measure of genetic differentiation than statistics based on the stepwise mutation model (Gaggiotti et al., 1999; Slatkin, 1995).

Since simple-sequence repeats are largely restricted to noncoding regions of the genome, microsatellite variation is generally considered to be selectively neutral (Schlötterer and Wiehe, 1999). The validity of this assumption was evaluated for the markers used in this study by comparing observed F_{ST} values with a null distribution of values generated by a coalescent-based simulation model. Specifically, the model of Beaumont and Nichols (1996) was used to generate the expected neutral distribution of F_{ST} as a function of heterozygosity. Coalescent simulations were performed using a symmetrical 100-island model of population structure, with sample sizes of 30 diploid individuals (= median of actual sample sizes). Two separate sets of simulations were performed in which mutational dynamics conformed to either the infinite alleles model or the stepwise mutation model. In order to generate a wide range of heterozygosity values, simulations were based on two different mutation rates ($N_e\mu = 0.1$ and 1.0, where $N_e\mu$ is the mutation rate scaled to effective population size). Results of the simulations revealed no evidence for departures from neutral expectations at any locus, regardless of the underlying mutation model ($P > 0.05$ for every locus $\times$ model combination). When single-locus F_{ST} values were plotted as a function of heterozygosity, observed points were well within the 0.025 and 0.975 quantiles of the expected neutral distribution.

Testing for Evidence of Spatially Varying Selection as a Cause of Phenotypic Differentiation

Clinal variation in quantitative traits is often attributed to the effects of spatially varying selection across an environmental gradient. However, identical patterns can be produced by the interplay between purely stochastic processes. For example, clinal variation in allelic frequencies at genes underlying a particular trait can result from gene flow between partially isolated populations that have diverged via drift, or admixture between two or more genetically differentiated founding populations. One means of inferring the role of selection in the maintenance of clinal variation is to compare relative levels of between-population divergence in quantitative traits and neutral DNA markers. Such comparisons can be used to test whether the observed level of trait divergence exceeds neutral expectations (Lande, 1992; Lynch, 1994; Rogers and Harpending, 1983; Whitlock, 1999). When a species is distributed across an environmental selection gradient, a joint analysis of phenotypic divergence and isolation by distance for neutral DNA markers can elucidate the spatial scale at which adaptation to local environmental conditions can evolve in response to spatially varying selection (Slatkin, 1973).

If the geographic patterning of additive genetic variance underlying a particular trait is exclusively attributable to migration–drift equilibrium (and if there are no departures from allelic or genotypic equilibria within populations), variance components can be defined as $\sigma_b{}^2 = 2F_{ST}\sigma_o{}^2$, $\sigma_w 2 = (1 - F_{ST})\sigma_o{}^2$, and $\sigma_t{}^2 = (1 + F_{ST})\sigma_o{}^2$, where $\sigma_w{}^2$, $\sigma_b{}^2$, and $\sigma_t{}^2$ represent the within-population, between-population, and total genetic variances in trait values, respectively, and $\sigma_o{}^2$ represents the total variance in the trait under panmixia (Lande, 1992; Rogers and Harpending, 1983; Wright, 1951). Thus, a dimensionless measure of differentiation for quantitative traits, analogous to Wright's (1951) F_{ST}, can be defined as

$$Q_{ST} = \sigma_b{}^2/(\sigma_b{}^2 + 2\sigma_w{}^2). \quad (4)$$

The partitioning of phenotypic variance within and between populations of *C. sphinx* was assessed using a two-way ANOVA on PC1 scores, with sex included as a fixed-effects factor. Using the method-of-moments approach outlined in Storz (2002), variance components were estimated by equating observed mean squares to their expectations. Accordingly, Q_{ST} for body size of *C. sphinx* was calculated for each pairwise combination of populations using moment-based estimates of the within- and between-population variance components (Var(w) and Var(b), respectively) and within-population phenotypic variance (×0.5) was used as a proxy for additive genetic variance (equivalent to assuming that body size is characterized by a narrow-sense heritability of $h^2 = 1/2$).

In peninsular India, the geographic pattern of variation in external morphology of *C. sphinx* conforms to Bergmann's rule, as indicated by a steep, monotonic cline of increasing body size from South to North (figure 14.6; Storz et al., 2001a). If clinal size variation of *C. sphinx* is simply attributable to isolation by distance, the positive association between Q_{ST} and separation distance should disappear when the effects of neutral genetic divergence (as measured by F_{ST} for microsatellites) are held constant. By contrast, the null hypothesis of isolation by distance would be rejected if the increase in pairwise Q_{ST} as a function of geographic/environmental distance remained statistically significant after controlling for pairwise F_{ST}. A significant partial regression of pairwise Q_{ST} on distance would indicate that migration–drift equilibrium is not a sufficient explanation for the latitudinal pattern of clinal size variation in *C. sphinx*.

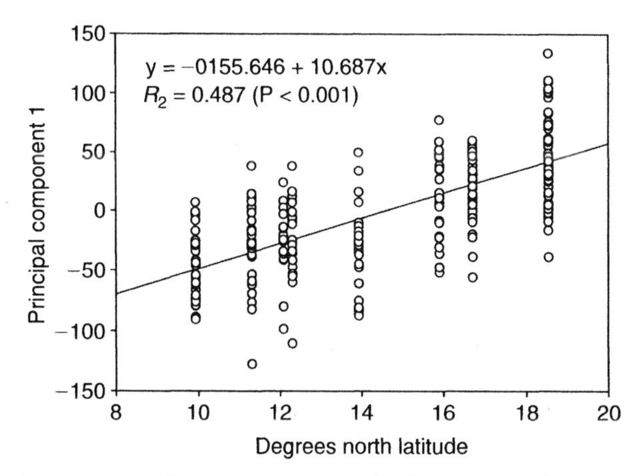

Figure 14.6 Least-squares linear regression of body size (as indexed by PC1 scores) against latitude for *Cynopterus sphinx* ($n = 251$) sampled from peninsular India.

The association between body-size variation and geographic/environmental distance was tested using pairwise and partial Mantel tests (Manly, 1997). Using a stepwise multiple regression procedure, a matrix of arcsin $\sqrt{Q_{ST}}$ (**Q**) was related to a matrix of arcsin $\sqrt{F_{ST}}$ (F) and a matrix of pairwise measures of geographic or environmental distance (**D**). For the three respective matrices, let q_{ij}, f_{ij}, and d_{ij} denote distances between localities i and j. The following model was then evaluated:

$$q_{ij} = \beta_0 + \beta_1 f_{ij} + \beta_2 d_{ij} + \varepsilon_{ij}, \quad (5)$$

where β_2 measures the association between q_{ij} and d_{ij} while controlling for the effects of f_{ij}, and ε_{ij} represents an independent error term. Statistical significance of the association between the dependent variable matrix **Q** and the two independent variable matrices (**F** and **D**) was assessed by means of a randomization test (Manly, 1997). In addition to the tests based on linear measures of geographic distance, partial Mantel tests were performed using multivariate measures of pairwise environmental distance (see Storz, 2002). In all pairwise and partial Mantel tests, the matrix element representing the Thithimathi–Mysore comparison was excluded since these two localities were closely situated at the same latitude.

Across the entire transect, the average pairwise estimate of Q_{ST} was 13.5-fold larger than that of F_{ST} (0.323 vs. 0.024). In contrast to the levels of population subdivision observed at microsatellite loci, pairwise Q_{ST} exhibited a remarkably steep increase as a positive function of distance (figure 14.7). Pairwise Mantel tests revealed highly significant correlations between Q_{ST} and the following variables: F_{ST}, geographic distance, and an environmental distance metric based on a temperature/equability vector (PC1-T; table 14.2). The correlation between Q_{ST} and F_{ST} is primarily attributable to the large number of cases where both measures were close to zero. Regression

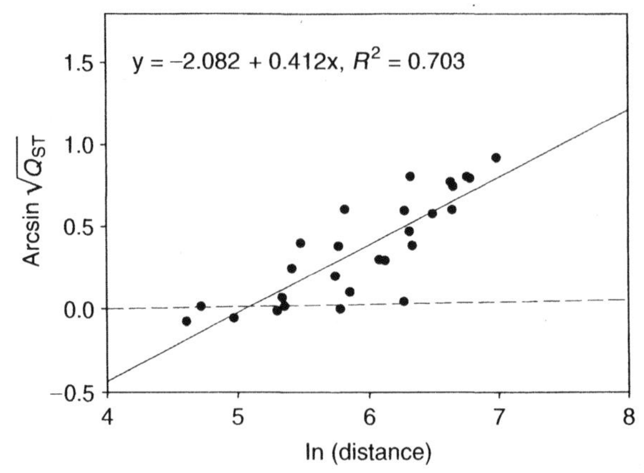

Figure 14.7 Least-squares linear regression of arcsin ($\sqrt{Q_{ST}}$ against ln-distance for each pairwise combination of populations. The dashed line denotes the linear regression line for arcsin $\sqrt{F_{ST}}$ versus distance (note the difference in scale of the *y*-axis compared with figure 14.4). Statistical significance of the regression coefficient was assessed using a matrix randomization test.

coefficients for the matrices of geographic and environmental distance remained highly significant in the partial Mantel tests (table 14.2). In other words, the increase in Q_{ST} as a positive function of geographic and environmental distance remained statistically significant even when the effects of neutral divergence (F_{ST}) were partialed out. Regression coefficients for matrices of geographic and environmental distance also remained significant when randomizations were restricted to spatially defined groups of populations along the transect (Storz, 2002). Thus, we can be confident that the regression coefficients reflect real associations and cannot be explained as artifacts of spatial autocorrelation (Oden and Sokal, 1992; Raufaste and Rousset, 2001).

In conclusion, results of the partial Mantel tests confirmed that migration–drift equilibrium is not a sufficient explanation for the latitudinal pattern of clinal size variation in *C. sphinx*. Between-population divergence in body size increased with environmental distance across a South-to-North gradient of decreasing temperature, decreasing relative humidity, and increasing seasonality (Storz et al. 2001a; Storz, 2002). The geographic patterning of pairwise Q_{ST} is most likely attributable to spatially varying selection and/or the direct influence of latitudinally ordered environmental effects.

Inferences about the adaptive basis of clinal variation are strengthened when the environmental component of phenotypic variation can be identified and statistically removed (Coyne and Beecham, 1987; Huey et al., 2000; Long and Singh, 1995; Mousseau and Roff, 1989, 1995). This can be accomplished by comparing different populations using common-garden or reciprocal transplant experiments (Mousseau, 1999). Unfortunately, this approach is not readily tractable for the study of bat populations. To examine the effects of different assumptions about the genetic basis of body size

Table 14.2 Results of Pairwise (A) and Partial (B) Mantel Tests of Causal Hypotheses Regarding the Pattern of Clinal Size Variation in *C. sphinx*

	Independent variable matrices			
	Arcsin $\sqrt{F_{ST}}$	**ln-distance**	**PC1-T distance**	**PC1-R distance**
(A) Pairwise MCTs				
arcsin $\sqrt{Q_{ST}}$	**0.634 (P = 0.0003)**	**0.838 (P = 0.0001)**	**0.463 (P = 0.0007)**	−0.082 (P = 0.6970)
(B) Partial MCTs				
arcsin $\sqrt{Q_{ST}}$	**0.303 (P = 0.0020)**	**0.694 (P = 0.0001)**	–	–
arcsin $\sqrt{Q_{ST}}$	**0.743 (P = 0.0001)**	–	**0.599 (P = 0.0004)**	–

Partial Mantel tests were performed in a stepwise regression procedure for variables that showed a significant degree of assocciation in pairwise tests. A matrix of arcsin $\sqrt{F_{ST}}$ was included as an independent variable in each of the partial regression analyses to control for the effects of neutral genetic divergence. Tests were performed on independent variable matrices of pairwise ln-distance and two separate pairwise measures of environmental distance (indexed by PC1-T and PC1-R vectors). PC1-T and PC1-R are multivariate axes that summarize latitudinal variation in temperature and precipitation, respectively. Standardized regression coefficients and associated P values that remained statistically significant after Bonferroni correction are in bold. P values for one-sided tests are expressed as the proportion of 10,000 randomizations that yielded values greater than or equal to observed t values. Since the pairwise MCT revealed no significant matrix correlation between arcsin $\sqrt{Q_{ST}}$ and PC1-R distance, the latter variable was not included in the partial MCT with the PC1-T distance matrix.

variation in *C. sphinx*, Q_{ST} was recalculated over a range of values for Var(w) (assuming h^2 in the range 0.50–0.85) and Var(b) (assuming that 50–85% of the between-population variance was attributable to a nonheritable environmental component). Results of this exercise indicate that over the full range of biologically plausible values for Var(w) and Var(b), the linear regression slope for arcsin $\sqrt{Q_{ST}}$ versus ln-distance remained significantly steeper than that for arcsin $\sqrt{F_{ST}}$ versus ln-distance. When it was assumed that $h^2 = 0.85$ and that 85% of the between-population variance in body size was attributable to environmental effects (%Var(b) genetic = 15), partial Mantel tests between the recalculated Q_{ST} matrix and each of the geographic and PC1-T distance matrices remained statistically significant ($P < 0.05$). Thus, the environmental component of the between-population variance in body size would have to be extraordinarily large to accept the null hypothesis of neutral phenotypic divergence.

In studies of geographic variation, inferences about the interplay between different evolutionary forces are greatly enhanced when patterns of genetically based trait variation are considered in conjunction with estimates of neutral genetic divergence (e.g., Prout and Barker, 1993; Spitze, 1993). As demonstrated in our analysis of clinal size variation in *C. sphinx*, a matrix of pairwise estimates of neutral genetic divergence can be used as an independent variable in a partial Mantel test when the observed pattern of phenotypic divergence is tested against causal hypotheses. The chief merit of the approach based on Q_{ST} versus F_{ST} contrasts is that results can be interpreted within the framework of the neutral theory of phenotypic evolution (reviewed by Lynch, 1994).

Potential Causes of Latitudinal Variation in Phenotypic Optima

If the observed pattern of clinal size variation in *C. sphinx* has an adaptive genetic basis, what are the underlying causes of spatially varying selection? In pteropodid bats, basal metabolic rate is highly size-dependent and medium-sized frugivores such as *C. sphinx* are typically characterized by precise regulation of body temperature (McNab, 1989; McNab and Bonaccorso, 1995). Because the energetics of temperature regulation have important consequences for fecundity, gestation period, and rates of postnatal growth in bats (McNab, 1982), the ecologically optimal body size of nonmigratory species may be expected to vary across broad-scale climatic gradients. Although the underlying causes remain to be elucidated, it seems clear that spatially varying selection has played an important role in shaping latitudinal size variation in *C. sphinx*.

Geographic Variation, Clines, and Subspecies

There is considerable uncertainty surrounding the taxonomic status of the many named forms of *C. sphinx* in the Indomalayan Region (reviewed by Storz and Kunz, 1999). Larger specimens from northern India were originally referred to *C. sphinx gangeticus*, while generally smaller specimens from the south were referred to *C. sphinx sphinx* (Andersen, 1912; Hill, 1983). Results of the analysis reported here indicate that the nominal subspecies *C. s. gangeticus* and *C. s. sphinx* represent two ends of a continuous spectrum of clinal size variation. The combined analysis of genetic and phenotypic variation indicates that subspecies boundaries in peninsular India represent arbitrary subdivisions of a continuous gradation of morphological characters. Imposing a formal nomenclature on this gradation of form does nothing to elucidate the underlying causes of geographic variation. It remains to be seen whether other examples of clinal variation in bat populations represent cases of selection–migration equilibrium across an environmental gradient, isolation by distance, or secondary intergradation across a contact zone between phylogenetically distinct forms.

LITERATURE CITED

Andersen, K. 1912. Catalogue of the Chiroptera in the collection of the British Museum, Vol. 1: Megachiroptera. British Museum of Natural History, London.

Baker, R.J. 1981. Chromosome flow between chromosomally characterized taxa of a volant mammal, *Uroderma bilobatum* (Chiroptera: Phyllostomidae). Evolution, 35: 296–305.

Barton, N.H. 1982. The structure of the hybrid zone in *Uroderma bilobatum*. Evolution, 36: 863–866.

Barton, N.H. 1990. The genetic consequences of dispersal. Pp. 37–59. In: Animal Dispersal: Small Mammals as a Model (N.C. Stenseth and W.Z. Lidicker, eds.). Chapman and Hall, New York.

Barton, N.H., and K.S. Gale. 1993. Genetic analysis of hybrid zones. Pp. 13–45. In: Hybrid Zones and the Evolutionary Process (R.G. Harrison, ed.). Oxford University Press, Oxford.

Beaumont, M.A., and R.A. Nichols. 1996. Evaluating loci for use in the genetic analysis of population structure. Proceedings of the Royal Society of London, Series B, 263: 1619–1626.

Bouteiller, C., and N. Perrin. 2000. Individual reproductive success and effective population size in the greater white-toothed shrew *Crocidura russula*. Proceedings of the Royal Society of London, Series B, 267: 701–705.

Bush, G.L., S.M. Case, A.C. Wilson, and J.L. Patton. 1977. Rapid speciation and chromosomal evolution in mammals. Proceedings of the Naional Academy of Sciences of the USA, 74: 3942–3946.

Coyne, J.A. 1984. Correlation between heterozygosity and rate of chromosome evolution in animals. The American Naturalist, 123: 725–729.

Coyne, J.A., and E. Beecham. 1987. Heritability of two morphological characters within and among natural populations of *Drosophila melanogaster*. Genetics, 117: 727–737.

Daly, J.C., and J.L. Patton. 1990. Dispersal, gene flow, and allelic diversity between local populations of *Thomomys bottae* pocket gophers in the coastal ranges of California. Evolution, 44: 1283–1294.

Endler, J.A. 1977. Geographic Variation, Speciation, and Clines. Princeton University Press, Princeton, NJ.

Gaggiotti, O.E., O. Lange, K. Rassmann, and C. Gliddon. 1999. A comparison of two indirect methods for estimating average levels of gene flow using microsatellite data. Molecular Ecology, 8: 1513–1520.
García-Ramos, G., and M. Kirkpatrick. 1997. Genetic models of adaptation and gene flow in peripheral populations. Evolution, 51: 21–28.
Haldane, J.B.S. 1948. The theory of a cline. Journal of Genetics, 48: 277–284.
Hill, J.E. 1983. Bats (Mammalia: Chiroptera) from Indo-Australia. Bulletin of the British Museum (Natural History), 45: 103–208.
Hill, W.G. 1972. Effective size of populations with overlapping generations. Theoretical Population Biology, 3: 278–289.
Hill, W.G. 1979. A note on effective population size with overlapping generations. Genetics, 92: 317–322.
Huey, R.B., G.W. Gilchrist, M.L. Carlson, D. Berrigan, and L. Serra. 2000. Rapid evolution of a geographic cline in size in an introduced fly. Science, 287: 308–309.
Hutchison, D.W., and A.R. Templeton. 1999. Correlation of pairwise genetic and geographic distance measures: inferring the relative influences of gene flow and drift on the distribution of genetic variability. Evolution, 53: 1898–1914.
Kimura, M., and J.F. Crow. 1964. The number of alleles that can be maintained in a finite population. Genetics, 49:725–738.
Kimura, M., and G.H. Weiss. 1964. The stepping stone model of population structure and the decrease of genetic correlation with distance. Genetics, 49: 561–576.
Lande, R. 1992. Neutral theory of quantitative genetic variance in an island model with local extinction and colonization. Evolution, 46: 381–389.
Lidicker, W.Z., and J.L. Patton. 1987. Patterns of dispersal and genetic structure in populations of small rodents. Pp. 144–161. In: Mammalian Dispersal Patterns (B.D. Chepko-Sade and Z.T. Halpin, eds.). University of Chicago Press, Chicago.
Long, A.D., and R.S. Singh. 1995. Molecules versus morphology: the detection of selection acting on morphological characters along a cline in *Drosophila melanogaster*. Heredity, 74: 569–581.
Lynch, M. 1994. Neutral models of phenotypic evolution. Pp. 86–108. In: Ecological Genetics (L.A. Real, ed.). Princeton University Press, Princeton, NJ.
Manly, F.J.B. 1997. Randomization, Bootstrap and Monte Carlo Methods in Biology. Chapman and Hall, New York.
McCracken, G.F., and G.S. Wilkinson. 2000. Bat mating systems. Pp. 321–362. In: Reproductive Biology of Bats (E.G. Crichton and P.H. Krutzsch, eds.). Academic Press, New York.
McNab, B.K. 1982. Evolutionary alternatives in the physiological ecology of bats. Pp. 151–200. In: Ecology of Bats (T.H. Kunz, ed.). Plenum Press, New York.
McNab, B.K. 1989. Temperature regulation and rate of metabolism in three Bornean bats. Journal of Mammalogy, 70: 153–161.
McNab, B.K., and F.J. Bonaccorso. 1995. The energetics of pteropodid bats. Pp. 111–122. In: Ecology, Evolution and Behaviour of Bats (P.A. Racey and S.M. Swift, eds.). Oxford University Press, Oxford.
Mousseau, T.A. 1999. Intra- and interpopulation genetic variation. Pp. 219–250. In: Adaptive Genetic Variation in the Wild (T.A. Mousseau, B. Sinervo, and J.A. Endler, eds.). Oxford University Press, Oxford.
Mousseau, T.A., and D.A. Roff. 1989. Adaptation to seasonality in a cricket: patterns of phenotypic and genotypic variance in body size and diapause expression along a cline in season length. Evolution, 43: 1483–1496.

Mousseau, T.A., and D.A. Roff. 1995. Genetic and environmental contributions to geographic variation in the ovipositor length of a cricket. Ecology, 76: 1473–1482.

Nunney, L. 1993. The influence of mating system and overlapping generations on effective population size. Evolution, 47: 1329–1341.

Nunney, L. 1996. The influence of variation in female fecundity on effective population size. Biological Journal of the Linnean Society, 59: 411–425.

Nunney, L., and D.R. Elam. 1994. Estimating the effective size of conserved populations. Conservation Biology, 8: 175–184.

Oden, N.L., and R.R. Sokal. 1992. An investigation of three-matrix permutation tests. Journal of Classification, 9: 275–290.

Patton, J.L., and J.H. Feder. 1981. Microspatial genetic heterogeneity in pocket gophers: non-random breeding and drift. Evolution, 43: 12–30.

Patton, J.L., and M.F. Smith. 1990. The evolutionary dynamics of the pocket gopher *Thomomys bottae*, with emphasis on California populations. University of California Press, Berkeley.

Patton, J.L., and S.Y. Yang. 1977. Genetic variation in *Thomomys bottae* pocket gophers: macrogeographic patterns. Evolution, 31: 697–720.

Prout, T., and J.S.F. Barker. 1993. *F* statistics in *Drosophila buzzatii*: selection, population size and inbreeding. Genetics, 134: 369–375.

Raufaste, N., and F. Rousset. 2001. Are partial Mantel tests adequate? Evolution, 55: 1703–1705.

Reynolds, J., B.S. Weir, and C.C. Cockerham. 1983. Estimation of the coancestry coefficient: basis for a short-term genetic distance. Genetics, 105: 767–779.

Rogers, A.R., and H.C. Harpending. 1983. Population structure and quantitative characters. Genetics, 105: 985–1002.

Schlötterer, C., and T. Wiehe. 1999. Microsatellites, a neutral marker to infer selective sweeps. Pp. 238–248. In: Microsatellites: Evolution and Applications (D.B. Goldstein and C. Schlötterer, eds.). Oxford University Press, Oxford.

Slatkin, M. 1973. Gene flow and selection in a cline. Genetics, 75: 733–756.

Slatkin, M. 1978. Spatial patterns in the distribution of polygenic characters. Journal of Theoretical Biology, 70: 213–228.

Slatkin, M. 1991. Inbreeding coefficients and coalescence times. Genetical Research, 58: 167–175.

Slatkin, M. 1993. Isolation by distance in equilibrium and nonequilibrium populations. Evolution, 47: 264–279.

Slatkin, M. 1995. A measure of population subdivision based on microsatellite allele frequencies. Genetics, 139: 457–462.

Spitze, K. 1993. Population structure in *Daphnia obtusa*: quantitative genetic and allozymic variation. Genetics, 135: 367–374.

Storz, J.F. 2002. Contrasting patterns of divergence in quantitative traits and neutral DNA markers: analysis of clinal variation. Molecular Ecology, 11: 2537–2552.

Storz, J.F., and M.A. Beaumont. 2002. Testing for genetic evidence of population expansion and contraction: an empirical analysis of microsatellite variation using a hierarchical Bayesian model. Evolution, 56: 154–166.

Storz, J.F., and T.H. Kunz. 1999. *Cynopterus sphinx*. Mammalian Species, 613: 1–8.

Storz, J.F., J. Balasingh, P.T. Nathan, K. Emmanuel, and T.H. Kunz. 2000a. Dispersion and site-fidelity in a tent-roosting population of the short-nosed fruit bat (*Cynopterus sphinx*) in southern India. Journal of Tropical Ecology, 16: 117–131.

Storz, J.F, H.R. Bhat, and T.H. Kunz. 2000b. Social structure of a polygynous tent-making bat, *Cynopterus sphinx* (Megachiroptera). Journal of Zoology (London), 251: 151–165.

Storz, J.F., J. Balasingh, H.R. Bhat, P.T. Nathan, A.A. Prakash, D.P. Swami Doss, and T.H. Kunz. 2001a. Clinal variation in body size and sexual dimorphism in an Indian fruit bat, *Cynopterus sphinx* (Chiroptera: Pteropodidae). Biological Journal of the Linnean Society, 72: 17–31.

Storz, J.F, H.R. Bhat, and T.H. Kunz. 2001b. Genetic consequences of polygyny and social structure in an Indian fruit bat, *Cynopterus sphinx*. I. Inbreeding, outbreeding, and population subdivision. Evolution, 55: 1215–1223.

Storz, J.F, H.R. Bhat, and T.H. Kunz. 2001c. Genetic consequences of polygyny and social structure in an Indian fruit bat, *Cynopterus sphinx*. II. Variance in male mating success and effective population size. Evolution, 55: 1224–1232.

Storz, J.F., U. Ramakrishnan, and S.C. Alberts. 2001d. Determinants of effective population size for loci with different modes of inheritance. Journal of Heredity, 92: 497–502.

Storz, J.F., U. Ramakrishnan, and S.C. Alberts. 2002. Genetic effective size of a wild primate population: influence of current and historical demography. Evolution, 56: 817–829.

Waite, T.A., and P.G. Parker. 1996. Dimensionless life histories and effective population size. Conservation Biology, 10: 1456–1462.

Weir, B.S. 1996. Genetic Data Analysis II. Sinauer Associates, Sunderland, MA.

Weir, B.S., and C.C. Cockerham. 1984. Estimating *F*-statistics for the analysis of population structure. Evolution, 38: 1358–1370.

Whitlock, M.C. 1999. Neutral additive genetic variance in a metapopulation. Genetical Research, 74: 215–221.

Wilkinson, G.S. 1987. Altruism and co-operation in bats. Pp. 299–323. In: Recent Advances in the Study of Bats (M.B. Fenton, P. Racey, and J.M.V. Rayner, eds.). Cambridge University Press, Cambridge.

Wright, S. 1938. Size of population and breeding structure in relation to evolution. Science, 87: 430–431.

Wright, S. 1951. The genetical structure of populations. Annals of Eugenics, 15: 323–354.

Wright, S. 1969. Evolution and the Genetics of Populations. Vol. 2. The Theory of Gene Frequencies. University of Chicago Press, Chicago.

15

Conflicts and Strategies in the Harem-Polygynous Mating System of the Sac-Winged Bat, *Saccopteryx bilineata*

Christian C. Voigt, Gerald Heckel, & Otto von Helversen

Daytime roosts of the neotropical sac-winged bat *Saccopteryx bilineata* (Emballonuridae) include up to 60 individuals and are located in buttress cavities of trees or in well-illuminated sections of caves or abandoned houses. Colonies are structured into harem groups, each of which includes a single male and a varying number of females. In daytime roosts, sac-winged bats do not aggregate into clusters; instead individuals are spaced at least 10 cm apart. Agonistic interactions among colony members are frequent. Males in general almost never win encounters with females, and harem males in particular cannot prevent females from switching between harems. Paternity studies confirmed that harem males are more successful in fathering offspring than nonharem males. Although harem males do not monopolize reproduction within their harem, reproductive success of males increases with harem size. Removal experiments with harem males show that nonharem males are probably queuing for access to harem territories. The importance of joining a queue as soon as possible in a male's life may have led to male philopatry in *S. bilineata*. We conclude that the mating pattern of *S. bilineata* does not resemble a classical harem system because female choice is probably more important than male competition. This has probably led to the evolution of elaborate acoustic and scent courtship displays by male sac-winged bats.

INTRODUCTION

Although polygynous mating systems are common among mammals (and also probably among bats), it was not until 1974 that a harem formation was first described for a tropical bat species (Bradbury and Emmons, 1974). This species was *Saccopteryx bilineata*, also known as the greater white-lined or sac-winged bat. Its conspicuous social behavior and general abundance in the Neotropics have stimulated extensive fieldwork starting with the studies by Bradbury and coworkers (Bradbury and Emmons, 1974; Bradbury and Vehrencamp, 1976, 1977). Their influential papers still present the baseline information about the natural history of *S. bilineata*. Since then, new laboratory techniques have been established, particularly in molecular genetics, and, consequently, new aspects of the sociobiology of *S. bilineata* can be explored. In this chapter, we summarize our present knowledge about the roosting ecology and the sociobiology of *S. bilineata* with a special focus on the conflicts and strategies in the mating system of this species.

NATURAL HISTORY

Saccopteryx bilineata is a member of the family Emballonuridae and a common insectivorous bat of the lowland regions in the Neotropics (usually below 800 m above sea level; Bradbury and Emmons, 1974). Sac-winged bats can easily be identified in the field by the two wavy white lines on the back and by the specific roosting posture—individuals cling with their feet to a vertical surface and support their body with folded forelimbs, which in males contain sac-like organs in the antebrachial wing membrane. Female sac-winged bats weigh on average 8.5 g, which is about 1 g more than males, and have only nonfunctional rudiments of the antebrachial wing sacs of males (Bradbury and Emmons, 1974).

Sac-winged bats inhabit evergreen rainforests, semideciduous forests, and even highly disturbed habitats or plantations. In some regions of the Neotropics, *S. bilineata* is one of the most abundant insectivorous bats present. Daytime roosts of *S. bilineata* are located in well-lit cavities of trees or at the basal portions of large buttressed trees. Colonies may include up to 60 individuals. Bradbury and Vehrencamp (1976) stated that, within their study area in Costa Rica, the distribution of colony sizes was bimodal, with small groups of about three to five individuals and large groups with more than 12 individuals as the most abundant colony sizes. The sex ratio in colonies of *S. bilineata* was approximately 2:1 in favor of females (Bradbury and Emmons, 1974).

Daytime roosts are used by *S. bilineata* throughout the year and given roosts may be occupied continuously for many decades (*c.* 15 years: Bradbury and Emmons, 1974; 20 years: O. Vargas, pers. comm.). Like most neotropical emballonurid species, sac-winged bats do not aggregate in clusters with close body contact. Instead, individuals are spaced at least 10 cm apart (Goodwin and Greenhall, 1961). The distance to a neighbor is rigorously maintained by agonistic interactions such as beating the folded wing toward an intruder. Individual spacing may have evolved in response to the predation risk that clustered groups potentially face at well-lit and exposed sites (Bradbury, 1977).

The temperature regime of the daytime roosts follows a diurnal pattern and, during an average sunny day, colony members spend most of the day close to thermoneutrality (lower critical temperature of the thermoneutral zone at 30 °C; Genoud and Bonnaccorso, 1986). Females seem to prefer the

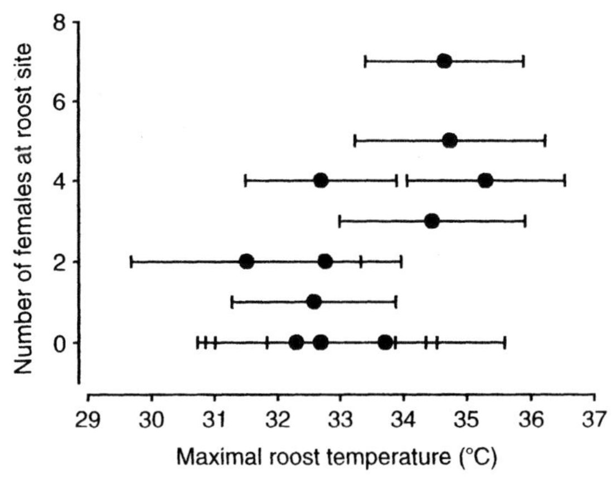

Figure 15.1 Mean number of females in a territory in relation to the maximum temperature at the same site (averaged over six consecutive, sunny days). Females seem to prefer roosting sites with high temperatures ($r_s = 0.63$, $P = 0.028$).

warmer areas of a daytime roost (figure 15.1), i.e., the number of females at a site is larger at places with higher temperatures (C.C. Voigt, unpublished observations). Thus, the temperature regime at a given roost and the thermal preference of this species could partly explain the formation of groups in colonies of the sac-winged bat.

In Costa Rican populations of *S. bilineata*, the mating season is restricted to approximately 4 weeks in December and January (Bradbury and Emmons, 1974; Bradbury and Vehrencamp, 1976; Voigt and von Helversen, 1999; Heckel et al., submitted). Copulations occur mostly during December and the first half of January. In our Costa Rican study population, the mating season corresponds with the end of the rainy season and parturition with its onset. It is unclear whether this pattern is also present in other geographical regions.

Testes are inconspicuous during most of the year and increase in size to an average of 0.2% of male body mass during the mating season (C.C. Voigt, unpublished observation). Conception and implantation of the embryo probably occurs shortly after copulation. Tannenbaum (1975) noted that fetuses are recognizable 8 weeks after the likely date of conception. This observation makes sperm storage in the female reproductive tract over a longer time period unlikely. However, given the fact that gestation lasts for approximately 6 months, it is possible, although not demonstrated, that embryonic development is delayed. Females give birth to a single offspring each year. Within a colony, births are synchronized and occur within a period of 2–3 weeks (Bradbury and Vehrencamp, 1976; Tannenbaum 1975; Heckel et al., 1999). Newborns are fully furred, weigh approximately 40% of the mother's body mass and have forearm lengths that are 60% of adult dimensions. Thus, newborns of *S. bilineata* are relatively well developed at birth compared with, for example, vespertilionid bats from temperate regions

(Hayssen and Kunz, 1996). First flight attempts by juveniles can be observed at an age of 10–14 days; they are probably weaned at an age of 2 months. About 75% of juvenile males remain in the maternal colony, sometimes as cryptic males in their natal harem territory (see the section on cryptic males below), whereas the majority of female juveniles disperse from the natal colony (Bradbury and Emmons, 1974; Bradbury and Vehrencamp, 1976). Tannenbaum (1975) reported that 84% of female offspring either left the study area or died, whereas only 16% of young females moved to known neighboring harems or colonies.

SOCIAL SYSTEM

The basic social units in colonies of *S. bilineata* are so-called harem groups that consist of a single adult male and a varying number of females (Bradbury and Emmons, 1974; Bradbury and Vehrencamp, 1976, 1977). Bradbury and coworkers have pointed out that it is probably more correct to describe the harems of *S. bilineata* as "one-male multiple-female groups" because the former term may imply exclusive access of harem males to all females of their harem. Like Bradbury and coworkers, and McCracken and Wilkinson (2002), we use the word "harem" here for reasons of simplicity without implying exclusivity of access of the harem male to his group of females.

Harem males defend the boundaries of their territories within which females roost. In a Costa Rican population, harem size averaged 1.5 females (Voigt et al., 2005). Within a given colony, harem sizes can change between the mating and lactation period. Larger colonies typically contain several harem territories. Nonharem or juvenile males roost either solitarily in the colony or adjacent to harem territories. Overall, the social system of *S. bilineata* has been characterized as stable in terms of male residency, but unstable in terms of female composition (McCracken and Wilkinson, 2002). In daytime roosts, the minimum space between individuals is rigorously enforced, except among mother–young pairs before weaning and males copulating with females. Among adult *S. bilineata*, allogrooming is absent and males of *S. bilineata* do not provide direct paternal care.

Bradbury and Vehrencamp, (1976) reported that *S. bilineata* forages in groups according to colony affiliation, and that the colony foraging site is partitioned into individual harem territories defended by harem males that include individual foraging patches of all current harem females. The foraging areas of female *S. bilineata* typically consist of two or more adjacent patches per individual, each 10–20 m in length. The foraging area of a harem male encompasses the foraging patches of all females in their harem. According to Bradbury and Vehrencamp (1976), females change foraging sites if they switch to a different colony or harem. Thus, details of roost site subdivision are probably mapped directly onto foraging distribution (Bradbury and Vehrencamp, 1976).

REPRODUCTIVE TACTICS OF FEMALES

Composition of Harem Groups

Tannenbaum (1975) reported that most harems consisted of resident females that could be observed during consecutive days at almost the same place within the territory. Newly arriving females were usually driven away by residents. However, integration into the harem group occurred if newcomers were persistent. We performed 12-hour video observations in a daytime roost during the mating season and found that harem females can be separated into two groups. Approximately 40% of females stayed throughout the day within the same territory, whereas the remaining females (nonresidents) switched between territories during the same day (figure 15.2). Thus, not all females can be exclusively assigned to a single harem territory, because they may use up to three different harem territories during the course of a day. Each time a female enters a territory, she is greeted by the harem holder with hovering displays. The number of hovering displays of males is not significantly different between resident and nonresident females (figure 15.3). Hence, males probably "greet" females coming into their harem irrespective of the strength of their affiliation to his or other harems.

Social Hierarchies among Females

Females interact with each other mostly in an aggressive way, and in most dyadic interactions the same female will be consistently the winner (Heckel et al., submitted). Females can be ranked according to the probability of winning an interaction and we refer to this rank in the remainder of the chapter when speaking about the social status of a female. Females with a high social status in a harem probably choose the best roosting sites within a

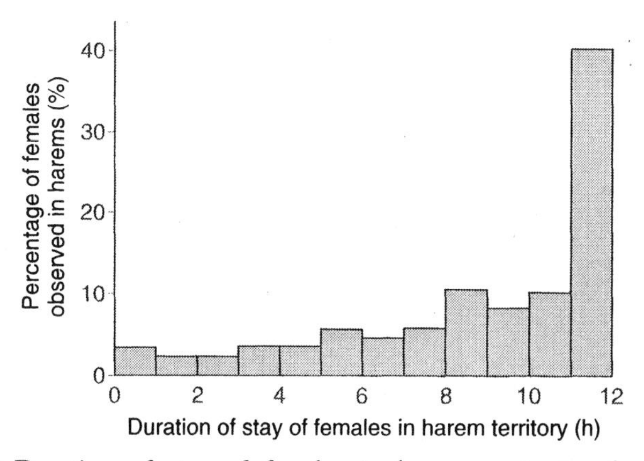

Figure 15.2 Duration of stay of females in harem territories during 12-hour continuous video observations in a daytime roost in Costa Rica. About 40% of the females remained in the same territory for the whole day, whereas 60% of the females switched between different territories.

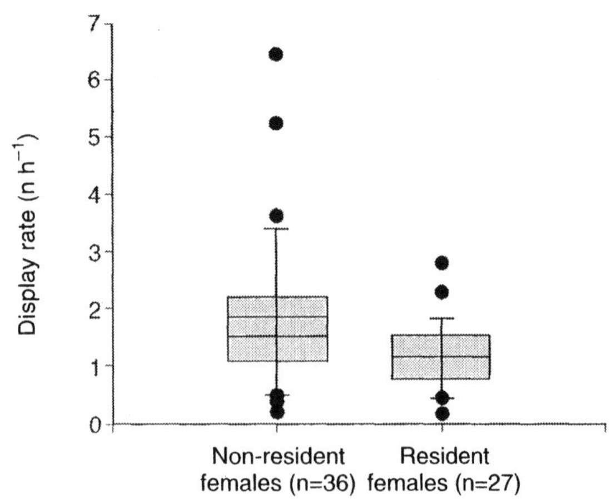

Figure 15.3 Rate of male hovering displays in front of resident and nonresident females (see figure 15.2 for definition of residency). The difference between the two groups was not significant (Mann–Whitney *U*-test: $U_{37,26} = 308$, $P > 0.05$; C.C. Voigt, unpublished observation). The box plots show the 25th to 75th percentile (box), the 10th and 90th percentiles (T), the mean value as a thick line, the median as a thin line, and outliers.

harem. Critical roost characteristics could offer protection against predators, microclimatic conditions, or the roost substrate quality. For instance, females of high social status usually roost in the higher parts of a territory and females of low social status in the lower parts (Heckel and von Helversen, 2003). A high roosting position could be beneficial because it is likely to provide protection against ground-dwelling predators.

Female Dominance

Females of a high social status were more dominant against the harem male than females of low status (figure 15.4). As a probable consequence, harem males more often courted subordinate than dominant females and attempted to copulate more often with subordinate than with dominant females (Heckel et al., submitted; figure 15.5). Taken together, subordinate females were more likely to be harassed by male *S. bilineata* and this could, in the long run, lower their fitness. Consequently, selection could have favored females with an increased aggressiveness toward males. Despite male harassment, females roost within harem territories and we suggest that female *Saccopteryx* gain a possible benefit from roosting within harems by being exposed "only" to a single male, namely the owner of the territory, instead of being harassed by all males of a colony outside a harem. Harems as shelters against sexual harassment have been suggested for other polygynous mating systems (Campagna et al., 1992; Linklater et al., 1999; Réale et al., 1996). In contrast to *Saccopteryx*, males of these species are larger than females, which increases the risk of females of being injured or killed by males. Thus far, we lack

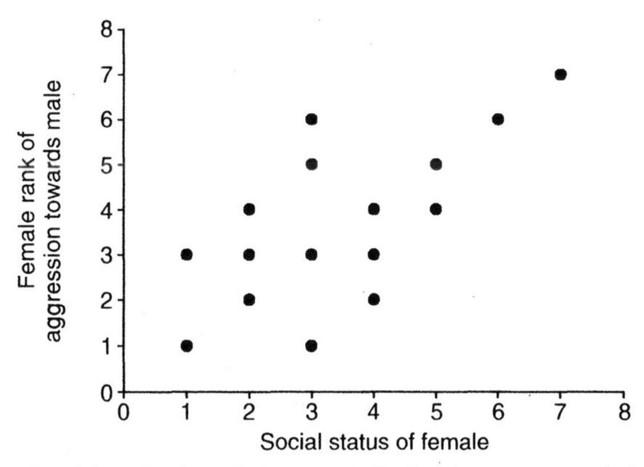

Figure 15.4 Rank of females in relation to their dominance toward the harem male (Heckel and von Helversen, 2003). Females of a high social status were more aggressive toward males compared with subordinate females ($r_s = 0.67$, $P < 0.05$). The highest social status within the female hierarchy was set to 1. Ranking was carried out following Jameson et al. (1999).

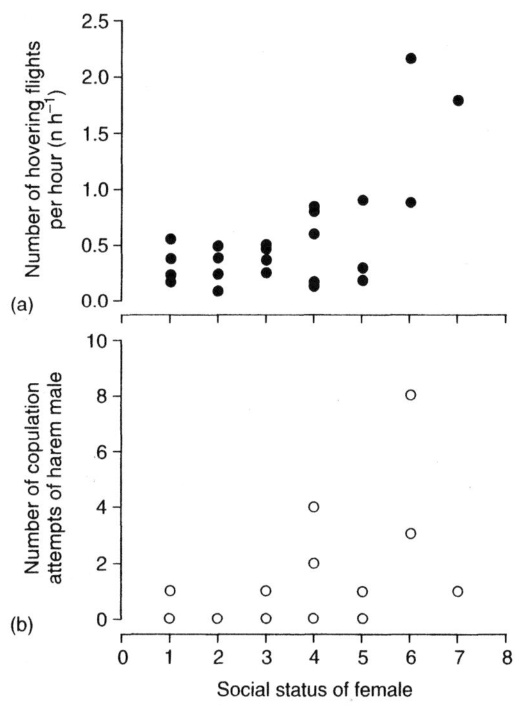

Figure 15.5 The number of hovers (a) and copulation attempts (b) of males in correlation with the social status of a female. Harem males hovered less often in front of dominant females than in front of subordinate females ($r_s = 0.46$, $P = 0.024$, $n = 24$). Harem males attempted to copulate more often with subordinate females than with dominant females ($r_s = 0.52$, $P = 0.019$, $n = 19$) (Heckel and von Helversen, 2003).

data on fitness costs of female *Saccopteryx* resulting from male sexual harassment.

Female Control over Copulation

Males perform numerous hovering flights in front of estrous females but only a fraction of these displays are followed by copulation attempts. Males usually approach the female from above and try to mount her with both wing sacs opened. Tannenbaum (1975) noted that 20% of all copulation attempts were successful. In most cases, females interrupted the approach of a male by barking, beating the forearm toward the male or by flying off. Obviously, for a successful copulation attempt a male requires the cooperation of the female. Other males frequently disturb copulations by invading the territory or attacking the pair (Tannenbaum, 1975; C.C. Voigt, unpublished observation). Harem males also prevent other males from copulating with the females in their harems.

During a 2-week period and a total of 37 hours of observation, Tannenbaum (1975) observed 13 presumed successful copulations and 63 unsuccessful copulations in two neighboring harems with five and six females respectively (figure 15.6). In two cases, one of the males attempted to copulate with a female of the neighboring harem. In both cases, the harem holder prevented the male from approaching the female. Although males monopolize copulations within their harems, they cannot prevent females from leaving. Thus, females could choose a mate other than the harem male by switching its territory and soliciting copulations from other males. All observed copulations lasted only a few seconds (Tannenbaum, 1975; C.C. Voigt, unpublished observation) and, in general, females terminated the

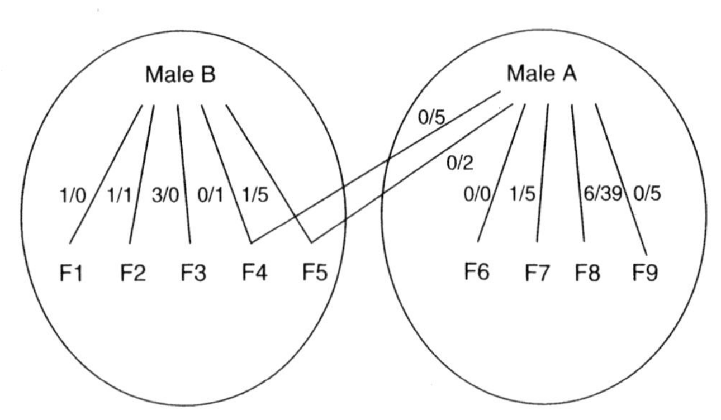

Figure 15.6 Observations of copulations in two neighboring harems within a 2-week period (modified after Tannenbaum, 1975). The numbers to the left of the slash indicate the total numbers of probably successful copulations and the numbers to the right of the slash the numbers of probably unsuccessful copulations. Because Tannenbaum observed the two harems for 37 hours within the 2-week period, some copulations may have remained unobserved. The male of harem A attempted copulations with two females of harem B. All copulation attempts of male A in territory B were interrupted by male B.

copulation by crawling or flying away from beneath the male. Thus, forced copulations seem unlikely.

REPRODUCTIVE TACTICS OF MALES

Reproductively active males of *S. bilineata* face two major constraints. Firstly, most female sac-winged bats are superior over males during agonistic interactions and, secondly, during daytime and probably also at night, harem males may be restricted in their movements to their territory because other males may usurp a deserted position (see below).

Courtship Displays

Harem males perform an elaborate variety of visual, vocal, and olfactory displays to attract and retain females in their harem.

Complex Songs Besides several types of simple calls that are uttered during the day, males emit songs with a complex composition and structure that are unusual for bats (Davidson and Wilkinson, 2002, 2003). Two types of male songs can be distinguished: territorial songs and courtship songs (Behr and von Helversen, 2004).

Territorial songs are the most conspicuous vocalizations of males in a colony. Territorial songs can be heard occasionally during the day, but mainly during the hours of highest activity in the colony at dawn when colony members return from their foraging trips, and at dusk before they leave to hunt (figure 15.7), a pattern also observed in carnivores (East and Hofer, 1991a).

Territorial songs last for up to 4 seconds and typically consist of 20 to 50 elements (figure 15.8). A song bout typically starts with short, upward frequency-modulated syllables, then merges into a series of up and down modulated ("roof-shaped") elements (or inverted-V elements *sensu* Bradbury and Emmons, 1974; Davidson and Wilkinson, 2002, 2003) and ends with more complex elements that are often preceded by a noisy buzz. A similar pattern, although at lower frequencies, has been described for the whooping bouts of spotted hyenas (East and Hofer, 1991b). In *Saccopteryx*, syllables last from about 10 milliseconds at the beginning to about 100 milliseconds at the end of a song. Bradbury and Emmons (1974) distinguished at least six different categories of syllables although they acknowledge that these are not discrete. Territorial songs that are emitted by males throughout the day often differed from the most complex structure of bouts (figure 15.8) by lacking complex end-syllables. These shorter songs have been named "short songs" by Bradbury and Emmons (1974) in contrast to the "long songs" emitted in the morning and evening. Behr and von Helversen (2004) suggested considering both as "territorial songs" because they are only uttered by males within their harem territory, often without the presence of females.

More complex than "territorial songs" are "courtship songs" that males emit toward individual females in their harem, which may last for minutes up

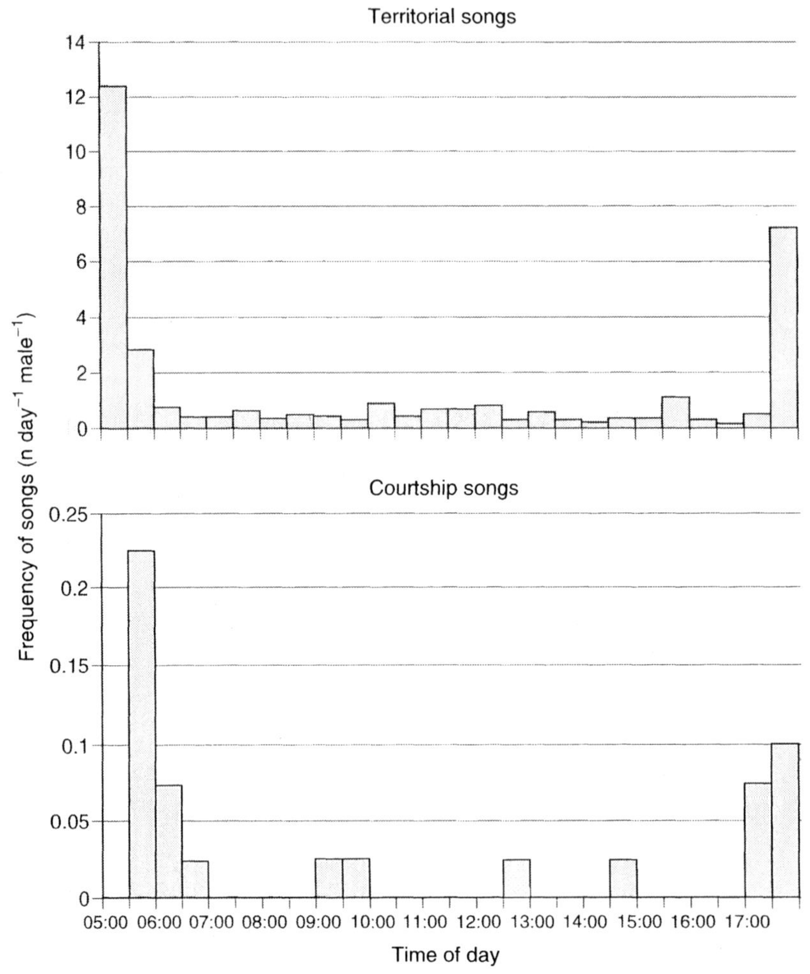

Figure 15.7 Activity pattern of courtship and territorial songs in daytime roosts. The number of songs per male is plotted against the half-hour intervals of the daytime. Territorial songs are most frequently performed at dawn before females have returned to the roost (females return on average around 0530 hours).

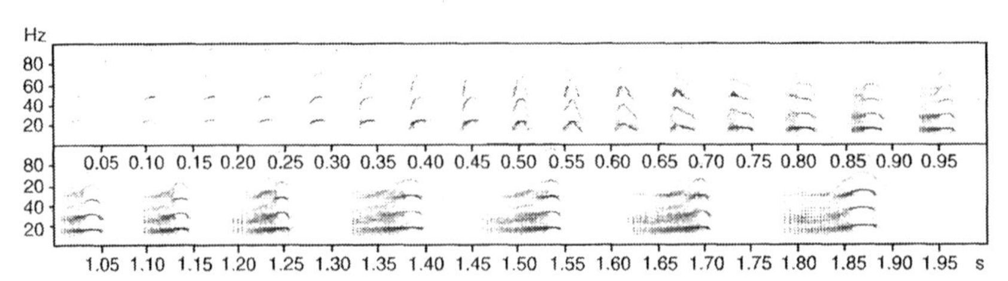

Figure 15.8 Example of a complete territorial song by *S. bilineata*. The frequency (0–100 kHz) is plotted against the time (s). During the first half of the territorial songs inverted-V elements are predominant, whereas these elements are altered at the end of the song by adding a noisy buzz to the beginning.

to an hour (Behr and von Helversen, 2004). These songs are emitted mostly during the hours of highest activity in the colony at dawn and dusk (figure 15.7). To human listeners, courtship songs are less intense than territorial songs because most of the sound energy of courtship songs is contained in frequencies above 20 kHz. The singing male always focuses on one of the females of his harem. He approaches and waves his head toward her. A harem male may repeat this consecutively with different females. Most often, females terminate the approaches of a courting male by striking the folded wing against him. Usually this causes the male to fly up and to start a hovering display in front of the female, then to land about 20–40 cm away from her and to begin a new attempt. Copulation attempts mostly followed this sequence of courting behaviors.

Courtship songs, like territorial songs, consist of a series of discrete call elements, typically lasting from 50 to 300 milliseconds (figure 15.9). Four main groups of calls are discernible: "trills," "quasi-CF-calls," "short tonal elements," and "multi-harmonic signals" (Behr and von Helversen, 2004). All these call types are variable so that one male may use up to 100 call types or more.

The calls most commonly found in courtship songs were trills. These are tonal elements with a rippled modulation of the basal frequency modulation (figure 15.9). In a quantitative analysis of more than 500 trills of three males of one colony Behr and von Helversen (2004) showed that different males used very different trill calls in their courtship songs (figure 15.10). When the dendrogram resulting from a cluster analysis was "cut off" at a similarity level of 90% (Wishart, 1969), 99 different clusters were identified with 40% of these clusters containing only "private" elements of one male, a result significantly different from a random expectation.

The enormous repertoire of courtship songs of different males (Davidson and Wilkinson, 2002; Behr and von Helversen, 2004) and the individuality of these repertoires (Davidson and Wilkinson, 2003; Behr and von Helversen, 2004) raises the question of the role that territorial and courtship songs may play in sexual selection, mainly for female choice of mates or for identification of individuals. Davidson and Wilkinson (2003) showed that males with a larger composite syllable repertoire had more females in their territories and that some acoustic features of a common

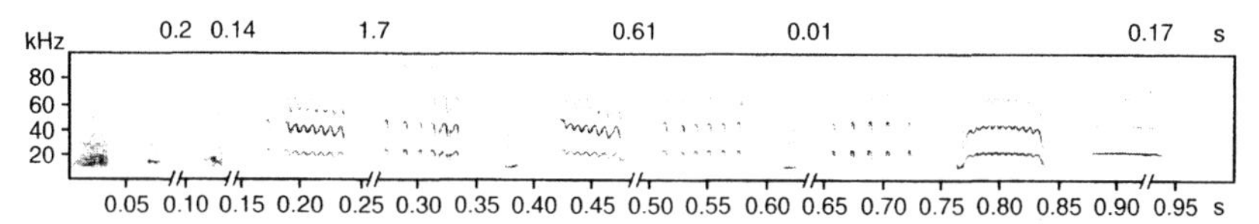

Figure 15.9 Example of a courtship song by *S. bilineata*. The frequency (0–100 kHz) is plotted against the time (s). Intermission periods are cut out from the sequence. The durations of intermission periods are given as time in seconds above the appropriate place (double slash on the *x*-axis). A multi-harmonic signal is shown at the beginning of the song; short tonal calls at 0.07, 0.12, 0.37, and 0.62 s; trill calls at 0.18, 0.28, 0.42, and 0.76 s; and a quasi-CF-call at the end.

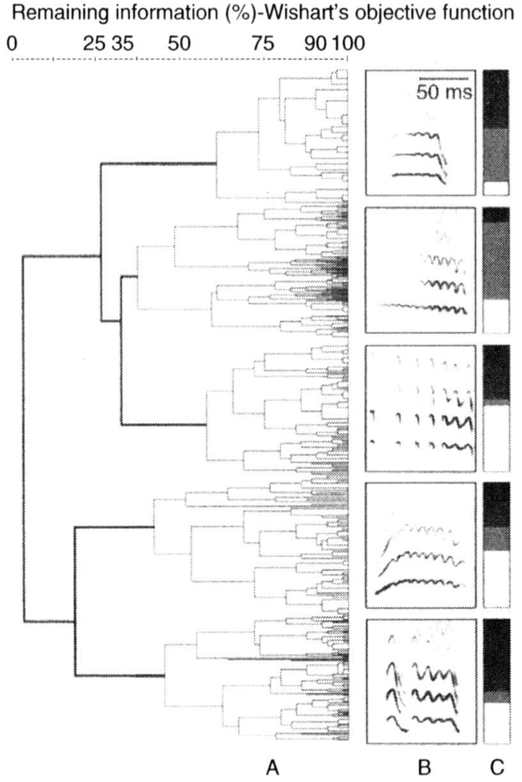

Figure 15.10 Dendrogram of 500 trills from three male *S. bilineata* (Behr and von Helversen, 2004). Similar calls are grouped on the left side of the graph and a typical example of the trill type is plotted to the right of the dendrogram. The cluster analysis extracted five distinct trill types when grouping was performed at a similarity level of 35% (Wishart, 1969). The bars to the right of the trill examples indicate how the three males (white, gray, and black) differed in the use of the given type of trill, with the full bar representing 100%.

call type such as the peak frequency, the duration of the inverted-V, the duration of so-called screeches, and the number of inverted-V calls, were significantly correlated with harem size. More experiments, particularly playback experiments, are needed to unravel the function of social calls and songs in *S. bilineata*.

Display of Perfume In addition to vocalizations, males also display scents to females. Male *S. bilineata* possess a sac-like organ in each of their front wing membranes (= antebrachium; figure 15.11) that can be opened and closed by two ligaments. Because these antebrachial sacs do not contain secreting cells (Starck, 1958; Scully et al., 2000), the wing sacs are considered organs for the storage and display of scents (Voigt and von Helversen 1999; Scully et al., 2000; Voigt, 2002). During each afternoon, male *S. bilineata* fill several body liquids into their holding sacs. This stereotypic and complex behavioral

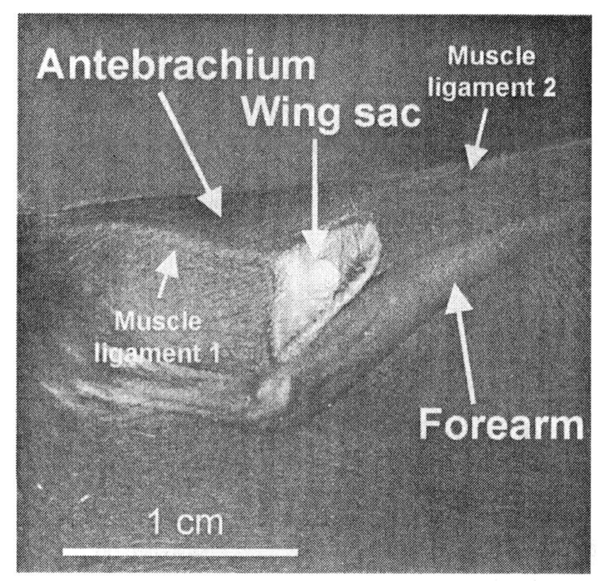

Figure 15.11 Dorsal view of the right antebrachial holding sac of a male *S. bilineata* (Voigt, 2002; see also Voigt and von Helversen, 1999). The right wing is extended to its full length but only the basal part is shown in the picture. Two muscle ligaments attach to the border of the wing sac. Contraction of ligament 1 opens the sac, whereas contraction of ligament 2 closes it. The inner epithelium of the sac is white to pink in adult males and structured by a main ridge and several smaller folds.

sequence consists of two phases (Voigt and von Helversen, 1999; Voigt, 2002). During phase 1, males take up urine with their mouth (figure 15.12) and clean the interior of their holding sacs extensively. Next, males transfer droplets of genital and gular secretion into the holding sacs (figure 15.13). Thus, the odoriferous liquid of the holding sacs is a mixture of urine, saliva, and secretions of the gular and genital region. Chemical analysis of the composite sac liquid has revealed large amounts of different fatty acids and at least three male-specific compounds of low molecular mass (C.C. Voigt, F. Schröder, S. Franke, and J. Meinwald, unpublished observations). All fatty acids identified thus far are also present in the holding sacs of subadult males or the nonfunctional rudiments of the holding sacs of females.

Males display the perfume of their holding sacs toward roosting conspecifics, preferably females. Two different types of perfume displays have been described in *S. bilineata*. One of these displays was named "salting" (Bradbury and Emmons, 1974) and constitutes a rapid fanning movement of the folded wing in the direction of another bat with wing sacs opened. The other type of display is a hovering flight during which males open their wing sacs periodically (figure 15.14; Bradbury and Emmons, 1974; Voigt and von Helversen, 1999). While hovering is performed only in front of individuals of a male's harem, salting is performed across harem boundaries. Hovering displays last for up to 14 seconds and can include

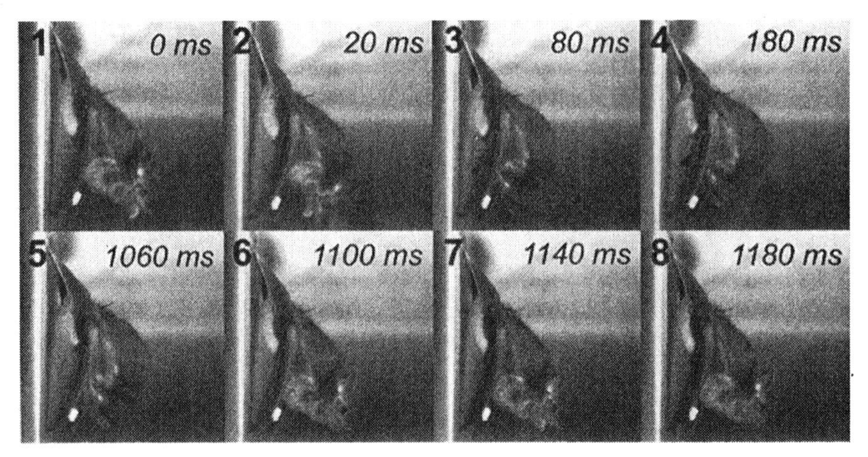

Figure 15.12 Uptake of urine from the penis by male *S. bilineata* (phase 1 of perfume-blending). Males bend down toward the genital region and take up droplets of urine from the penis. Afterwards, the wing sacs are licked and probably cleaned intensively (Voigt and von Helversen, 1999).

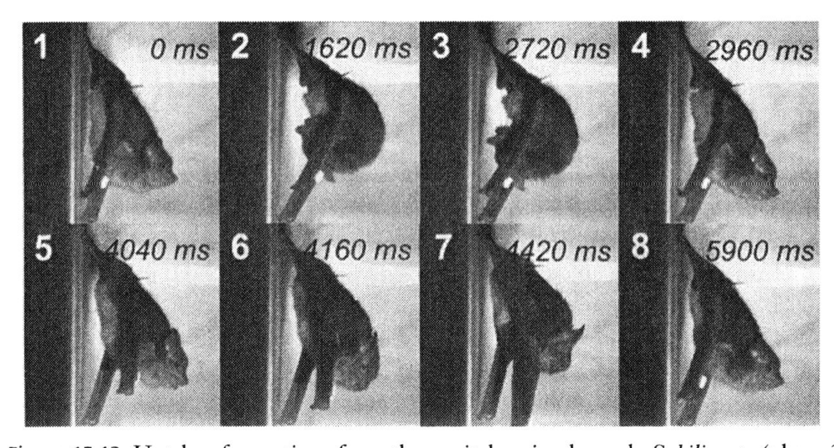

Figure 15.13 Uptake of secretions from the genital region by male *S. bilineata* (phase 2 of perfume-blending). Males bend down toward the genital region, press their gular region (region below the chin) against the penis (plate 2) and deposit a white secretion droplet on the gular region (compare plates 1 and 4). Afterwards, the droplet is smeared into one of the wing sacs by a quick movement of the head (plates 6 and 7) (Voigt and von Helversen, 1999).

up to 10 fanning movements. Females usually respond to a hovering male by emitting social calls or by waving their folded forearm toward the male, perhaps as a threatening gesture.

The frequency of perfume displays varies seasonally. Particularly intense courtship activity of males can be observed prior to and during the mating season. Within the same male, the pattern of perfume-blending does not vary

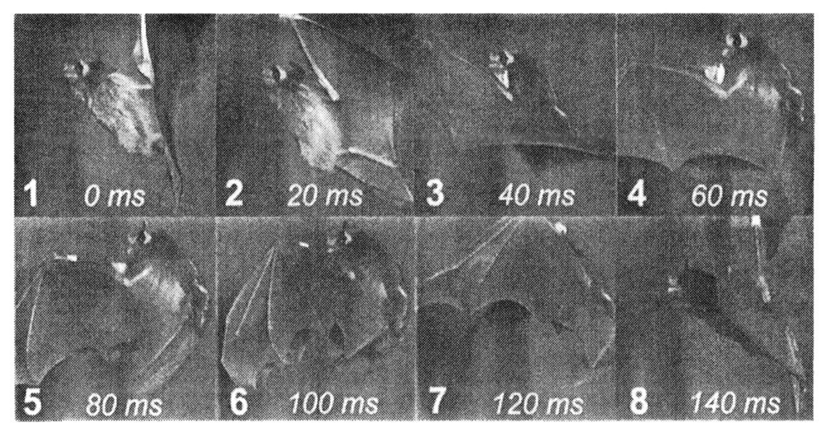

Figure 15.14 Hovering display of a male *Saccoperyx bilineata* in front of a female (not visible on this picture). Wing sacs snap open during approximately every seventh wing stroke. Scent molecules are probably drifted toward the roosting females. During the wing backstroke wing sacs are closed again (Voigt and von Helversen, 1999).

between the mating and nonmating season (Voigt, 2002). This observation is consistent with the notion that the display of fragrances serves functions in addition to courtship. Gular gland secretions, for example, are also used by male sac-winged bats for the scent-marking of territory boundaries (Bradbury and Emmons, 1974; Voigt and von Helversen, 1999) and the aerial display of gular gland secretions could demonstrate territory ownership.

REPRODUCTIVE VALUE OF HAREMS

Harem males vigorously defend their territory against male competitors and are eager to display songs and fragrances to females roosting within their territory. Overall, the number of hovering displays and flight maneuvers increases with increasing harem size (Voigt and von Helversen, 1999). Consequently, the energetic effort of harem maintenance also increases with increasing harem size (figure 15.15; Voigt et al., 2001). Given the fact that males spend a considerable amount of time and energy to prevent other males from entering their territory, it seems plausible that harems are of high value to the holder. A recent paternity study using DNA microsatellites supports this view: the reproductive success of harem males was, on average, higher than that of nonharem males: 1.6 offspring per year sired by harem males versus 0.7 offspring per year sired by nonharem males (figure 15.16; Heckel and von Helversen, 2002).

Nonetheless, this observation is consistent with the hypothesis that harem males do not monopolize reproduction within their territory because they cannot restrict movements of females outside their territory. Heckel et al. (1999) reported that harem males fathered only 30% of the offspring within their territories. Thus, the majority of offspring in the territories were sired by males other than the actual harem holders. Notwithstanding, reproductive

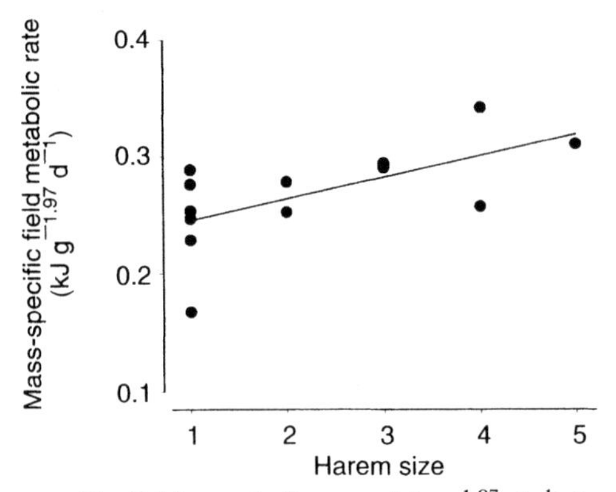

Figure 15.15 Mass-specific field metabolic rate (kJ g$^{-1.97}$ d^{-1}) in relation to harem size in *S. bilineata*. Field metabolic rate of males was significantly correlated with the harem size ($r^2 = 0.38$, $t_{12} = 2.6$, $P = 0.024$; Voigt et al., 2001).

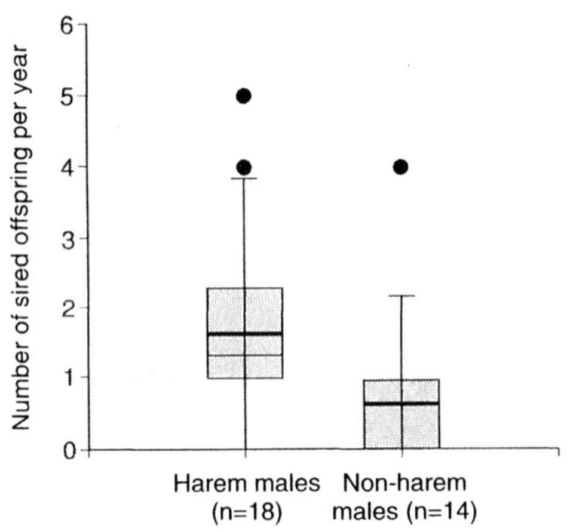

Figure 15.16 Average reproductive success (number of juveniles sired per year within the colony) of harem and nonharem males. Harem males sired on average 1.6 and nonharem males on average 0.7 offspring per reproductive season, which was significantly different according to a Mann–Whitney *U*-test (after Heckel and von Helversen, 2002). The box plots show the 25th to 75th percentile (box), the 10th and 90th percentile (T), the mean value as a thick line, the median as a thin line, and outliers.

success of harem males with females of their own harem increased with increasing harem size (figure 15.17a; Heckel and von Helversen, 2003). Thus, it is possible that harem males benefit in terms of number of sired offspring by defending a territory and investing effort in recruiting more females into their harem. By contrast, the total number of offspring sired by harem males outside their territory did not correlate with the number of females in the territories (figure 15.17b). This statistical result is interesting because our observations suggest that harem males are restricted in their movements and

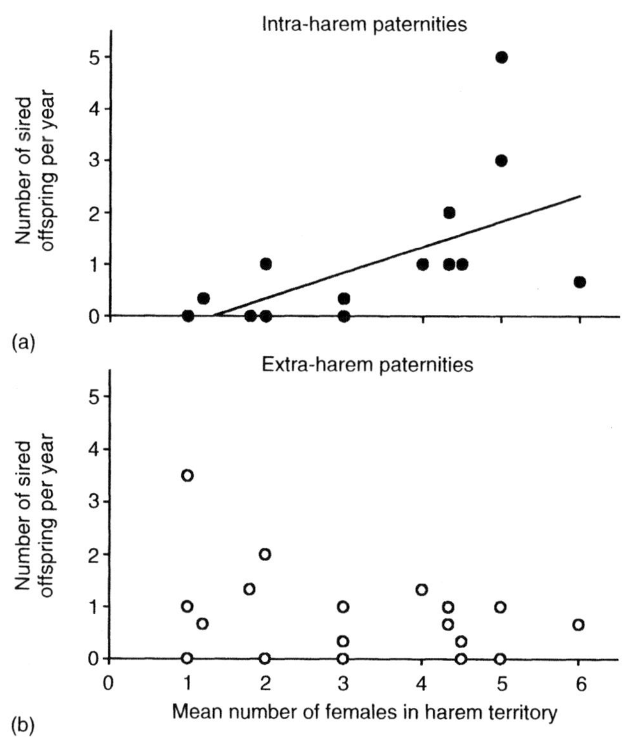

Figure 15.17 Relationship between reproductive success of harem males and the number of female *S. bilineata* in their corresponding territory during the lactation period (filled circles and regression line, number of offspring sired within the territory; open circles, number of offspring sired outside the territory). Despite large variance, reproductive success of males within their own harems was, on average, positively correlated with the number of females roosting in the territory during the lactation period ($r_s = 0.76$, $P < 0.001$; modified after Heckel, 2000). By contrast, the number of offspring fathered by harem males outside their territory was not related to harem size ($r_s = -0.33$, $P = 0.30$).

must remain in their territory during daytime when most copulations are likely to occur. When those males sire offspring outside their territory, either females visit the corresponding male during daytime or males and females meet for copulation during nightly foraging trips.

QUEUING OF NONHAREM MALES FOR TERRITORIES

Changes in harem possession are rare, and thus nonharem males may need to wait on average for 2 years (Voigt and Streich, 2003) before they can gain access to a territory or until they can take over a territory from a resident harem male. Thus far, we have found no morphological differences between harem males and nonharem males (Heckel and von Helversen, 2002; Voigt and Streich, 2003). Harem males are distinguished from nonharem males only by a longer tenure (Voigt and Streich, 2003). When harem males were

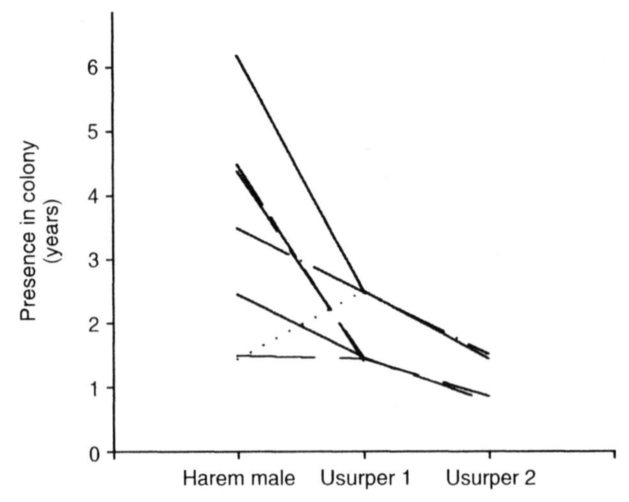

Figure 15.18 Comparison of the tenure in the colony (in years) of harem males and males that occupied the territory when harem males were removed (=usurper 1). When both harem males and usurper 1 were removed, a second male that was present in the colony for a shorter period is compared with the harem male and the corresponding first usurper that occupied the territory. The corresponding males of the same territory are linked with connecting lines. Five categories of colony presence could be distinguished in harem males when compared with two groups of usurping males.

removed temporarily, the nonharem male with the longest tenure at the site usurped the vacant territory. When both harem male and usurper of the first experiment were removed, a second nonharem male filled the vacancy. The second usurper had a shorter tenure in the colony than the harem male and the first usurper (figure 15.18). Nonharem males with a tenure position at a specific site in the colony did not participate in any usurpation attempts at different locations. Thus, it is likely that groups of nonharem males had site-specific hierarchies based on their respective tenure. This pattern is best described as queuing for access to harem territories. The fact that in *S. bilineata* only territories with large groups of females were usurped (Voigt and Streich, 2003) may reflect the fact that males evaluate the quality of a territory based on the potential reproductive success they could obtain. Similar strategic queuing behavior based on habitat or territory quality has been reported for birds (e.g., Beletsky and Orians, 1987; Ens et al., 1995).

In *S. bilineata*, a large proportion of male offspring remain in the colony in which they were born, whereas female offspring disperse to new colonies (see above). Subadult males most probably enter a queue during the first year of their life. Allozyme genetic assays revealed an allelic distribution among males in different colonies that was nonrandom in comparison with the outbred distribution amongst females, suggesting that males within *S. bilineata* colonies may be related as a consequence of male philopatry

(McCracken, 1984). Male philopatry may have evolved in *S. bilineata* in response to territory inheritance or the necessity to establish tenure as early as possible in a male's life. Philopatry may be favored if it is advantageous for males to enter a queue for a territory as early as possible in their life.

Cryptic Males

In her study on sac-winged bats in Panama, Tannenbaum (1975) described males younger than 1 year that spent the daytime silently in harems. During her 1-year field study, she noted 12 harem holders that disappeared from their territories. In five of these cases, cryptic males filled the vacancies. Two of the cryptic males replaced the harem male in their maternal territory and three outside of their maternal territory. Three of these five replacements occurred shortly before or during the mating season. Her observation suggests that males of 5 or 6 months of age are capable of maintaining harem ownership.

Similar observations of cryptic males have been made in Costa Rican populations (Voigt and Streich, 2003), where juvenile males stayed close to their mother in the maternal territory up to an age of 6–8 months. The corresponding harem holder usually ignored the presence of the juvenile male. During the removal experiments described above, the juvenile left the maternal area shortly after the harem was usurped by another male. In two cases, usurpers probably expelled cryptic males. Thus, it is possible that usurpers considered cryptic males as competitors. In two of seven removal experiments during which both harem males and the first usurper were excluded, a cryptic male remained the only male in the corresponding territory.

In contrast to all adult usurpers, these cryptic males never courted females. In line with this observation, Tannenbaum (1975) never observed cryptic males to steal copulations from harem females. However, at 6 months of age cryptic males have testes that are almost adult size (Voigt and Streich, 2003), and one incidence of a male fathering offspring at an age of 7 months has been demonstrated with DNA microsatellites (G. Heckel, unpubl. obs.). Full sexual maturity of juvenile males is probably achieved during the second mating season. The fact that juvenile *S. bilineata* do not participate in reproduction during their cryptic status in a harem is in contrast to the description of cryptic males in other bat species (e.g., *Nyctalus noctula* in Gebhard, 1997). On the other hand, the short-term disadvantage to cryptic males of postponing copulations may be outweighed by the possible long-term benefit of obtaining ownership of a nearby harem at a young age or the survival advantage associated with a prolonged period on familiar ground.

ACKNOWLEDGMENTS

We wish to express our sincere gratitude to Sonja Meister, Oliver Behr, and Sue Davidson for providing unpublished data. In addition, we would like to thank Sonja Meister and Oliver Behr for critical discussion and Heribert Hofer for comments on the manuscript. C.C.V. also wishes to express his thanks to Jack W. Bradbury for partial support of this study. We also thank the Organization for Tropical Studies for allowing us to conduct this project on their property and for providing the necessary infrastructure. In addition, we thank the Costa Rican authority MINAE, especially Javier Guevara, for support over many years. This work was funded by grants from the Deutsche Forschungsgemeinschaft to O.v.H and C.C.V.

LITERATURE CITED

Behr, O., and O. von Helversen. 2004. Bat serenades—complex courtship songs of the sac-winged bat (*Saccopteryx bilineata*). Behavioral Ecology and Sociobiology, 56: 106–115.

Beletsky, L.D., and G.H. Orians. 1987. Territoriality among male red-winged blackbirds. Behavioral Ecology and Sociobiology, 24: 309–319.

Bradbury, J.W. 1977. Social organization and communication. Pp. 1–72. In: Biology of Bats, Vol. 3 (W.A. Wimsatt, ed.). Academic Press, New York.

Bradbury, J.W., and L. Emmons. 1974. Social organization of some Trinidad bats. I. Emballonuridae. Zeitschrift für Tierpsychologie, 36: 137–183.

Bradbury, J.W., and S.L. Vehrencamp. 1976. Social organization and foraging in emballonurid bats. I. Field studies. Behavioral Ecology and Sociobiology, 1: 337–381.

Bradbury, J.W., and S.L. Vehrencamp. 1977. Social organization and foraging in emballonurid bats. III. Mating systems. Behavioral Ecology and Sociobiology, 2: 1–17.

Campagna, C., C. Bisoli, F. Quintana, F. Perez, and A. Vila. 1992. Group breeding in sea lions: pups survive better in colonies. Animal Behaviour, 43: 541–548.

Davidson, S.M., and G.S. Wilkinson. 2002. Geographic and individual variation in vocalization by male *Saccopteryx bilineata* (Chiroptera: Emballonuridae). Journal of Mammalogy, 83: 526–535.

Davidson, S.M., and G.S. Wilkinson. 2003. Function of male song in the greater white-lined bat, *Saccopteryx bilineata*. Animal Behaviour, 67: 883–891.

East, M., and H. Hofer. 1991a. Loud calling in a female-dominated mammalian society. I. Structure and composition of whooping bouts of spotted hyenas, *Crocuta crocuta*. Animal Behaviour, 42: 637–649.

East, M., and H. Hofer. 1991b. Loud calling in a female-dominated mammalian society. II. Behavioural contexts and functions of whooping of spotted hyaenas, *Crocuta crocuta*. Animal Behaviour, 42: 651–669.

Gebhard, J. 1997. Fledermäuse. Birkhäuser, Basel, Switzerland.

Genoud, M., and F.J. Bonaccorso. 1986. Temperature regulation, rate of metabolism, and roost temperature in the greater white-lined bat *Saccopteryx bilineata* Emballonuridae. Physiological Zoology, 591: 49–54.

Goodwin, G.C., and A.M. Greenhall. 1961. A review of the bats of Trinidad and Tobago. Bulletin of the American Museum of Natural History, 122: 187–302.

Hayssen, V., and T.H. Kunz. 1996. Allometry of litter mass in bats: comparisons with respect to maternal size, wing morphology, and phylogeny. Journal of Mammalogy, 77: 476–490.

Heckel, G. 2000. Mating System and Reproductive Tactics in the White-Lined Bat, *Saccopteryx bilineata*. Ph.D. Dissertation, University of Erlangen-Nürnberg, Erlangen, Germany.

Heckel, G., and O. von Helversen. 2002. Male tactics and reproductive success in the harem polygynous bat *Saccopteryx bilineata*. Behavioural Ecology, 13: 750–756.
Heckel, G., and O. von Helversen. 2003. Genetic mating system and the significance of harem associations in the bat *Saccopteryx bilineata*. Molecular Ecology, 12: 219–227.
Heckel, G., C.C. Voigt, F. Mayer, and O. von Helversen. 1999. Extra-harem paternity in the white-lined bat *Saccopteryx bilineata*. Behaviour, 136: 1173–1185.
Jameson, K.A., M.C. Appleby, and L. Freeman. 1999. Finding an appropriate order for a hierarchy based on probabilistic dominance. Animal Behaviour, 57: 991–998.
Linklater, W.L., E.Z. Cameron, E.O. Minot, and K.J. Stafford. 1999. Stallion harassment and the mating system of horses. Animal Behaviour, 58: 295–306.
McCracken, G.F. 1984. Social dispersion and genetic variation in two species of emballonurid bats. Zeitschrift für Tierpsychologie, 66: 55–69.
McCracken, G.F., and G.S. Wilkinson. 2002. Bat mating systems. Pp. 321–362. In: Reproductive Biology of Bats (E.G. Crichton and P.H. Krutzsch, eds.). Academic Press, New York.
Réale, D., P. Boussès, and J.-L. Chapuis. 1996. Female-biased mortality induced by male sexual harassment in a feral sheep population. Canadian Journal of Zoology, 74: 1812–1818,
Scully, W.M.R., M.B. Fenton, and A.S.M. Saleuddin. 2000. A histological examination of the holding sacs and glandular scent organs of some bat species (Emballonuridae, Hipposideridae, Phyllostomidae, Vespertilionidae, and Molossidae). Canadian Journal of Zoology, 78: 613–623.
Starck, D. 1958. Beitrag zur Kenntnis der Armtaschen und anderer Hautdrüsenorgane von *Saccopteryx bilineata* Temminck 1838 (Chiroptera, Emballonuridae). Gegenbaur morphologisches Jahrbuch, 99: 3–25.
Tannenbaum, R. 1975. Reproductive Strategies in the White-Lined Bat. Ph.D. Dissertation, Cornell University, Ithaca, NY.
Voigt, C.C. 2002. Individual variation of perfume-blending in male sac-winged bats. Animal Behaviour, 63: 31–36.
Voigt, C.C., and J.W. Streich. 2003. Queuing for harem access in colonies of the sac-winged bat. Animal Behaviour, 65: 149–156.
Voigt, C.C., and O. von Helversen. 1999. Storage and display of odor by male *Saccopteryx bilineata* (Chiroptera; Emballonuridae). Behavioral Ecology and Sociobiology, 47: 29–40.
Voigt, C.C., G. Heckel, and F. Mayer. 2005. Sexual selection favors small and symmetric males in the greater sac-winged bat *Saccopteryx bilineata* (Emballonuridae; Chiroptera). Behavioral Ecology and Sociobiology, 57: 457–464.
Voigt, C.C., O. von Helversen, R.H. Michener, and T.H. Kunz. 2001. The economics of harem maintenance in the sac-winged bat *Saccopteryx bilineata*. Behavioral Ecology and Sociobiology, 50: 31–36.
Wishart, D. 1969. An algorithm for hierarchical classifications. Biometrics, 25: 165–170.

16

Flexibility and Specificity in the Roosting Ecology of the Lesser Long-Eared Bat, *Nyctophilus geoffroyi*: A Common and Widespread Australian Species

Linda F. Lumsden & Andrew F. Bennett

An increasing body of literature suggests that tree-cavity roosting bats are selective in their use of roosts. By comparing aspects of roosting ecology for which there is a high level of specificity with those where roost use appears more flexible, we may gain a better understanding of the key influences on roost selection. Here we review eight studies on the roosting ecology of the lesser long-eared bat, *Nyctophilus geoffroyi*, a common and widespread vespertilionid in Australia. *Nyctophilus geoffroyi* is flexible in the distribution of roost sites, with roosts occurring in urban, rural, and a wide range of natural environments. While predominantly roosting in tree cavities, individuals (especially males) also use other natural and artificial roost sites. Tree roosts include a range of forms and orientations, but roosts located under bark and in fissures are common, frequently with a northerly orientation. A strong preference is shown by both sexes for roosts in dead trees, and entrance dimensions of roosts are consistently narrow (2.5–2.8 cm). Males predominantly roost solitarily, while females form larger colonies, especially while breeding. In some studies, females displayed a significant preference for roosts in larger trees, especially for maternity roosts, but this pattern was not consistent in all areas. In all studies, individuals shifted roosts frequently (every 1–2 days), and these roosts occurred within restricted areas. We suggest that key influences on roosting ecology for this species are the thermal requirements of roosts, the risk of predation, social organization within populations, and the pattern of roost availability in the landscape.

INTRODUCTION

Diurnal roost sites are critical components of the ecological requirements of bats. Many species roost in cavities within trees, but knowledge of how they select such roosts is a relatively recent field of study. An increasing body of literature suggests that tree-cavity roosting bats are highly selective in their choice of roosts (reviewed by Kunz and Lumsden, 2003). It has been proposed by Vonhof and Barclay (1996) that selection pressures experienced by tree-cavity roosting bats in temperate regions should generally be similar, and hence that all species should select similar roosts. While a range of species conform to this prediction (Kunz and Lumsden, 2003), a number of studies on sympatric species have identified differences in the types of tree-cavity roosts that they select, despite access to the same roosting opportunities (Foster and Kurta, 1999; Lumsden et al., 2002b).

One way to investigate patterns and influences on roosting ecology is to examine a single species throughout its geographic range, to determine how consistent it is in roost selection despite the differing environments and climatic regimes in which it lives. By comparing various aspects of the roosting ecology of a species, including those that show a high level of specificity and those where roost use appears more flexible, we may gain a better understanding of the key influences on roost selection.

The lesser long-eared bat, *Nyctophilus geoffroyi* (figure 16.1), is a common vespertilionid that is endemic to Australia. It is a small (7 g body mass) species that forages with a slow, maneuverable flight pattern, capturing prey by aerial hawking and, at times, gleaning insects from vegetation (Brigham et al., 1997; Grant, 1991; O'Neill and Taylor, 1986). The diet consists primarily of

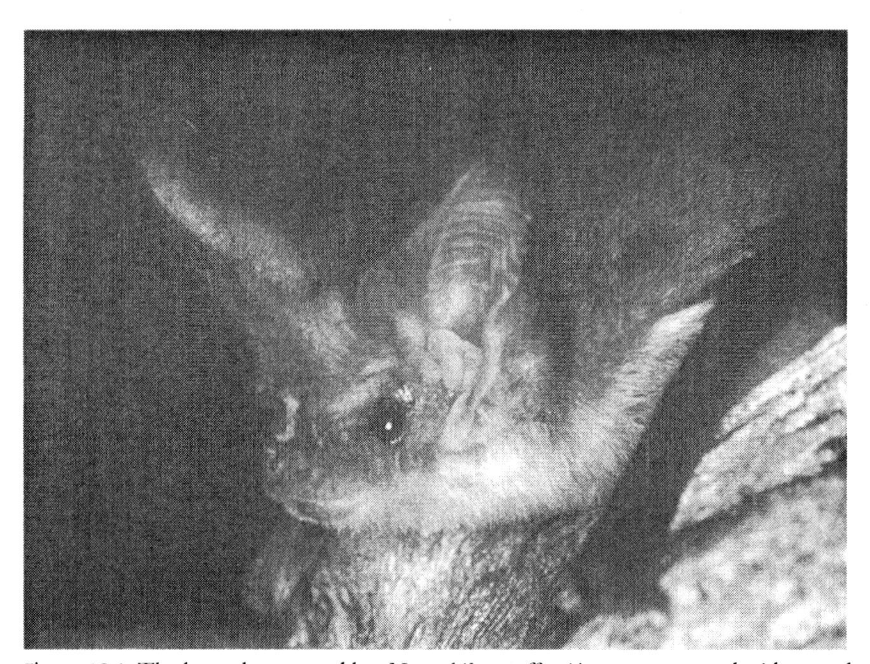

Figure 16.1 The lesser long-eared bat *Nyctophilus geoffroyi* is a common and widespread species in Australia.

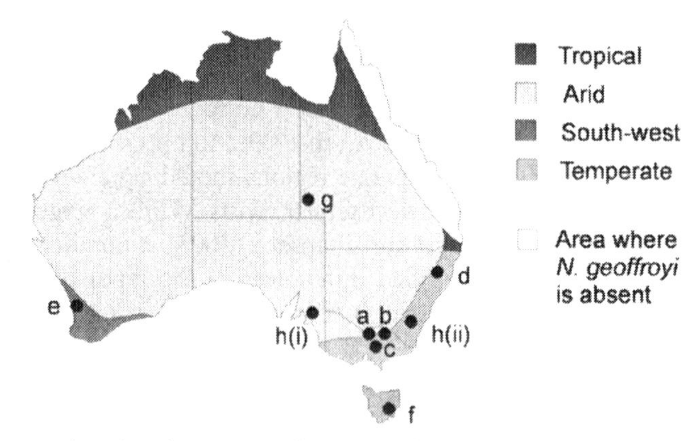

Figure 16.2 The distribution of the lesser long-eared bat *Nyctophilus geoffroyi* (Churchill, 1998) and biogeographic regions in Australia, and the location of study areas where roosting ecology has been investigated. Study area locations: a: Northern Plains, Victoria (Barmah study); b: Northern Plains, Victoria (farmland study); c: Strathbogie Ranges, Victoria; d: Armidale, northern New South Wales; e: southwestern Western Australia; f: Tasmania; g: Simpson Desert; h(i): Adelaide, South Australia; and h(ii): Canberra, Australian Capital Territory.

Lepidoptera, Orthoptera, Coleoptera, Hemiptera, and Hymenoptera (Churchill, 1998; L.F. Lumsden, unpublished data).

Nyctophilus geoffroyi has an extensive geographic range, being absent only from a narrow strip along the northeastern coast of Australia (figure 16.2). A wide range of environments are inhabited, from arid regions in central Australia to tropical forests and woodlands in the north, and temperate forests, rainforests, and alpine areas in the south (Churchill, 1998). This species has adapted to living in both urban and rural areas, and is often one of the more common bats in these modified environments (Lumsden and Bennett, 1995). A wide variety of roost types has been reported (see below), which led Maddock and Tidemann (1995) to suggest that *N. geoffroyi* is a generalist that shows little selectivity for roost sites. However, several recent studies have revealed a high level of specificity in roost site selection (Lumsden et al., 2002a,b; Schedvin et al., in prep.). Here we review eight studies on the roosting ecology of *N. geoffroyi* in order to identify those characteristics that appear to be flexible or specific, and use this analysis as a basis for exploring possible influences on roost site selection.

ROOSTING ECOLOGY

Nyctophilus geoffroyi uses a wide variety of roost types. Roosts are predominantly in tree cavities, but it has also been recorded within mud nests of the fairy martin (*Hirundo ariel*), between layers in clothing hung outside, sacks hung over rafters, canvas awnings, fence posts, under rocks on the ground, and even inside a pendulum cuckoo clock (Lumsden and Bennett,

1995; Schulz, 1998). Buildings are also used for roosting, predominantly in urban environments. Although all roosts for *N. geoffroyi* reported by Tidemann and Flavel (1987) were within buildings, this may overemphasize the use of these structures, as their study was primarily in urban areas and based mainly on roosts reported by the general public. In rural settings, buildings are used less frequently, despite being readily available, suggesting that where there are suitable tree cavities these are used preferentially (Lumsden et al., 2002b). Unlike some Northern Hemisphere species, which use tree cavities in summer but hibernate underground (Mayle, 1990; Schober and Grimmberger, 1989; van Zyll de Jong, 1985), *N. geoffroyi* roosts in trees year-round, despite occupying areas where temperatures fall below 0 °C in winter. There are, however, occasional records of individuals within the mouths of caves or rock crevices, or being found (often dead) further into complex cave systems (Maddock, 1972; Parker, 1973; Turbill, 1999).

Eight studies have investigated the roosting ecology of *N. geoffroyi*, together resulting in the location of 343 roosts. These studies are the source material for this review (table 16.1), with the locations of study sites shown in figure 16.2. Most studies are from temperate forests and woodlands in southern Australia, including southwestern Western Australia (Hosken, 1996), Tasmania (Taylor and Savva, 1988), the Strathbogie Ranges of central Victoria (A. Schedvin, pers. comm.), and the Northern Tablelands of New South Wales (Turbill, 1999). The most extensive studies have been conducted in Victoria, with 139 roosts located during an investigation of the use of remnant vegetation in farmland and a nearby floodplain forest (Barmah forest) in the Northern Plains (Lumsden, 2004; Lumsden et al., 1998, 2002a,b), and 86 roosts were located in a follow-up study investigating the roosting ecology of *N. geoffroyi* in farmland isolated from extensive forests (Lumsden, 2004). One study has been conducted in urban areas (Tidemann and Flavel, 1987). Only limited information is available on the roosting ecology of *N. geoffroyi* from arid regions of central Australia (Williams, 2001), and there are no published studies from tropical environments of northern Australia. Anecdotal information is included where available; however, unless otherwise stated, information presented is based on these eight studies.

A number of variables from each study are summarized in table 16.1, although not all studies collected data on every variable listed. The variables include aspects of the roost tree, the roost cavity within the tree, the landscape context of roosts, and roosting behavior. Some studies have presented data for both sexes, and these have been shown separately in table 16.1 where there are marked differences. The data have been categorized based on the sex of the individual carrying the radiotransmitter that led to the location of the roost.

Characteristics of Roost Trees

As expected from the extensive distribution and range of habitats used by *N. geoffroyi*, roost sites have been recorded in a wide variety of tree species.

Table 16.1 A Comparison of Roost Characteristics and Roosting Behavior between Eight Studies on the Roosting Ecology of the Lesser Long-Eared Bat *Nyctophilus geoffroyi* (values are means ± SD, or percentages as indicated)

	Northern Plains, Vic (Barmah)	**Northern Plains, Vic (farmland)**	**Strathbogies, Vic**	**North NSW**	**SW WA**	**Tasmania**	**Simpson Desert**	**SA & ACT**
Source	Lumsden, 2004; Lumsden et al., 1998, 2002a,b	Lumsden, 2004	Schedvin et al., in prep.	Turbill, 1999	Hosken, 1996	Taylor and Savva, 1988	Williams, 2001	Tidemann and Flavel, 1987
No. of roosts located	139	86	27	34	16	7	6	28
No. of individuals tracked	45	34	14	13	5	5	4	–
Environment	Remnants in farmland and extensive woodland	Remnants in farmland	Forest	Woodland	Woodland	Forest	Arid woodland	Urban
Type of roost								
Tree or timber remains	96%	91%	100%	97%	100%	100%	100%	–
Building	3%	9%	–	–	–	–	–	100%
Rock structure	1%	–	–	3%	–	–	–	–
Proportion of roosts in dead trees	79%	47%	37%	64%	75%	14%	67%	–
Proportion of roosts in dead limbs (on live or dead trees)	89%	88%	56%	73%	75%	–	100%	–
Tree size (diameter, cm)	–	–	–	–	57[a]	–	14.2 ± 4.2[a]	–
Males	42.4 ± 28.9	63.6 ± 28.5	57.0 ± 46.7	–	–	–	–	–
Females (all)	79.4 ± 48.6	70.1 ± 39.8	90.1 ± 53.8	–	–	–	–	–
Maternity roosts	105.1 ± 48.9	66.9 ± 34.4	83.4 ± 51.5	–	–	–	–	–

Type of cavity within tree[b]	–	–	–	–	–	–	–	
Under bark	46%	35%	–	61%	100%	50%	17%	–
Hole	11%	30%	41%	18%	–	–	–	–
Spout	12%	13%	23%	–	–	–	33%	–
Fissure	31%	22%	36%	21%	–	25%	50%	–
Bole	–	–	–	–	–	25%	–	–
Roost entrance dimensions (cm) (based on width of entrance)	2.5 ± 1.1	2.8 ± 1.3	2.5 ± 1.0	–	–	3	2.1 ± 1.3	3.8 ± 4.6
Predominant roost entrance orientation	North	All directions	North and west	North-west	North and west	–	North and west	East
Height of roost (m)	–	–	–	–	1.9 ± 1.4[a]	–	2.5 ± 1.9[a]	–
Males	3.3 ± 2.9	4.3 ± 2.6	14.5 ± 13.4	2.5 ± 0.3	–	–	–	–
Females	8.7 ± 6.2	5.4 ± 3.2	13.4 ± 10.5	2.1 ± 0.4	–	–	–	3.8 ± 1.2
Distance between roost site and capture point (km)	–	–	–	–	0.9–1.2 maximum[a]	–	0.6 ± 0.3[a]	–
Males	1.9 ± 2.9	0.7 ± 0.8	–	0.4 ± 0.2	–	–	–	–
Females	6.7 ± 2.9	1.4 ± 2.5	–	0.5 ± 0.2[c]	–	–	–	–
Mean no. individuals in roost								
Males	1.5 ± 2.2	2.3 ± 2.6	32.5 ± 12.0	1	1 ± 0	–	1 ± 0	1 ± 0
Females	7.0 ± 10.7	9.7 ± 7.5	12.7 ± 8.2	9.5	1 ± 0	12.7 ± 10.0	3.5 ± 0.7	19.6 ± 9.6
Mean no. of days spent in the one roost	1.8 ± 1.6	2.5 ± 2.1	–	1.5 ± 0.5	1.1 ± 0.2	–	1.5 ± 0.6	–
Mean distance between consecutive roosts (m)	322 ± 400	489 ± 784	180 ± 170	97 ± 74	194 ± 57	–	250 ± 212	–

[a]Data for both sexes combined.
[b]All roosts were within buildings.
[c]This figure is the mean for lactating females; the one nonbreeding female that was tracked roosted 10.3 km from where it was caught.

There is some evidence that particular tree species are favored: A. Schedvin et al. (pers. comm.) found that all roosts were in eucalypt species with rough bark (e.g., narrow-leaf peppermint, *Eucalyptus radiata*, red stringybark, *E. macroryncha*), compared with those with smooth bark, which were preferentially used by other species of bats (e.g., *Falsistrellus tasmaniensis*). In addition, use of tree species may vary with weather conditions. For example, Hosken (1996) reported that individuals shifted between dead banksias (*Banksia* spp.) and live melaleuca (*Melaleuca* spp.) trees in response to storms, and attributed this movement to thermal characteristics of the trees. It is unlikely that *N. geoffroyi* selects roosts based on tree species per se, rather that particular tree species provide different types of cavities with different microclimates and physical dimensions. This has also been reported for species in North America that have similar patterns of roost site selection and roosting ecology to *N. geoffroyi* (e.g., *Lasionycteris noctivagans*: Mattson et al., 1996; *Myotis septentrionalis* and *M. sodalis*: Foster and Kurta, 1999; Kurta et al., 1996).

A common finding was that many of the roosts of *N. geoffroyi* were in dead trees (table 16.1, figure 16.3). In addition, where roosts were in live trees, they were predominantly located within a dead section of the tree. In most studies, over three quarters of roosts were in dead timber. Where the availability of potential roost trees was assessed, dead trees were used significantly more often than expected. For example, in the Northern Plains of Victoria, only 19% of trees in the vicinity of roosts were dead, but 79% of roosts were in dead trees (Lumsden et al., 2002b). Eucalypts form cavities while still alive and healthy (Mackowski, 1984), and abundant cavities were available in live trees in these areas, which were used extensively by sympatric species of bats (e.g., *Chalinolobus gouldi*: Lumsden et al., 2002b; *F. tasmaniensis*: A. Schedvin, pers. comm.). Although little information is available from arid environments in Australia, the same pattern of roosting in dead timber may be emulated (Williams, 2001; Wood Jones, 1923–1925).

The size of roost trees varies depending on the form of the locally available trees. For example, in the wet forests of the Strathbogies where trees are typically large, the diameter of trees used was often greater than 80 cm. In contrast, in the arid woodlands of the Simpson Desert where trees are typically small, the mean diameter of roost trees was only 12 cm (table 16.1). Where roost trees have been compared with available trees, roost trees tend to be larger in diameter (Lumsden et al., 2002b). There appears to be intraspecific variation in the size of trees used as roosts, with males often selecting roost trees of smaller diameter than females, and breeding females selecting larger trees than nonbreeding females (table 16.1). For example, in one study area (Northern Plains, Barmah), maternity roosts were located in significantly larger trees than those used by nonbreeding females (105.1 ± 48.9 vs. 57.4 ± 36.3 cm, $P < 0.001$; Lumsden et al., 2002b). However, in two other studies this was not the case (table 16.1).

Figure 16.3 Roost sites selected by *Nyctophilus geoffroyi* were frequently in dead trees, such as this isolated tree in farmland in northern Victoria. The roost entrance is indicated by the arrow.

Characteristics of Roost Cavities

Approximately half of the roosts in most studies were located under exfoliating bark on either live or dead trees (table 16.1). An exception to this pattern was from forests in the Strathbogie Ranges, where no roosts were located under bark, despite its availability. Extensive use of cavities under bark is a feature common to all species in the genus *Nyctophilus* (Churchill, 1998; Lunney et al., 1988, 1995), and this is the main group of bats in Australia that use this roosting resource. Fissures, which often form when dead timber cracks, were consistently used by *N. geoffroyi* in most studies. Fissures and cavities under bark usually constituted about three quarters of all roosts (table 16.1). Other types of cavities available as potential roost sites included those in the trunk and main branches, spouts (hollow broken-off branches), and cavities in the boles at the base of trees. Sympatric bat species used one or more of these alternate roost types extensively (Lumsden et al., 2002b; A. Schedvin, pers. comm.), but *N. geoffroyi* made comparatively less use of these types of cavities.

The use of roosts under bark varies seasonally. In some areas, nonbreeding females roost under bark more frequently than do breeding females. For example, in the Northern Plains (Barmah) study, no maternity roosts were located under bark despite 50% of roosts used by females outside the breeding season were located there (Lumsden et al., 2002b). However, in nearby farmland where only remnants of woodland persist and the overall availability of potential roost sites is lower, 11% of maternity roosts were under bark (Lumsden, 2004). In addition, Turbill (1999) located two maternity roosts in substantial cavities under thick bark on large, dead pine trees (*Pinus* sp.), and McKean and Hall (1964) reported a maternity roost under the bark of a live rough-barked eucalypt.

Of all roost characteristics, the most consistent and specific feature of roosts used by *N. geoffroyi* is the dimension of the entrance to the roost. Entrances to cavities used by *N. geoffroyi* are often long and thin, especially when under bark and in fissures (e.g., 2.5 cm wide × 1.5 m long; figure 16.4), and so both the length and width of entrances are usually measured. The width is the dimension most likely to constrain other species from entering the roost, and consequently this measure is used to compare between studies and between species. The three studies with extensive data on entrance dimensions to tree cavity roosts had remarkably similar results (table 16.1), reporting means of 2.5 ± 1.0 to 2.8 ± 1.3 cm. These entrance dimensions approximate the average body width of *N. geoffroyi*. Although these studies did not compare the cavities used with those available, larger cavities were widespread in these forests as shown by other species selecting cavities with larger openings (e.g., mean of 10 cm diameter for *C. gouldii*: Lumsden et al., 2002b; mean of 8 cm diameter for *F. tasmaniensis*: A. Schedvin, pers. comm.). Roosts within buildings recorded by Tidemann and Flavel (1987) were somewhat larger than those in tree cavities, with a mean width of 3.8 cm.

Roost entrances face a wide range of orientations; however, in most studies the predominant directions faced were north and west. North and west-facing roosts optimize exposure to solar radiation, which is likely to result in energetic savings for bats (McNab, 1982). In the Northern Hemisphere, roosts often face south for the same reason (e.g., Kalcounis and Brigham, 1998; Mattson et al., 1996; Vonhof and Barclay, 1997). This pattern was also found in hot desert regions, where bats may have been expected to avoid excessive temperatures. Williams (2001) recorded temperatures up to 36 °C within a *N. geoffroyi* roost under bark on the northwest side of a dead tree in the Simpson Desert. An additional benefit of roosting on the northern side of

Figure 16.4 Small entrance dimensions were a consistent feature of *Nyctophilus geoffroyi* roosts, which were often under exfoliating bark or within a narrow fissure.

a tree in southern Australia is greater protection from cold and wet weather conditions, which are predominantly from the south and west.

A consistent feature of roosts in woodland vegetation was that they were often low on the trunk, some less than 2 m above ground (figure 16.5). In contrast, the taller forests of the Strathbogie Ranges had roosts that were higher, with a mean of 14 m above the ground. This is likely to be influenced by the lack of roosts under bark in this area, as exfoliating bark is often located low on the main trunk of trees. In one study (Northern Plains, Barmah) males roosted significantly lower to the ground than did females (Lumsden et al., 2002b), but this was not consistent across all studies (table 16.1).

Figure 16.5 Roosts used by male *Nyctophilus geoffroyi* were often low to the ground, such as under bark on this fallen tree. The arrow indicates the location of the roost.

The height of the roost proportional to the height of the roost tree was comparable across studies, with roost heights between 38% and 50% of the height of the tree. That is, roosts are generally located in the mid- to lower sections of trees rather than in the upper canopy, which reflects the typical location of the main roost types of exfoliating bark and fissures.

Landscape Context of Roost Sites

The slow maneuverable flight pattern of *N. geoffroyi* while foraging has led to the suggestion that this species does not move large distances between roost sites and foraging areas (Hosken, 1996). However, Lumsden et al. (2002a) found that substantial distances were covered, with some individuals roosting up to 12 km from where they foraged. This study area was located near the interface between farmland and extensive forest, and individuals appeared to select optimal areas for roosting (within the extensive forest) and optimal areas for foraging (in remnant woodland in farmland), readily commuting between the two. The sexes behaved markedly differently in this respect, with females commuting significantly greater distances than males (table 16.1). In other studies with more uniform environments (e.g., continuous forest, or remnant vegetation in farmland), shorter distances were traversed between roost sites and capture points (capture points are used to represent a foraging area of each individual: see Lumsden et al., 2002a) (table 16.1). Lumsden et al. (2002a), using the wing morphology equation of Jones et al. (1995), predicted that the foraging range for *N. geoffroyi* should be 2.2 km. In general, the results from most studies conformed to this prediction (table 16.1), although a small number of individuals from many of the studies moved greater distances. It seems that in environments where foraging and roosting resources are uniformly distributed, there is little need to commute large distances; however, where there is an advantage in doing so, individuals are capable of regularly undertaking greater flight distances.

Nyctophilus geoffroyi often concentrates its roosts within particular parts of the habitat. Taylor and Savva (1988) found that all roosts were located in mature forest, despite individuals having been trapped initially in regrowth forest. A. Schedvin (pers. comm.) found *N. geoffroyi* to predominantly roost on drier ridges within the forest, and that these areas had a greater density of cavity-bearing trees than that available in randomly located sites. Likewise, roost sites were not randomly distributed within the extensive Barmah floodplain forest in northern Victoria. *Nyctophilus geoffroyi* preferentially roosted in areas of the forest that had significantly higher densities of dead cavity-bearing trees than were generally available. In the same area, the sympatric species *C. gouldii* selected forested areas that had higher densities of live trees with cavities (Lumsden et al., 2002a).

Roosting Behavior

Colony sizes of *N. geoffroyi* are typically small, with means of fewer than 10 individuals in most studies (table 16.1). Females, especially when forming maternity colonies, typically roosted in larger groups than did males. Males frequently roosted singly, while a lower proportion of females (predominantly nonbreeding individuals) were solitary. Similar patterns have been recorded for other tree-cavity roosting species (e.g., Law and Anderson, 2000; Mattson et al., 1996; Tidemann and Flavel, 1987). There are, however, exceptions to this general pattern. Two roosts found by A. Schedvin (pers. comm.) contained 28 and 41 individuals, respectively. The individuals carrying radiotransmitters in these roosts were males, but the sex of other individuals was not known. All females in the study by Hosken (1996) roosted solitarily; however, this investigation was conducted during autumn and winter, and colony sizes in maternity roosts were not available. Colonies that form in buildings are often larger than those in tree cavities: a mixed-sex colony of some 200 individuals was reported from a building in South Australia (Reardon and Flavel, 1987).

Nyctophilus geoffroyi regularly shifts roost sites. New roosts are often used on a daily basis although some individuals may return to the same roost for up to 14 consecutive days (Lumsden et al., 1998). The mean number of consecutive days in the same roost was consistent between studies, ranging from 1.1 to 2.6 days (table 16.1). This pattern conforms with many other species of bats that roost in tree cavities and under bark (Kunz and Lumsden, 2003; Lewis, 1995). Maternity colonies of *N. geoffroyi* moved their roost as frequently as did nonbreeding colonies (Lumsden, 2004; Turbill, 1999), which contrasts with the findings for some other species (e.g., Kurta et al., 1996; Mattson et al., 1996; Vonhof and Barclay, 1996). For example, lactating females shifted roosts every 1.5 ± 0.7 days on average in the Northern Plains (Lumsden, 2004). This involved the female returning to the roost after the first foraging bout and carrying the young in flight to the subsequent roost. As this species typically has twins, this flight was usually undertaken twice.

When individuals move to a new roost, it is usually to one nearby (e.g., <300 m; table 16.1). In this respect, *N. geoffroyi* follows a pattern typical of many tree-cavity roosting bats, in being faithful to a roost area but moving roost sites within that area on a regular basis (Kunz and Lumsden, 2003). There was, however, some variation between studies, based on the density of available roost sites. In continuous forest of the Strathbogie Ranges and Barmah forest, mean distances between successive roosts were approximately 175 m. However, where roosts were in remnant vegetation in a matrix of cleared farmland devoid of trees, roost trees were on average more than 400 m apart.

KEY FACTORS INFLUENCING ROOSTING ECOLOGY

Nyctophilus geoffroyi occurs in a wide variety of environments and occupies a broad range of roost structures. Despite this variation, several features of its roosting ecology are consistent and specific across southern Australia. Roosts are typically under bark or in fissures in dead trees, and have a northerly or westerly orientation with narrow entrance dimensions.

The thermal condition of a bat's environment has a major influence on energy expenditure (McNab, 1982), with a range of species known to select roosts that have a specific internal microclimate (e.g., Hall, 1982; Kerth et al., 2001; Sedgeley, 2001; Vonhof and Barclay, 1997). In temperate regions of Australia, preferred roost cavities are insulated against temperature extremes. Roosts used by *N. geoffroyi* often have a smaller range in temperature when compared with external ambient conditions, with cooler temperatures during the day and warmer temperatures at night (Hosken, 1996; Turbill, 1999).

Males, nonbreeding females and reproductive females are likely to have different thermal requirements, and hence may select different roosts. Maternity sites are often warmer than roosts used during the nonbreeding period, to enable rapid development of the young before and after birth (McNab, 1982). Adult males and nonreproductive females may select cooler roost sites to facilitate entry into torpor, thus minimizing energy expenditure (Hamilton and Barclay, 1994; Hosken and Withers, 1999; Kerth et al., 2001). Turbill (1999) found that lactating female *N. geoffroyi* entered torpor on 75% of roost days monitored, while males at the same time of the year entered torpor every day. Females used shallower bouts of torpor with a mean skin temperature of 21 °C while in torpor, compared with a skin temperature of 11 °C for males.

Physical characteristics of the roost cavity and roost tree are likely to contribute to the microclimate in roosts used by *N. geoffroyi*. A north-facing orientation is more protected from prevailing weather conditions and receives greater solar radiation, contributing to warming of the roost and keeping it drier during rainy periods. Large trees and dead timber are likely to influence the degree of insulation of the roost and hence the microclimate. Tree trunks of smaller diameter (e.g., <50 cm diameter) are reported to offer less insulation against extreme temperatures than those of larger ones (e.g., >70 cm diameter; Alder, 1994; Sluiter et al., 1973). The type of roost will also influence thermal conditions within the cavity. For example, Turbill (1999) reported a significant difference in the amount of time that individual *N. geoffroyi* spent in torpor in roosts under bark, in fissures, and in cavities. Bats that roosted under thin bark aroused from torpor more frequently than did those roosting under thicker bark or in cavities, which offered more insulation.

The most consistent characteristic of roosts occupied by *N. geoffroyi* was the small size of the entrance. *Nyctophilus geoffroyi* also appears to prefer internal cavities that are not much larger than their body size. If the size of the internal cavity is a critical component in the roosting ecology of this species because of the microclimate that it provides, then this is likely to influence entrance dimensions, because small internal cavities are likely to have small entrances. If, however, the size of the entrance is a more critical component (to reduce the risk of predation and competition), this may be a strong influence on other aspects of roost selection. For example, roosts under bark and in fissures are those most likely to have narrow entrances, with entrances to spouts and cavities often being larger. In addition, exfoliating bark and fissures are more likely to occur on dead, rather than live trees.

A narrow entrance to a roost is likely to reduce the risk of both predation and competition (Tidemann and Flavel, 1987). There is a range of other species in the forests and woodlands of southern Australia that may compete for tree cavities, such as possums, gliders, marsupial carnivores, and other species of bats, cavity-nesting birds and feral bees. Most of these species would be excluded by an entrance dimension of 2.5 cm width. However, the frequency of shifts between roosts and the use of many roosts by individual *N. geoffroyi* suggest that roosts may not be in such short supply that competition is a strong influence. Rather, the consistency of roost dimensions across different study areas and its close match with body size may be a response to the risk of predation.

Potential predators include cats, marsupial carnivores, owls, and goannas (Lumsden and Bennett, 1995), most of which would be excluded by a narrow entrance dimension of 2.5 cm. Bats are especially vulnerable to predation from within the roost when they are in torpor, or when nonvolant young are left in the roost while adult females forage. There is also a risk of predation from diurnal and nocturnal birds as bats exit their roost at dusk. Shifting roost sites frequently has been suggested as a strategy to reduce such predation, although a number of other factors such as social interactions, reducing parasite loads, and gaining familiarity with a range of roost microclimates may also influence this behavior (Lewis, 1995). *Nyctophilus geoffroyi* shifted roost sites regularly in most studies, suggesting that this is a consistent feature of its roosting ecology. Lumsden et al. (2002a) suggested that the low position of roosts used by *N. geoffroyi* may also subject them to a greater risk of predation than species of bats that roost higher in the tree. However, there are no quantitative data on the frequency or intensity of predation on *N. geoffroyi* to test this hypothesis. Likewise, there are no empirical data to compare the relative influence of predation risk versus other possible reasons for regularly shifting to alternate roosts.

Little is known of the social organization and mating system of *N. geoffroyi* (Lumsden and Bennett, 1995). However, aspects of roosting ecology such as colony size and differences in colony size between sexes, and regular shifting of roosts are all likely to be influenced by social organization. Most males appear to roost solitarily and this occurs throughout the year, including during the mating season. The mating system is not known, but some form of polygyny (McCracken and Wilkinson, 2000) is most likely. Hosken (1998) inferred sperm competition within a captive colony of *N. geoffroyi*, based on one male fathering all offspring despite each female mating with both males present.

FURTHER INVESTIGATIONS

Even common and widespread species such as *N. geoffroyi* can show specificity in their selection of roosts and in roosting behavior. Some characteristics of roosts are consistently selected (e.g., entrance size, dead timber), whereas others vary depending on what is locally available within the particular environment (e.g., height of roost). The preliminary study by Williams (2001) suggests that findings from temperate regions may also be applicable in arid regions. However, further studies are required from the arid areas of central Australia and from the tropical north, to test the generality of the trends reported here for *N. geoffroyi*, in different environments. To determine the relative importance of factors influencing roosting ecology, there is a need for quantitative information on the microclimatic properties of roosts (and of cavities not used as roosts), on the rate of predation of bats when in or leaving roosts, on social organization within populations, and on the implications for reproductive success and survival rates.

ACKNOWLEDGMENTS

We thank Natasha Schedvin, Christopher Turbill, and Amy Williams for access to unpublished data; Martin Schulz for commenting on an earlier draft; and Tom Kunz for inviting us to present this review.

LITERATURE CITED

Alder, H. 1994. Erste Erfahrungen mit dem Data Logger: Ereigniszählung vor Baumhöhlenquartieren von Wasserfledermäusen, *Myotis daubentoni*, bei gleichzeitiger Messung mikroklimatischer Werte. Mitteilungen der Naturforschenden Gesellschaft Schaffhausen, 39: 119–133.

Brigham, R.M., R.L. Francis, and S. Hamdorf. 1997. Microhabitat use by two species of *Nyctophilus* bats: a test of ecomorphology theory. Australian Journal of Zoology, 45: 553–560.

Churchill, S. 1998. Australian Bats. Reed New Holland, Sydney.

Foster, R.W., and A. Kurta. 1999. Roosting ecology of the northern bat (*Myotis septentrionalis*) and comparisons with the endangered Indiana bat (*Myotis sodalis*). Journal of Mammalogy, 80: 659–672.

Grant, J.D.A. 1991. Prey location by two Australian long-eared bats, *Nyctophilus gouldi* and *N. geoffroyi*. Australian Journal of Zoology, 39: 45–56.

Hall, L.S. 1982. The effect of cave microclimate on winter roosting behaviour in the bat, *Miniopterus schreibersii blepotis*. Australian Journal of Ecology, 7: 129–136.

Hamilton, I.M., and R.M.R. Barclay. 1994. Patterns of daily torpor and day-roost selection by male and female big brown bats (*Eptesicus fuscus*). Canadian Journal of Zoology, 72: 744–749.

Hosken, D.J. 1996. Roost selection by the lesser long-eared bat, *Nyctophilus geoffroyi*, and the greater long-eared bat, *N. major* (Chiroptera: Vespertilionidae) in *Banksia* woodlands. Journal of the Royal Society of Western Australia, 79: 211–216.

Hosken, D.J. 1998. Sperm fertility and skewed paternity during sperm competition in the Australian long-eared bat *Nyctophilus geoffroyi* (Chiroptera: Vespertilionidae). Journal of Zoology (London), 245: 93–100.

Hosken, D.J., and P.C. Withers. 1999. Metabolic physiology of euthermic and torpid lesser long-eared bats, *Nyctophilus geoffroyi* (Chiroptera: Vespertilionidae). Journal of Mammalogy, 80: 42–52.

Jones, G., P.L. Duverge, and R.D. Ransome. 1995. Conservation biology of an endangered species: field studies of greater horseshoe bats. Symposia of the Zoological Society of London, 67: 309–324.

Kalcounis, M.C., and R.M. Brigham. 1998. Secondary use of aspen cavities by tree-roosting big brown bats. Journal of Wildlife Management, 62: 603–611.

Kerth, G., K. Weissmann, and B. König. 2001. Day-roost selection in female Bechstein's bats (*Myotis bechsteinii*): a field experiment to determine the influence of roost temperature. Oecologia, 126: 1–9.

Kunz, T.H., and L.F. Lumsden. 2003. Ecology of cavity and foliage roosting bats. Pp. 3–89. In: Bat Ecology (T.H. Kunz and M.B. Fenton, eds.). University of Chicago Press, Chicago.

Kurta, A., K.J. Williams, and R. Mies. 1996. Ecological, behavioural, and thermal observations of a peripheral population of Indiana bats (*Myotis sodalis*). Pp. 102–117. In: Bats and Forests Symposium, October 19–21, 1995, Victoria, British Columbia, Canada (R.M.R. Barclay and R.M. Brigham, eds.). British Colombia Ministry of Forests, Victoria, BC, Canada.

Law, B.S., and J. Anderson. 2000. Roost preferences and foraging ranges of the eastern forest bat *Vespadelus pumilus* under two disturbance histories in northern New South Wales, Australia. Austral Ecology, 25: 352–367.

Lewis, S.E. 1995. Roost fidelity of bats: a review. Journal of Mammalogy, 76: 481–496.

Lumsden, L.F. 2004. The Ecology and Conservation of Insectivorous Bats in Rural Landscapes. Ph.D. Thesis, School of Ecology and Environment, Deakin University, Burwood, Victoria, Australia.

Lumsden, L.F., and A.F. Bennett. 1995. Lesser long-eared bat *Nyctophilus geoffroyi*. pp. 184–186. In: Mammals of Victoria: Distribution, Ecology and Conservation (P.W. Menkhorst, ed.). Oxford University Press, Melbourne, Australia.

Lumsden, L.F., A.F. Bennett, and J.E. Silins. 1998. The roosting behavior of two species of vespertilionids in southern Australia. Bat Research News, 39: 80–81.

Lumsden, L.F., A.F. Bennett, and J.E. Silins. 2002a. Location of roosts of the lesser long-eared bat *Nyctophilus geoffroyi* and Gould's wattled bat *Chalinolobus gouldii* in a fragmented landscape in south-eastern Australia. Biological Conservation, 106: 237–249.

Lumsden, L.F., A.F. Bennett, and J.E. Silins. 2002b. Selection of roost sites by the lesser long-eared bat (*Nyctophilus geoffroyi*) and Gould's wattled bat (*Chalinolobus gouldii*) in south-eastern Australia. Journal of Zoology (London), 257: 207–218.

Lunney, D., J. Barker, D. Priddel, and M. O'Connell. 1988. Roost selection by Gould's long-eared bat, *Nyctophilus gouldi* Tomes (Chiroptera: Vespertilionidae), in logged forests on the south coast of New South Wales. Australian Wildlife Research, 15: 375–384.

Lunney, D., J. Barker, T. Leary, D. Priddel, R. Wheeler, P. O'Connor, and B. Law. 1995. Roost selection by the north Queensland long-eared bat *Nyctophilus bifax* in littoral rainforest in the Iluka World Heritage Area, New South Wales. Australian Journal of Ecology, 20: 532–537.

Mackowski, C.M. 1984. The ontogeny of hollows in Blackbutt (*Eucalyptus pilularis*) and its relevance to the management of forests for possums, gliders and timber. Pp. 553–567. In: Possums and Gliders (A.P. Smith and I.D. Hume, eds.). Surrey Beatty and Sons, Chipping Norton, UK.

Maddock, T.H. 1972. The lesser long-eared bat, *Nyctophilus geoffroyi* Leach: cave-dweller or occasional visitor? South Australian Naturalist, 46: 63–64.

Maddock, T.H., and C.R. Tidemann. 1995. Lesser long-eared bat *Nyctophilus geoffroyi*. Pp. 502–503. In: The Mammals of Australia (R. Strahan, ed.). Reed Books, Chatswood, New South Wales, Australia.

Mattson, T.A., S.W. Buskirk, and N.L. Stanton. 1996. Roost sites of the silver-haired bat (*Lasionycteris noctivagans*) in the Black Hills, South Dakota. Great Basin Naturalist, 56: 247–253.

Mayle, B.A. 1990. A biological basis for bat conservation in British woodlands: A review. Mammal Review, 20: 159–195.

McCracken, G.F., and G.S. Wilkinson. 2000. Bat mating systems. Pp. 321–362. In: Reproductive Biology of Bats (E.G. Crichton and P.H. Krutzsch, eds.). Academic Press, San Diego.

McKean, J.L., and L.S. Hall. 1964. Notes on Microchiropteran bats. Victorian Naturalist, 81: 36–37.

McNab, B.K. 1982. Evolutionary alternatives in the physiological ecology of bats. Pp. 151–200. In: Ecology of Bats (T.H. Kunz, ed.). Plenum Press, New York.

O'Neill, M.G., and R.J. Taylor. 1986. Observations on the flight patterns and foraging behaviour of Tasmanian bats. Australian Wildlife Research, 13: 427–432.

Parker, S.A. 1973. An annotated checklist of the native land mammals of the Northern Territory. Records of the South Australian Museum, 16: 1–57.

Reardon, T.B., and S.C. Flavel. 1987. A Guide to the Bats of South Australia. South Australian Museum in Association with Field Naturalists' Society of South Australia (Inc.), Adelaide, Australia.

Schober, W., and E. Grimmberger. 1989. A Guide to Bats of Britain and Europe. Hamlyn, London.

Schulz, M. 1998. Bats and other fauna in disused Fairy Martin *Hirundo ariel* nests. Emu, 98: 184–191.

Sedgeley, J.A. 2001. Quality of cavity microclimate as a factor influencing selection of maternity roosts by a tree-dwelling bat, *Chalinolobus tuberculatus*, in New Zealand. Journal of Applied Ecology, 38: 425–438.

Sluiter, J.W., A.M. Voûte, and P.F. van Heerdt. 1973. Hibernation of *Nyctalus noctula*. Periodicum Biologorum, 75: 181–188.

Taylor, R.J., and N.M. Savva. 1988. Use of roost sites by four species of bats in state forest in south-eastern Tasmania. Australian Wildlife Research, 15: 637–645.

Tidemann, C.R., and S.C. Flavel. 1987. Factors affecting choice of diurnal roost sites by tree-hole bats (Microchiroptera) in south-eastern Australia. Australian Wildlife Research, 14: 459–473.

Turbill, C. 1999. Thermal Biology and Roost Selection of Long-eared Bats, *Nyctophilus geoffroyi*. B.Sc. (Hons.) Thesis, University of New England, Armidale, Australia.

Vonhof, M.J., and R.M.R. Barclay. 1996. Roost-site selection and roosting ecology of forest-dwelling bats in southern British Columbia. Canadian Journal of Zoology, 74: 1797–1805.

Vonhof, M.J., and R.M.R. Barclay. 1997. Use of tree stumps as roosts by the western long-eared bat. Journal of Wildlife Management, 61: 674–684.

Williams, A. 2001. The Ecology of Insectivorous Bats in the Simpson Desert, South-western Queensland. B.Sc. (Hons.) Thesis, School of Biological Sciences, University of Sydney, Sydney.

Wood Jones, F. 1923–1925. The Mammals of South Australia, Parts I–III. Government Printer, Adelaide, Australia.

Zyll de Jong, C.G. van. 1985. Handbook of Canadian Mammals. 2. Bats. National Museums of Canada, Ottawa.

17

Causes and Consequences of Tree-Cavity Roosting in a Temperate Bat, *Chalinolobus tuberculatus*, from New Zealand

Colin F.J. O'Donnell & Jane A. Sedgeley

Among the 79 taxa of Microchiroptera in Australasia, frequency of tree-cavity roosting increases as mean annual temperature decreases and latitude increases. This gradient suggests there may be significant thermal benefits to tree-cavity roosting in cold climates. We explore the causes and consequences of tree-cavity roosting during summer months in *Chalinolobus tuberculatus*, a species that occurs at the southern limit (highest latitude) of this gradient. Five geographically distinct populations are compared. *C. tuberculatus* selected the oldest and largest trees for maternity roosting and avoided roosting under bark and in caves and buildings, despite the abundance of these sites. It also selected small, well-insulated cavities that accrue significant energy conservation benefits compared with other potential roosts (the "thermal hypothesis"). Reproductive females selected roosts that reach maximum temperatures late in the day and retain high temperatures through the night, thus benefiting nonvolant young. Productivity and survival were significantly higher in populations that selected well-insulated roosts. We propose that selection favors smaller, rather than larger, roosting group sizes in this cold, temperate climate. Smaller groups of bats that use relatively small, well-insulated cavities have higher survival rates than larger groups that use larger, less insulated cavities. *C. tuberculatus* formed behaviorally, though not geographically, isolated subgroups. All colonies exhibited extreme roost-site lability on a daily basis, but strong long-term philopatry among pools exceeding 100 roosts. Most roosts were used once per year but date of reuse was similar each year. Strict temporal philopatry suggests that bats do not switch roosts in response to daily variability in weather conditions. The thermal hypothesis suggests that development of grouping behavior may be an incidental response to physiological constraints on thermoregulation and reproduction. Nevertheless, social interdependence would increase the probability that clusters are large enough on any one day to be thermally beneficial and individuals could improve the reproductive success of other relatives within the group. We conclude by outlining hypotheses that could test the general applicability of findings to tree-cavity roosting bats.

INTRODUCTION

The Importance of Tree-Cavity Roosts

Cavities in trees are important for many bird and mammal species for shelter and as sites for breeding (Sedgeley, 2001). The quality of a roost site can profoundly affect its occupants by directly influencing fitness (Du Plessis and Williams, 1994; Li and Martin, 1991; Zahn, 1999). Consequently, the availability of tree cavities is thought to be a limiting factor for cavity-using species (Newton, 1994). Tree-cavity roosting is a common strategy among Microchiroptera (Kunz and Lumsden, 2003). High levels of selectivity for roosts relative to available sites have been demonstrated in many tree-dwelling bat species in a variety of habitats (e.g., Boonman, 2000; Brigham et al., 1998; Kalcounis and Brigham, 1998; Law and Anderson, 2000; Lumsden et al., 2002; Sedgeley and O'Donnell, 1999a,b; Vonhof and Barclay, 1996, 1997). As most species are secondary cavity users (they do not create or structurally modify cavities; Kunz and Lumsden, 2003), selection of trees and cavities with physical properties that provide high-quality roost sites is critical.

Factors that may influence the quality of a roost cavity include its provision of protection from predators and weather, its location relative to food sources, and its internal microclimate (Sedgeley, 2001). For example, the thermal regime of the roost can be a main factor influencing energy expenditure in bats (McNab, 1982). Energy demands of breeding females are highest during pregnancy and lactation (Kurta et al., 1990; Speakman and Racey, 1987), but individuals tend to minimize use of torpor because its can result in extended gestation and depressed postnatal growth (Hamilton and Barclay, 1994; Kunz, 1987; Kunz and Hood, 2000; Racey and Swift, 1981). Roosts with a physical structure that results in a warm and stable temperature should allow breeding females to reduce energy expenditure while remaining homeothermic for longer periods. Therefore, roost selection may be one of the most important mechanisms by which breeding females reduce their energy expenditure during reproduction with little cost to reproductive success (Racey, 1982; Sedgeley, 2001; Zahn, 1999). Moreover, tree-cavity roosting behavior has important implications for the productivity and survival of Microchiroptera, and understanding causes and consequences of this behavior is important for the conservation of many species.

We predict there may be significant thermal benefits to tree-cavity roosting in cold climates. New Zealand bats inhabit the coldest temperate rainforests and highest latitudes in the Australasian region. Climate at these latitudes is harsher than in Northern Hemisphere areas at similar latitudes (figure 17.1). For example, at 45° S, annual temperature and rainfall are almost identical to the west of Scotland at 55–57° N (O'Donnell, 2002a). The New Zealand bat fauna is limited to three species: the long-tailed bat (*Chalinolobus tuberculatus*, Vespertilionidae), lesser short-tailed bat (*Mystacina tuberculata*, Mystacinidae), and greater short-tailed bat (*M. robusta*) (Daniel, 1990). Both *M. tuberculata* and *C. tuberculatus* roost primarily in tree cavities despite an apparent abundance of caves and buildings in many parts of the country (Daniel and Williams, 1984; O'Donnell et al., 1999; Sedgeley and O'Donnell, 1999a). Little is known about roosting requirements of *M. robusta*, which is thought to be extinct.

Objectives

We propose that *C. tuberculatus*, which occurs at the southern limit of bat distribution in Australasia, provides a model for exploring the causes and consequences of tree-cavity roosting, particularly in temperate regions. The objectives of this review are to: (1) assess patterns of roost site selection by bats over a latitudinal gradient in Australasia to determine whether there is general evidence for selection of tree cavities being related to thermal benefits; (2) summarize results from intensive research on roosting behavior and population structure of *C. tuberculatus* in New Zealand; (3) examine the extent to which the hypotheses developed are applicable to patterns of roost selection in Australasia generally; (4) and develop general hypotheses to explain the causes and consequences of tree-cavity roosting in temperate Microchiroptera.

This study provides a contrast to the growing literature on bats from temperate coniferous and deciduous forests from the Northern Hemisphere.

Study Species

Chalinolobus tuberculatus is a moderately small (8–11 g) insectivorous bat. It is threatened with extinction because of habitat loss, human disturbance at roost sites, and predation by exotic mammals (O'Donnell, 2000a, 2001a). A long-term study of *C. tuberculatus* in the Eglinton Valley, in South Island, New Zealand commenced in summer 1992–1993. From 1993–1994 to the present (10 years), studies have concentrated on examining roosting ecology, habitat use, breeding, population structure, and movements. The study area is subject to cold temperate conditions (figure 17.1) with prolonged periods below 0 °C in winter.

Roosting ecology has been investigated in four other localities ranging across 10° of latitude (table 17.1, figure 17.2). The areas represent a broad range of habitat types present through the range of *C. tuberculatus*. Unlike the Eglinton Valley, which comprised mountainous rainforests, other sites were in areas that were mosaics of modified forest and developed farmland. Techniques adopted in each study were similar. Methods are explained by O'Donnell and Sedgeley (1999), O'Donnell (2000b,c, 2001b), Sedgeley and O'Donnell (1996, 1999a,b), and Sedgeley (2001). Primary techniques were radiotracking and sampling the composition of bat colonies at their roost sites using harp traps. This review specifically examines roosting behavior during the austral summer (October–March), and the use of communal roosting sites rather than those used by solitary bats.

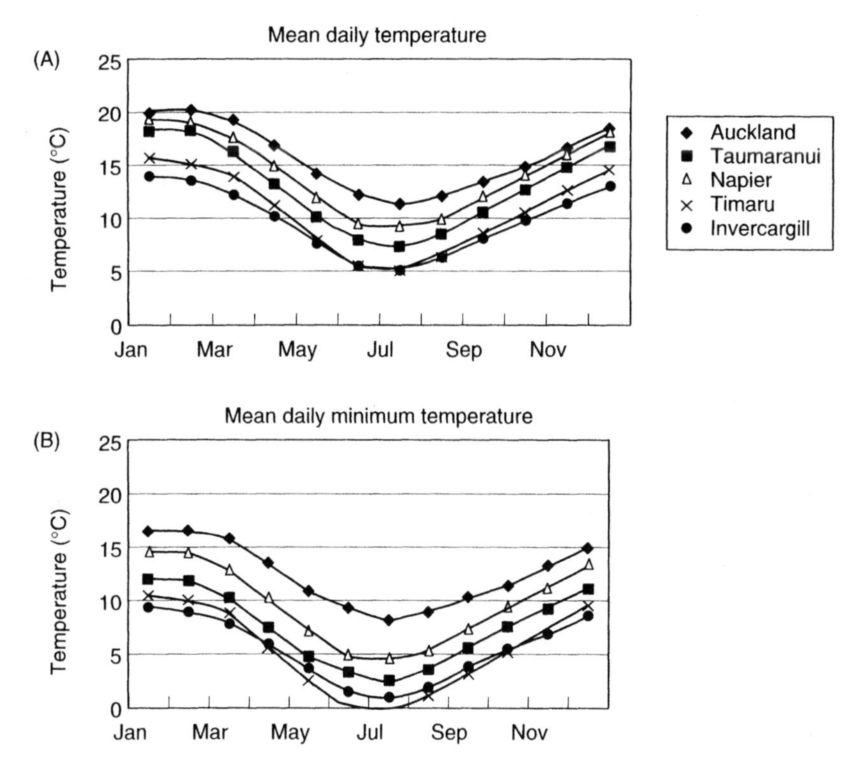

Figure 17.1 (A) Mean daily temperature (°C) and (B) mean daily minimum temperature at New Zealand Meteorological Service climate stations closest to the study areas mentioned in the text.

BIOGEOGRAPHY OF TREE-CAVITY ROOSTING IN AUSTRALASIA

Microchiroptera in Australasia are represented by 79 currently recognized taxa from seven families (Daniel, 1990; Duncan et al., 1999). Thirty-four taxa (43%) use tree cavities as primary roosts. Twenty-one (48%) of 44 Vespertilionidae taxa, 8 (72%) of 11 Mollosidae, 2 (100%) Mystacinidae, 2 (25%) of 8 Emballanuridae, and 1 (11%) of 9 Hipposideridae roost primarily in tree cavities. No Megadermatidae or Rhinolophidae roost in tree cavities.

The number of bat taxa declines with increasing latitude, from 41 taxa at 10–15° S to only three taxa at 45–50° S. However, frequency of tree-cavity roosting increases as mean annual temperature decreases and latitude increases (figure 17.3). Significantly more tropical than temperate taxa roost in caves and mines and foliage ($P < 0.02$) and more temperate species roost in tree cavities ($P < 0.001$), although there is no difference between frequency of roosting in buildings, under bark, or in bird nests ($P > 0.26$) (figure 17.4). Members of the most diverse family, Vespertilionidae, are equally distributed between tropical (17 taxa) and temperate (18 taxa) zones (9 taxa occur in both zones). Significantly more tropical vespertilionids

Table 17.1 Study Areas for *Chalinolobus tuberculatus* Research on Roosting Behavior

Study Area	Latitude (°S)	Habitat Type	Study Period	No. of Bats Radiotracked	No. of Bats Banded	References
Waitakere Ranges	36	*Agathis australis* dominant forest remnants and regenerating forest	Dec.–Mar. 1999–2001[a]	11	26	Alexander, 2001; S. Chapman, unpublished data
Western King Country	38	Podocarp–hardwood forest remnants dominated by *Beilschmiedia* surrounded by developed farmland	Oct.–Mar. 1998–2001[a]	12	533	C. O'Donnell, unpublished data
Puketitiri	39	Podocarp–hardwood–beech forest remnants and surrounded by developed farmland, plantations, and shrublands	1993–1995	19	240	Gillingham, 1996
Hanging Rock	44	Developed farmland, *Salix* woodland, and remnant *Kunzea* shrublands	Jan.–May 1996, 1999–2001[a]	37	163	Griffiths, 1996; O'Donnell, 2000d; O'Donnell and Sedgeley, 2004
Eglinton Valley	45	Extensive mixed *Nothofagus* forest	Oct.–Apr. 1992–2001[a]	107	789	O'Donnell and Sedgeley, 1999; Sedgeley and O'Donnell, 1999a,b; Sedgeley, 2001

[a]Ongoing study.

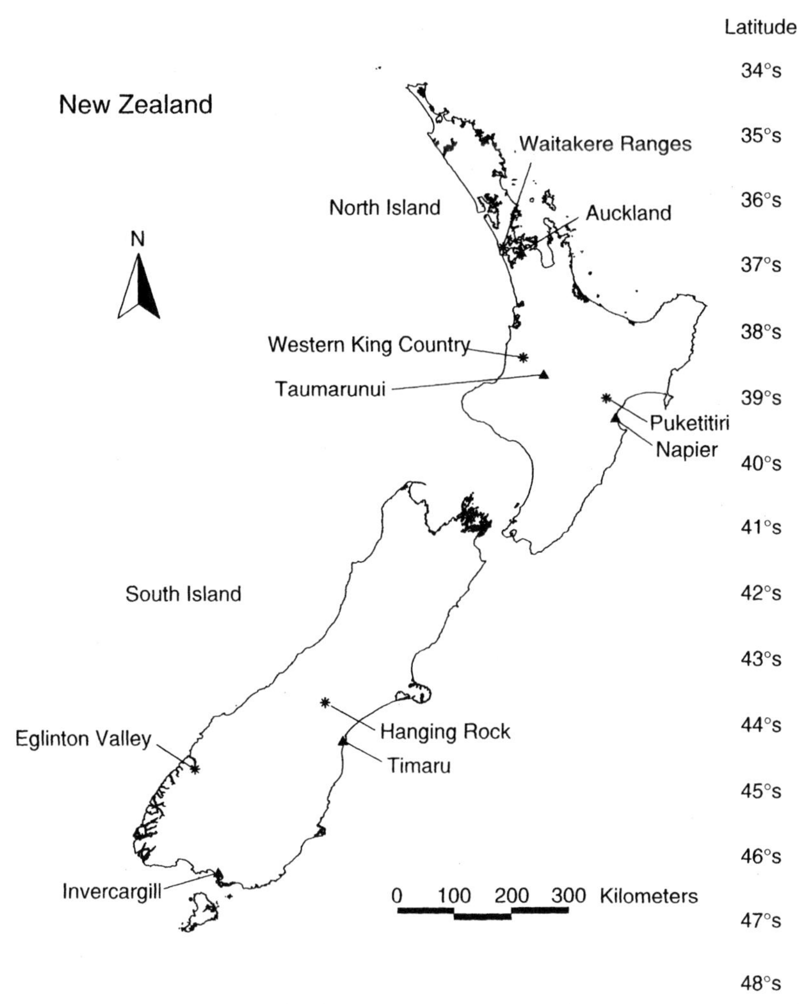

Figure 17.2 Map of New Zealand denoting study areas (∗) and climate stations (▲) mentioned in the text.

select foliage roosts ($P < 0.05$) whereas more temperate vespertilionids select tree cavities ($P < 0.01$). There is no significant difference in the use of caves, buildings, bark, and nest roosts ($P > 0.2$) between vespertilionids occupying the two zones.

ROOSTING PATTERNS OF *C. TUBERCULATUS* IN NEW ZEALAND

Frequency of Tree-Cavity Roosting

Chalinolobus tuberculatus roosts primarily in tree cavities in all study areas (75–100% of roosts; table 17.2). Previous studies that did not use radio-tracking underestimated the importance of tree-cavity roosts. For example,

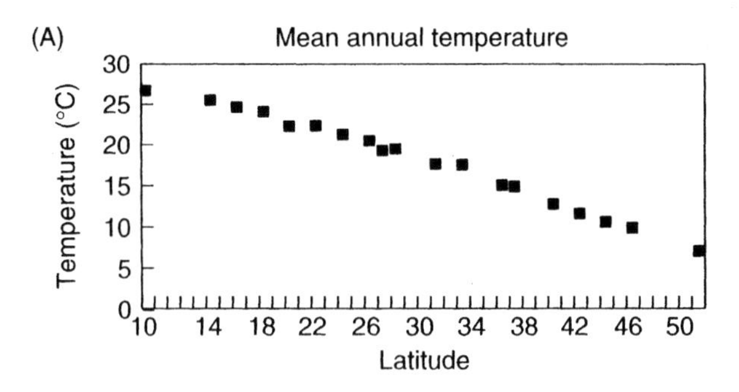

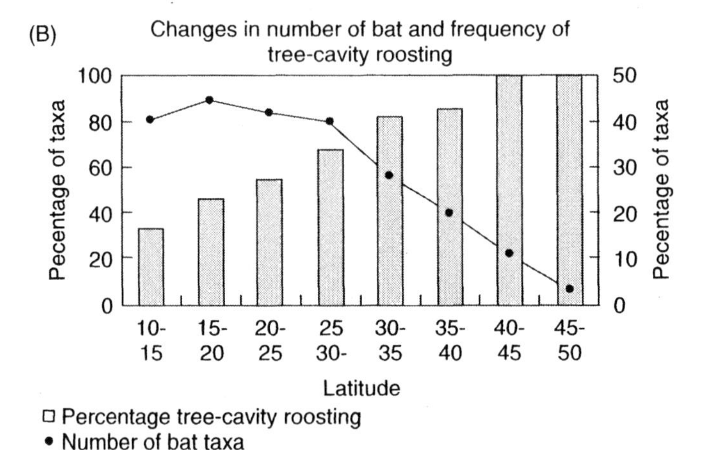

Figure 17.3 Relationship between latitude and (A) mean annual temperature (°C) in Australasia, and (B) number of taxa of Microchiroptera and percentage of taxa that roost primarily in tree cavities. Taxa were assigned to either tropical and/or temperate regions based on their occurrence in each 5° latitudinal zone in a gradient from 10° to 50° S. Mean temperatures are taken from Australian and New Zealand standard meteorological stations along the eastern seaboards of both countries.

at Hanging Rock, Daniel and Williams (1981) recorded anecdotal sightings of *C. tuberculatus* between 1957 and 1981 and all records of roost sites were in farm buildings. However, among the 67 roosts found in the same area by Griffiths (1996) and O'Donnell (2000d), none was in buildings or caves; 75% were in trees and 25% in limestone rock crevices.

In contrast to tree roosting, frequency of cave roosting was low despite areas of high cave abundance in some study areas. Over 830 caves more than 250 m long are listed on the New Zealand Speleological Society Database. Most are visited regularly by speleologists and if they see bats they often report them (Daniel and Williams, 1984; Department of Conservation National Bat database). However, use of caves by *C. tuberculatus* was confirmed only in the Western King Country study area. Of 221 large caves listed in this area, *C. tuberculatus* was reported from only six (2.7%). Our radiotracking study detected 23 roosts but only two (8%) were in caves

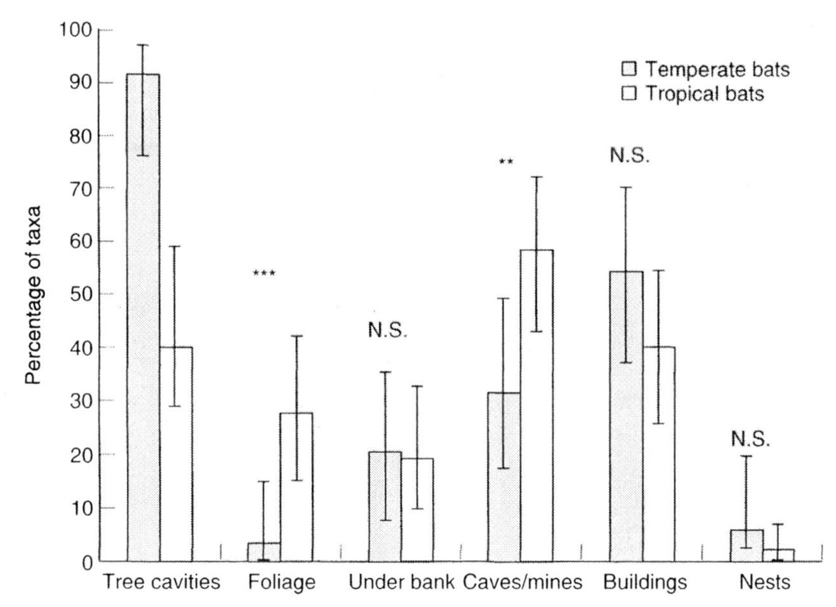

Figure 17.4 Percentage of tropical and temperate Microchiroptera occurring in Australasia using different roost types as primary roosts. Taxa were classed as roosting primarily (1) in tree cavities, (2) in foliage, epiphytes, and trunk surfaces, (3) in under bark, (4) in cliffs, caves, and mines, (5) in buildings, (6) in bird nests (source: 51 papers cited in Churchill (1998), Duncan et al., (1999), and O'Donnell (2001a)). Tropical and temperate zones were defined as north and south of the Tropic of Capricorn, respectively. Primary roosts were defined as those known to be used exclusively for roosting or were categorized qualitatively in the literature as "common," "usual" or "important" roosts. Classification of tree cavities as primary roosts did not preclude coding taxa as using more than one primary roost type (e.g., tree cavities, buildings, and caves were important roosts for *Chalinolobus pictatus*; Thomson et al., 1999). Differences in the frequency of use of roost types between tropical and temperate bats were compared using Fisher's Exact Test (***$P < 0.001$, **$P < 0.01$, N.S. not significant). The quantitative importance of each roost type was poorly known for the majority of taxa or taxa may have been misclassified. Because of potential recording errors, binomial confidence intervals were calculated (Mainland et al., 1956).

(table 17.2). Although radiotracking may underestimate use of caves because radio signals are difficult to detect underground, recent paleoecological studies support the contention that roosting by *C. tuberculatus* in caves is rare. Of more than 300 recent excavations of caves, none has recorded *C. tuberculatus* bones except in fossilized owl pellets ($n = 2$; $< 0.07\%$), whereas the bones of other small animals of similar size were common (Worthy and Holdaway, 1996, 2001).

Roost Site Selection

Bats inhabiting both unmodified rainforest and modified agricultural landscapes were highly selective of trees despite forest types and species of tree differing among the five study areas (tables 17.1, 17.2). *Chalinolobus tuberculatus* roosted in a variety of tree species, but the ranges of species used were small compared with those available (table 17.2). For example, in the Hanging Rock area, roosts were located in six (16%) of the 37 tree species available (O'Donnell, 2000d), and most roosts were located in just three species (*Salix fragilis*, *Kunzea ericioides*, *Cordyline australis*). *Chalinolobus tuberculatus* in the Eglinton Valley actively selected red beech (*Nothofagus fusca*) and standing dead trees for roosting (table 17.2), and avoided silver and mountain beech (*N. menziesii*, *N. solandri*) (Sedgeley and O'Donnell, 1999a). Roosts were close to forest edge providing bats with short commuting times to preferred foraging areas (O'Donnell, 2000c).

Most roosts were located in indigenous tree species. All study areas except the Eglinton Valley contained both introduced and indigenous species, but only 18 of the 100 roosts located in these areas were in exotic species. In the Hanging Rock study area, 27 species (73%) were exotic but significantly more roosts were in indigenous trees (47%) than expected on the basis of their availability ($\chi^2 = 21.69$, d.f. $= 1$, $P < 0.001$).

Chalinolobus tuberculatus roosted in some of the largest and oldest trees at low altitude and on fertile sites within the study areas (table 17.2). In the Eglinton Valley, roosts were significantly larger (mean $\pm$ SD $= 104.9 \pm 36.2$ cm diameter at breast height, dbh) than available trees (63.0 ± 36.4 cm dbh) ($t = 16.78$, d.f. $= 719$, $P < 0.0001$). The majority of trees (75.5%) had trunk diameters greater than 80 cm dbh, whereas only 25.0% of available trees were of comparable size (figure 2 in Sedgeley and O'Donnell, 1999a). Although no data on frequency of size classes of available trees were collected in the Waitakere, Puketitiri, and Western King Country areas, researchers commented that roosts were in large trees (Alexander, 2001; Gillingham, 1996). At Hanging Rock, roost trees were significantly smaller than trees in other areas (Tukey tests, $P < 0.05$), although communal roost trees (73.1 ± 51.4 cm dbh) were still significantly larger than available trees (49.3 ± 37.4 cm dbh) ($t = 3.12$, d.f. $= 384$, $P < 0.002$).

In the Eglinton Valley, the smallest trees selected were 100–200 years old and the majority were 450–600 years old (Ogden, 1978; Wardle, 1984). The *Agathis australis* used in the Waitakere study area were likely > 1000 years old (Ahmed and Ogden, 1987) and podocarps (*Dacrydium cuppressinum*, *D. dacrydioides*, *Prumnopytys ferruginea*, *P. taxifolia*) used in the Waitakere, Western King Country, and Puketitiri areas were likely 500 to > 1000 years old (Lusk and Ogden, 1992; Norton et al., 1988). The oldest and largest trees in the forest were those most likely to develop cavities and to provide cavities with qualities *C. tuberculatus* selected for (Sedgeley and O'Donnell, 1999a,b). Roosts at Hanging Rock were again an exception. Most were in introduced *Salix fragilis* likely < 100 years old. However, *S. fragilis* is very fast growing and reaches maturity and a large size at a relatively young age (R. Hall, pers. comm.). They represent some of the oldest trees in that landscape.

Table 17.2 Frequency of Tree-Cavity Roosting, Size of Trees Used, and Percentage Occurrence of Communal *C. tuberculatus* Roosts in Different Tree Species in Five Study Areas (see table 17.1 for descriptions and source data)

	Waitakere Ranges ($n = 20$)	**Western King Country ($n = 23$)**	**Puketitiri ($n = 13$)**	**Hanging Rock ($n = 44$)**	**Eglinton Valley ($n = 246$)**
Frequency of tree cavity roosting (%)	100	92[a]	100	77[a]	100
Stem diameter (mean dbh ± SD)	186.2 ± 54.5	79.4 ± 50.8	91.8 ± 28.2	57.8 ± 40.9	104.9 ± 36.2
Tree species used	3	9	5	8	4
Tree species available	> 25	> 25	> 25	37	8
Tree species:					
*Acacia dealbata**	–	–	–	2.3	–
Agathis australis	85.0	–	–	–	–
*Pinus radiata**	–	8.7	–	2.3	–
*Populus nigra**	–	–	–	2.3	–
Prumnopytys ferruginea	–	–	53.8	–	–
Prumnopytys taxifolia	–	–	7.7	–	–
*Salix fragilis**	–	–	–	22.7	–
Standing dead tree	–	17.4	–	25.0	22.4

dbh, diameter at breast height.
*Introduced species.
[a]Remainder of roosts located in limestone bluffs or caves.

Detailed analysis of cavity dimensions is currently only available from the Eglinton Valley. Cavities selected as roost sites were not a random subset of available cavities. All roosts were located in knothole cavities, were relatively high off the ground, and had small to medium-sized entrances (mean ± SD area = 100 ± 85 cm^2) that, compared with available sites, were uncluttered by external vegetation. Internally, roost cavities were relatively small with a vertical interior orientation, and were dry (Sedgeley and O'Donnell, 1999b). Bats in most other study areas also occupied tree cavities with small entrances (*c.* 60 × 60 mm) that were high off the ground (10–35 m) and likely had similar physical characteristics to the roosts in the Eglinton Valley. These roosts had the appearance of being small cavities rather than large hollow trunks. However, the Hanging Rock roosts were more variable, and roosts in willows often had large entrances exposed to the weather (Sedgeley and O'Donnell, 2004).

Thermal Characteristics of Cavity Roosts

Internal roost temperature was only measured in the Eglinton Valley and at Hanging Rock. Sedgeley (2001) showed that in the Eglinton Valley, the structural characteristics of roost cavities resulted in a microclimate that conveyed significant thermoregulatory advantages to roost occupants. Bat roosts had stable microclimates, displaying small ranges in temperature and humidity compared with external ambient conditions and available knothole cavities and trunk hollows that were not used by bats. Temperatures inside roost cavities were lower than ambient temperatures in the day and were warmer (and peaked) at night. Maximum temperatures occurred significantly later in the day in roosts and were held for significantly longer (figure 17.5A). Humidity in cavities was constantly high. Mean temperatures in more spacious hollow trunks, which were common in the Eglinton Valley, but not used by *C. tuberculatus*, were cooler and less stable than temperatures inside roosts. Results indicated *C. tuberculatus* select maternity roost sites with microclimatic conditions that accrue significant energetic benefits to reproductive females during the day and to nonvolant young, which were left alone for most of the night (O'Donnell, 2002b; Sedgeley, 2001). Predicted energy savings for adult bats using roost cavities compared with available knotholes were 1.1–3.3% and 3.4–7.3%, compared with hollows. Much greater energy savings would occur at night and benefit nonvolant young, but energy savings for nonvolant young have not yet been quantified.

Roosts in the Hanging Rock area that were > 80 cm dbh had stable temperature profiles similar to those in the Eglinton Valley (figure 17.5B). However, trees of this size were rare, and the majority had smaller stem diameters that were poorly insulated (figure 17.5C). Thus, microclimate followed the external ambient conditions more closely.

Structure of Roosting Groups

Bats in all study areas formed summer colonies dominated by reproductive females and their young (Alexander, 2001; Gillingham, 1996;

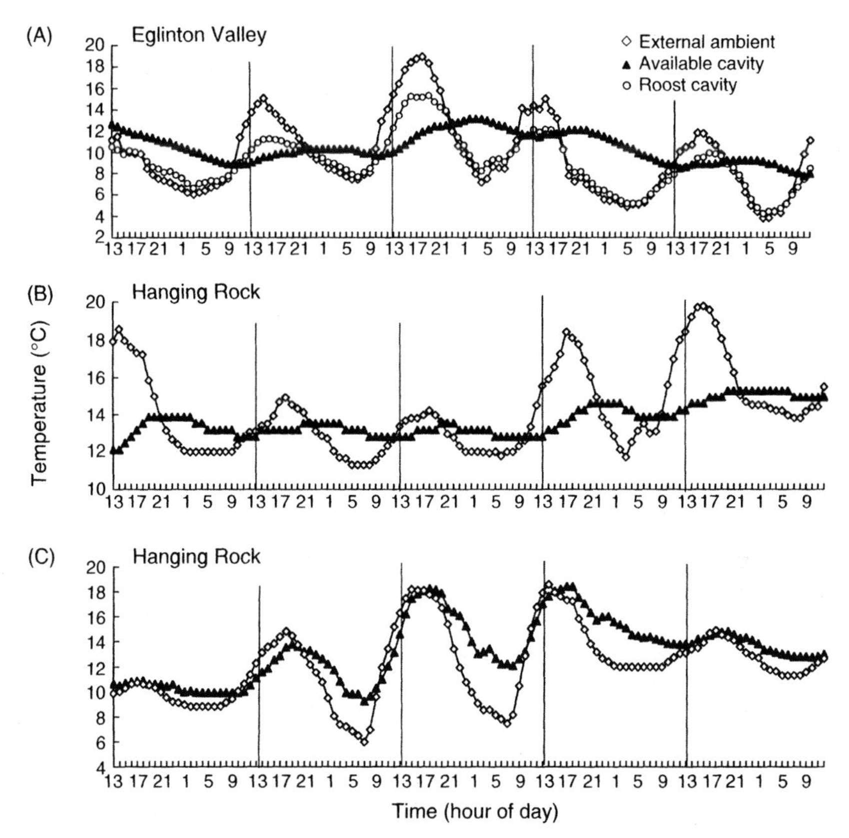

Figure 17.5 Comparison between external ambient temperature (°C) and temperature inside cavities used by bats. Three typical examples are shown: (A) roost temperature compared with ambient temperature and available cavities with similar structural characteristics but unused by bats in the Eglinton Valley; (B) a well-insulated roost cavity at Hanging Rock; and (C) a poorly insulated cavity from Hanging Rock.

O'Donnell, 2000d; O'Donnell and Sedgeley, 1999). Individual roosting aggregations were generally small compared with cave and building roosts of other species (Waitakere: mean ± SE = 11.6 ± 1.5, $n = 24$; Puketitiri: 70.2 ± 14.9, $n = 15$; Hanging Rock: 8.8 ± 0.9, $n = 63$; Eglinton Valley: 34.7 ± 13.2, $n = 178$). In the period before volant young were present, captures of bats at roosts in the Eglinton Valley comprised 62.8% reproductive females, 22.1% nonreproductive females, and 15.1% adult males. Comparable data on social structure from marked populations are now being collected in the Hanging Rock and Western King Country areas.

Composition of 95 communal groups was sampled as bats emerged at dusk in the Eglinton Valley (O'Donnell, 2000b). Long-term nonrandom associations among individuals were found by a cluster analysis that revealed three distinct social groups. On average, social groups contained (mean ± SD) 72.0 ± 26.0, 99.3 ± 19.0, and 131.7 ± 16.5 marked individuals per year. Number of bats emerging from individual communal roosts at night underestimated total size of each social group. Members of a social group did not necessarily roost together on any one night. They roosted in smaller groups spread over several roost sites, thus representing subsets of the larger social group. These smaller roosting groups were mobile; each subset moved to new sites virtually every day, and the combination of individuals within each subset also changed daily. The social groups were cryptic because collective foraging ranges of three groups overlapped and bats belonging to each group were spread over many roosts each day. However, roosting occurred in three geographically distinct adjacent areas. Only 1.6% of individuals switched between social groups. Nonreproductive females and males switched between social groups more often than reproductive females. However, individuals switched only once or twice during the study, and then only for a night. Juveniles of both sexes were associated with their natal group as 1-year-olds and later when breeding.

Results from 2 years at Hanging Rock suggest that there were two distinct social groups numbering 53.5 and 29 bats (average minimum number alive per year). As in the Eglinton Valley, foraging ranges of bats overlap, but the group roosting areas are distinct, though occupying adjacent sections of the Opihi River. Only three individuals have so far been detected roosting outside their natal groups.

Roost Switching and Reuse of Roosts

Bats in all populations moved to new roost sites virtually every day (70% of sites were only occupied for 1 day). Residency times of communal summer roosts averaged (± SE) 2.0 ± 0.4 days (range 1–4 days) in the Waitakeres, 1.2 ± 0.1 days (range 1–2 days) in the Western King Country, 1.7 ± 0.3 days (range 1–5 days) in Puketitiri, 1.6 ± 0.2 days (range 1–8 days) at Hanging Rock, and 1.7 ± 0.2 days (range 1–5 days) in the Eglinton Valley. Bats usually abandoned roosts simultaneously as a group (>90% of roosts) in all study areas. Residency varied significantly among sexes and stages of the breeding season in the Eglinton Valley. Lactating females changed roosts every 1.1 ± 0.2 days, postlactating females every 1.2 ± 0.3 days, pregnant females every 1.3 ± 0.2 days, and nonreproductive females every 1.3 ± 0.3 days. Males (3.0 ± 4.0 days per roost) and juveniles (1.8 ± 0.8 days per roost) occupied sites for significantly longer. At Hanging Rock, residency was longer in winter (3.6 days, Griffiths, 1996).

Bats did not cycle among a small pool of roosts, as is the case with other Microchiroptera (e.g., Brigham, 1991; Rieger, 1996), but used hundreds of different trees. O'Donnell and Sedgeley (1999) recorded roosts in 301 trees after radiotracking 58 bats over just three summers. Bats seldom reused a site within the same season, but had reused 37% of roosts by the end of the third summer. The sample size of roosts increased to 444 after radiotracking 107 bats over eight summers. We generally radiotracked only one individual from each social group at a time (O'Donnell, 2000b). Thus, rates of reuse were indicative only of patterns of roost use for each social group.

Nevertheless, reuse increased steadily over the years. Of the three social groups identified, by year 8 we detected a minimum reuse of 34.9% of roosts used by group 1 bats, 51.6% of roosts used by group 2, and 28.0% of roosts used by group 3. The period *C. tuberculatus* occupies communal roosts spans the austral summer (*c.* 180 days). Thus, each social group likely had a core pool of >150 trees used as communal roosts (based on the residency times of 1.1–1.7 days per roost and the knowledge that bats move to new roosts rather than cycling among a small pool of sites).

There was strong temporal synchrony in the reuse of the same roosts from year to year. Roosts were reused at the same time each summer. Date of reuse of 13.4% of roosts was within 0–3 days of the date of their first recorded use, 36.6% were reused within 9 days of the date of first use, and 62% reused within 20 days. Thus, date of reuse was not random with respect to date that first use was detected ($\chi^2 = 231.81$, d.f. $= 11$, $P < 0.0001$; after classification of date of reuse in 12 blocks covering 10-day periods, December–March, $n = 136$ cases of reuse). These results suggest that bats do not switch roosts in response to daily variability in weather conditions.

Consequences for Productivity and Survival

In the Eglinton Valley, reproductive history of 789 marked bats was studied over seven summers (O'Donnell, 2002a). Annual productivity was high (mean $\pm$ SE $= 0.91 \pm 0.07$ juveniles per female weaned). However, probability of survival to 1 year old varied from 0.26 to 0.88 (mean $\pm$ SE $= 0.53 \pm 0.06$). In contrast, at Hanging Rock, productivity and survival were significantly lower. Productivity was low with 0.22–0.24 young per parous female reaching volancy ($n = 3$ summers, 163 bats). Probability of surviving the first year was only 0.23.

The majority of sites where *C. tuberculatus* occurs in New Zealand are associated with unmodified indigenous rainforest (O'Donnell, 2000a). Higher productivity and survival occur in the Eglinton Valley where bats are using large-diameter, well-insulated native tree species. At Hanging Rock, productivity and survival are significantly lower and bats frequently use poorly insulated roosts in exotic tree species. We suggest that the lower survival at Hanging Rock reflects use of poor-quality roost cavities. Ninety-seven percent of the indigenous forest cover has been cleared over the last *c.* 150 years at Hanging Rock (O'Donnell, 2000d) and thus large-diameter, well-insulated trees are rare. However, other factors may explain differences in productivity and survival of bats between areas, including varying quality of foraging habitat, differential pressure from introduced predators, quality of maternity roost sites, differential overwinter mortality, quality of winter roost sites, and extent bats use torpor in winter. Although many of these factors have not been quantified, some are similar between study areas (e.g., predation risk), or less favorable in the colder Eglinton study area (e.g., food availability and temperature).

The occurrence of differential survival among social groups within the Eglinton Valley further supports the hypothesis that survival of bats is related to roost quality. Annual productivity did not differ among the three social groups investigated. However, probability of survival of juveniles to 1 year of age was inversely proportional to mean number of reproductive females in the groups (figure 6 in O'Donnell, 2002a). The smallest group (group 1) had the highest juvenile survival estimate (mean $\pm$ SE $= 0.70 \pm 0.07$), while the largest group (group 3) had the lowest estimate (0.30 ± 0.02). Reproductive females and juveniles in group 1 had significantly higher body condition. Bats in group 3, which live in a colder part of the Eglinton Valley (O'Donnell 2002a), may be forced to use cavities of poorer quality. The energetic costs of clustering are higher in the cold-temperate climates because heat loss is greater in larger clusters compared with smaller groups (Sedgeley, 2001; Zahn, 1999).

Currently, data on tree cavity selection are available for the two southernmost and coldest study areas. If our thermal hypothesis is correct, we predict that bats would be less selective in their choice of cavities in warmer parts of the range, as suggested by the increase in variance in tree characteristics recorded in the northern study areas (Bartlett's tests, $P < 0.05$).

Consequences for Social Structure

We suggest that roost site selection and roosting social behavior are linked, and both have implications for the evolution of size of social groups and reproductive fitness. Development of grouping behavior may be adaptive, or an incidental response to an organism's physiological needs (Jamieson and Craig, 1987). We further suggest that an optimum group size in *C. tuberculatus* developed as a consequence of use of optimal cavities. If thermal benefits are driving cavity roosting patterns, the development of highly structured social groups may have evolved incidentally. If the small cavities used by *C. tuberculatus* provide an optimum temperature regime suitable for raising young, then development of small structured groups may be in response to physiological limits on thermoregulation and reproduction. If *C. tuberculatus* uses observational learning, and shares information at roosts (Wilkinson, 1992), learning may be a mechanism by which group structure is maintained. In the Eglinton Valley, we estimated that *C. tuberculatus* used more than 450 different trees for roosts, but it exhibited extreme temporal philopatry, using each tree at a similar time each year. If nonvolant young learn the location of roosting sites, repetitive roost switching may imprint the group's pool of favored roosts on young.

Social breeding groups are composed of close relatives in many mammal species (Dobson et al., 1998; Kerth et al., 2000). In bats, formation of small clusters can be thermally advantageous (Roverud and Chappell, 1991), potentially increasing fitness (Hamilton, 1964) of colony members that could in turn improve the reproductive success of others within the group. The thermal benefit would be most pronounced at night when adult females leave the roost to forage. Juvenile *C. tuberculatus* return to their natal groups as 1-year-olds, females only rear young within those groups, and males return to groups when reproductively active. These characteristics suggest that group members may be related. Because they switch roosts often, social interdependence would increase the probability that clusters would be large enough on any one day to provide a thermal benefit. Analysis of genetic structure of social groups would resolve whether bats within groups are related (e.g., Kerth et al., 2000).

HYPOTHESES FOR FUTURE INVESTIGATION

The patterns of tree-cavity roosting during summer months identified in these New Zealand studies suggest that there may be significant thermal benefits to tree-cavity roosting in cold climates compared with roosting in open foliage or caves. Because *C. tuberculatus* shows a roosting pattern of increasing use of tree cavities along a latitudinal gradient, including the coldest latitudes, it would be a good species in which to examine the costs and benefits of tree-cavity roosting. This review raises questions about the extent to which roosting patterns of *C. tuberculatus* are typical of temperate rainforest bats. Examination of the roosting ecology of a range of Australasian Microchiroptera should aim to elucidate whether dependence on tree cavities, high roost site lability, demic structuring in populations, thermal benefits of cavity roosting, and cryptic subgroups are characteristic of cavity-roosting bats in the southern temperate region.

Several recent studies indicate similar roosting behavior in other Australasian species. For example, a number of taxa are highly labile in their use of roost sites, often moving every day, and select roosting trees and cavities with specific characteristics (e.g., Herr and Klomp, 1999; Law and Anderson, 2000; Lumsden et al., 2002; Lunney et al., 1995). Use of cavities with stable microclimates has been recorded elsewhere in Australasia (Hosken 1996; Turbill, 1999) and in the Northern Hemisphere (Alder, 1994; Kalcounis and Brigham, 1998; Vonhof and Barclay, 1997). Despite being highly selective in choice of roosting cavity, frequent shifting may be normal in bats that are not forced into regular or continual use of caves and buildings, where few roost sites are available (Lewis, 1995).

Apart from the studies of O'Donnell (2000b), we are unaware of any published studies in which social structure of Australasian bat has been investigated. However, in the Northern Hemisphere nonrandom associations of individuals within roosting aggregations at traditional sites suggest that well-structured subgroups occur regularly in different species (Entwistle et al., 1997; Kerth and König, 1999; Park et al., 1998; Rossiter et al., 2000; Wilkinson, 1985).

In summary, the results for *C. tuberculatus* suggest a number of hypotheses requiring investigation to test the general applicability to other bat taxa: (1) greater frequency of tree-cavity roosting as latitude increases reflects increased thermal benefits to bats during summer, particularly as these benefits affect survival; (2) an associated hypothesis is that thermal efficiency of roosting in caves and buildings during summer would decline;

(3) dependence on thermal benefits accrued in tree cavities will decline in a gradient as mean temperature increases with latitude; (4) productivity and survival of bats that use high-quality, thermally beneficial cavities as roosts will be significantly higher than in bats that use low-quality roosts; (5) tree-cavity dwelling bats are more likely to be adversely affected by reduction in abundance of high-quality roosts in colder temperate zones compared with warmer zones; (6) tree selection preferences will decrease as dependence on thermally beneficial roosts declines; (7) productivity and survival can be improved through management of roost site quality, for example, by erecting thermally beneficial roost boxes or by restoring indigenous forest habitat; and (8) highly structured groups develop as a consequence of roosting in small, thermally beneficial cavities.

ACKNOWLEDGMENTS

We thank Jane Alexander and Simon Chapman for access to unpublished data, Lindy Lumsden for data on Australian bats, Trevor Worthy for access to NZ cave databases, and Craig Willis and an anonymous referee for improvements to the manuscript.

LITERATURE CITED

Ahmed, M., and J. Ogden. 1987. Population dynamics of the emergent conifer *Agathis australis* (D. Don) Lindl. (kauri) in New Zealand. I. Population structures and tree growth rates in mature stands. New Zealand Journal of Botany, 25: 217–229.

Alder, H. 1994. Erste Erfahrungen mit dem Data Logger: Ereigniszählung vor Baumhöhlenquartieren von Wasserfledermäusen, *Myotis daubentoni*, bei gleichzeitiger Messung mikroklimatischer Werte. Mitteilungen der Naturforschenden Gesellschaft Schaffhausen, 39: 119–133.

Alexander, J. 2001. Ecology of long-tailed bats *Chalinolobus tuberculatus* (Forster, 1844) in the Waitakere Ranges: implications for monitoring. M.Sc. Thesis, Lincoln University, Lincoln, New Zealand.

Boonman, M. 2000. Roost selection by noctules (*Nyctalus noctula*) and Daubenton's bats (*Myotis daubentonii*). Journal of Zoology (London), 251: 385–389.

Brigham, R.M. 1991. Flexibility in foraging and roosting behaviour by the big brown bat (*Eptesicus fuscus*). Canadian Journal of Zoology, 69: 117–121.

Brigham, R.M., M.J. Vonhof, R.M.R. Barclay, and J.C. Gwilliam, 1998. Roosting behavior and roost-site preferences of forest-dwelling California bats (*Myotis californicus*). Journal of Mammalogy, 78: 1231–1239.

Churchill, S. 1998. Australian Bats. Reed New Holland, Sydney.

Daniel, M.J. 1990. Order Chiroptera. Pp. 114–137. In: The Handbook of New Zealand Mammals (C.M. King, ed). Oxford University Press, Auckland.

Daniel, M.J., and G.R. Williams. 1981. Long-tailed bats (*Chalinolobus tuberculatus*) hibernating in farm buildings near Geraldine, South Canterbury. New Zealand Journal of Zoology, 8: 425–430.

Daniel, M.J., and G.R. Williams. 1984. A survey of the distribution, seasonal activity and roost sites of New Zealand bats. New Zealand Journal of Ecology, 7: 9–25.

Dobson, F.S., R.K. Cheeser, J.L. Hoogland, D.W. Sugg, and D.W. Foltz. 1998. Breeding groups and gene dynamics in a socially structured population of prairie dogs. Journal of Mammalogy, 79: 671–801.

Duncan, A., G.B. Baker, and N. Montgomery (eds.). 1999. The Action Plan for Australian Bats. Natural Heritage Trust, Canberra.

Du Plessis, M.A., and J.B. Williams. 1994. Communal cavity roosting in green woodhoopoes: consequences for energy expenditure and the seasonal pattern of mortality. The Auk, 111: 292–299.

Entwistle, A.C., P.A. Racey, and J.R. Speakman. 1997. Roost selection by the brown long-eared bat *Plecotus auritus*. Journal of Applied Ecology, 34: 399–408.

Gillingham, N.J. 1996. The behaviour and ecology of long-tailed bats (*Chalinolobus tuberculatus* Gray) in the central North Island. M.Sc. Thesis, Massey University, Palmerston North, New Zealand.

Griffiths, R. 1996. Aspects of the ecology of a long-tailed bat, *Chalinolobus tuberculatus* (Gray, 1843), population in a highly fragmented habitat. M.Sc. Thesis, Lincoln University, Lincoln, New Zealand.

Hamilton, I.M., and R.M.R. Barclay. 1994. Patterns of daily torpor and day-roost selection by male and female big brown bats (*Eptesicus fuscus*). Canadian Journal of Zoology, 72: 744–748.

Hamilton, W.D. 1964. The genetic evolution of social behaviour. Journal of Theoretical Biology, 7: 1–52.

Herr, A., and N.I. Klomp. 1999. Preliminary investigation of roosting habitat preferences of the large forest bat *Vespadelus darlingtoni* (Chiroptera, Vespertilionidae). Pacific Conservation Biology, 5: 208–213.

Hosken, D.J. 1996. Roost selection by the lesser long-eared bat, *Nyctophilus geoffroyi*, and the greater long-eared bat, *N. major* (Chiroptera, Vespertilionidae) in *Banksia* woodlands. Journal of the Royal Society of Western Australia, 79: 211–216.

Jamieson, I.G., and J.L. Craig. 1987. Critique of helping behavior in birds: a departure from functional explanations. Pp. 79–98. In: Perspectives in Ethology, Vol. 7 (P. Bateson, and P. Klopfer, eds.). Plenum Press, New York.

Kalcounis, M., and R.M. Brigham 1998. Secondary use of aspen cavities by tree-roosting big brown bats. Journal of Wildlife Management, 62: 603–611.

Kerth, G., and B. König. 1999. Fission, fusion and nonrandom associations in female Bechstein's bats (*Myotis bechsteinii*). Behaviour, 136: 1187–1202.

Kerth, G., F. Mayer, and B. König. 2000. Mitochondrial DNA (mtDNA) reveals that female Bechstein's bats live in closed societies. Molecular Ecology, 9: 793–800.

Kunz, T.H. 1987. Post-natal growth and energetics of suckling bats. Pp. 395–420. In: Recent Advances in the Study of Bats (M.B. Fenton, P. Racey, and J.M.V. Rayner, eds.). Cambridge University Press, Cambridge.

Kunz, T.H., and W.H. Hood. 2000. Parental care and postnatal growth in the Chiroptera. Pp. 415–468. In: Reproductive Biology of Bats (E.G. Crichton and P.H. Krutzsch, eds.). Academic Press, San Diego.

Kunz, T.H., and L.F. Lumsden. 2003. Ecology of cavity and foliage roosting bats. Pp. 3–89. In: Bat Ecology (T.H. Kunz and M.B. Fenton, eds.). University of Chicago Press, Chicago.

Kurta, A., T.H. Kunz, and K.A. Nagy. 1990. Energetics and water flux of free-ranging big brown bats (*Eptesicus fuscus*) during pregnancy and lactation. Journal of Mammalogy, 71: 59–65.

Law, B.S., and J. Anderson, 2000. Roost preferences and foraging ranges of the eastern forest bat *Vespadelus pumilus* under two disturbance histories in northern New South Wales, Australia. Austral Ecology, 25: 352–367.

Lewis, S.E. 1995. Roost fidelity in bats: a review. Journal of Mammalogy, 76: 481–496.

Li, P., and T.E. Martin. 1991. Nest-site and nesting success of cavity-nesting birds in high elevation forest drainages. The Auk, 108: 405–418.

Lumsden, L.F., A.F. Bennett, and J.E. Silins. 2002. Roost site selection by the lesser long-eared bat *Nyctophilus geoffroyi* and Gould's wattled bat *Chalinolobus gouldii*. Journal of Zoology (London), 257: 207–218.

Lunney, D., J. Barker, T. Leary, D. Priddel, R. Wheeler, P. O'Connor, and B. Law, 1995. Roost selection by the north Queensland long-eared bat *Nyctophilus bifax* in littoral rainforest in the Iluka World Heritage Area, New South Wales. Australian Journal of Ecology, 20: 532–537.

Lusk, C., and J. Ogden 1992. Age structure and dynamics of a podocarp-broadleaf forest in Tongariro National Park, New Zealand. Journal of Ecology, 80: 379–393.

Mainland, D., L. Herrera, and M.I. Sutcliffe. 1956. Statistical tables for use with binomial samples: contingency tests, confidence limits, and sample size estimates. Department of Medical Statistics, New York University College of Medicine, New York.

McNab, B.K. 1982. Evolutionary alternatives in the physiological ecology of bats. Pp. 151–200. In: Ecology of Bats (T.H. Kunz, ed.). Plenum Press, New York.

Newton, I. 1994. The role of nest sites in limiting the numbers of hole-nesting birds: a review. Biological Conservation, 70: 265–276.

Norton, D.A., J.W. Herbert, and A.E. Beveridge. 1988. The ecology of *Dacrydium cupressinum*: a review. New Zealand Journal of Botany, 26: 37–62.

O'Donnell, C.F.J. 2000a. Conservation status and causes of decline of the threatened New Zealand long-tailed bat *Chalinolobus tuberculatus* (Chiroptera: Vespertilionidae). Mammal Review, 30: 89–106.

O'Donnell, C.F.J. 2000b. Cryptic local populations in a temperate rainforest bat *Chalinolobus tuberculatus* in New Zealand. Animal Conservation, 3: 287–297.

O'Donnell, C.F.J. 2000c. Influence of season, habitat, temperature, and invertebrate availability on nocturnal activity by the New Zealand long-tailed bat (*Chalinolobus tuberculatus*). New Zealand Journal of Zoology, 27: 207–221.

O'Donnell, C.F.J. 2000d. Distribution, status and conservation of long-tailed bat (*Chalinolobus tuberculatus*) communities in Canterbury, New Zealand. Environment Canterbury Report U00/38. Environment Canterbury, Christchurch, New Zealand.

O'Donnell, C.F.J. 2001a. Advances in New Zealand Mammalogy 1990–2001: long-tailed bat. Journal of the Royal Society of New Zealand, 31: 43–57.

O'Donnell, C.F.J. 2001b. Home range and use of space by *Chalinolobus tuberculatus*, a temperate rainforest bat from New Zealand. Journal of Zoology (London), 253: 253–264.

O'Donnell, C.F.J. 2002a. Timing of breeding, productivity and survival of long-tailed bats *Chalinolobus tuberculatus* (Chiroptera: Vespertilionidae) in cold-temperate rainforest in New Zealand. Journal of Zoology (London), 257: 311–323.

O'Donnell, C.F.J. 2002b. Influence of sex and reproductive status on nocturnal activity and night roosting by the New Zealand long-tailed bat *Chalinolobus tuberculatus*. Journal of Mammalogy, 83: 794–803.

O'Donnell, C.F.J., and J.A. Sedgeley. 1999. Use of roosts by the long-tailed bat, *Chalinolobus tuberculatus*, in temperate rainforest in New Zealand. Journal of Mammalogy, 80: 913–923.

O'Donnell, C.F.J., J. Christie, C. Corben, J.A. Sedgeley, and W. Simpson. 1999. Rediscovery of short-tailed bats (*Mystacina* sp.) in Fiordland, New Zealand: preliminary observations of taxonomy, echolocation calls, population size, home range, and habitat use. New Zealand Journal of Ecology, 23: 21–30.

Ogden, J. 1978. On the diameter growth rates of red beech (*Nothofagus fusca*) in different parts of New Zealand. New Zealand Journal of Ecology, 1: 16–18.

Park, K.J., E. Masters, and J.D. Altringham. 1998. Social structure of three sympatric bat species (Vespertilionidae). Journal of Zoology (London), 244: 379–389.

Racey, P.A. 1982. Ecology of bat reproduction. Pp. 57–104. In: Ecology of Bats (T.H. Kunz, ed.). Plenum Press, New York.

Racey, P.A., and S.M. Swift. 1981. Variations in gestation length in a colony of pipistrelle bats (*Pipistrellus pipistrellus*) from year to year. Journal of Reproduction and Fertility, 61: 123–129.

Rieger, I. 1996. Wie nutzen Wasserfledermäuse, *Myotis daubentonii* (Kuhl, 1817), ihre Tagesquartiere? Zeitschrift für Saugetierkunde, 61: 202–214.

Rossiter, S.J., G. Jones, R.D. Ransome, and E.M. Barratt. 2000. Parentage, reproductive success and breeding behaviour in the greater horseshoe bat (*Rhinolophus ferrumequinum*). Proceedings of the Royal Society of London, Series B, 267:545–551.

Roverud, R.C., and M.A. Chappell. 1991. Energetic and thermoregulatory behavior in the neotropical bat *Noctilio albiventris*. Physiological Zoology, 64: 1527–1540.

Sedgeley, J.A. 2001. Quality of cavity micro-climate as a factor influencing maternity roost selection by a tree-dwelling bat, *Chalinolobus tuberculatus*, in New Zealand. Journal of Applied Ecology, 38: 425–438.

Sedgeley, J.A., and C.F.J. O'Donnell. 1996. Harp-trapping bats at tree roosts in tall forest and an assessment of the potential for disturbance. Bat Research News, 37: 110–115.

Sedgeley, J.A., and C.F.J. O'Donnell. 1999a. Roost selection by the long-tailed bat, *Chalinolobus tuberculatus*, in temperate New Zealand rainforest and its implications for the conservation of bats in managed forests. Biological Conservation, 88: 261–276.

Sedgeley, J.A., and C.F.J. O'Donnell. 1999b. Factors influencing the selection of roost cavities by a temperate rainforest bat (Vespertilionidae: *Chalinolobus tuberculatus*) in New Zealand. Journal of Zoology (London), 249: 437–446.

Sedgeley, J.A., and C.F.J. O'Donnell. 2004. Roost use by long-tailed bats in South Canterbury: examining predictions of roost-site selection in a highly fragmented landscape. New Zealand Journal of Ecology, 28: 1–18.

Speakman, J.R., and P.A. Racey. 1987. The energetics of pregnancy and lactation in the brown long-eared bat, *Plecotus auritus*. Pp. 367–393. In: Recent Advances in the Study of Bats (M.B. Fenton, P. Racey and J.M.V. Rayner, eds.). Cambridge University Press, Cambridge.

Thomson, B., M. Schulz, C. Clague, M. Ellis, and A. Young. 1999. Little pied bat. Pp. 51–52. In: The Action Plan for Australian Bats (A. Duncan, G.B. Baker, and N. Montgomery, eds.). Natural Heritage Trust, Canberra.

Turbill, C. 1999. Thermal biology and roost selection of long-eared bats, *Nyctophilus geoffroyi*. B.Sc. Honours Thesis, University of New England, New South Wales, Australia.

Vonhof, M.J., and R.M.R. Barclay. 1996. Roost-site selection and roosting ecology of forest-dwelling bats in southern British Columbia. Canadian Journal of Zoology, 74: 1797–1805.

Vonhof, M.J., and R.M.R. Barclay. 1997. Use of tree stumps as roosts by the western long-eared bat. Journal of Wildlife Management, 61: 674–684.

Wardle, J.A. 1984. The New Zealand Beeches. Ecology, Utilization and Management. New Zealand Forest Service, Christchurch, New Zealand.

Wilkinson, G.S. 1985. The social organization of the common vampire bat. I. Pattern and cause of association. Behavioral Ecology and Sociobiology, 17: 111–121.

Wilkinson, G.S. 1992. Information transfer at evening bat colonies. Animal Behaviour, 44: 501–518.

Worthy, T.H., and R.N. Holdaway. 1996. Diet and biology of the laughing owl *Sceloglaux albifacies* (Aves: Strigidae) on Takaka Hill, Nelson, New Zealand. Journal of Zoology (London), 239: 545–572.

Worthy, T.H., and R.N. Holdaway. 2001. The Lost World of the Moa: Prehistoric Life of New Zealand. Indiana University Press, Bloomington, IN.

Zahn, A. 1999. Reproductive success, colony size and roost temperature in attic-dwelling bat *Myotis myotis*. Journal of Zoology (London), 247: 275–280.

Species Index

Author Index

Subject Index